Greening of Industry Networks Studies

Volume 15

Series Editors

Diego A. Vazquez-Brust, Faculty of Business and Law, University of Portsmouth, Portsmouth, Hampshire, UK

Joseph Sarkis, Business School, Worcester Polytechnic Institute, Worcester, MA, USA

This series aims to improve our understanding of how shifts in industrial regimes, trade, and technology are not only creating significant environmental and social impacts and inequities around the world but also opportunities for sustainable economic growth. It focuses on issues of industrial development, environment and society, and sustainable economy, dealing with the UN Sustainable Development Goals.

The series publishes high-quality peer-reviewed monographs, edited volumes and graduate textbooks related to the theme of sustainable economic development. The topics covered include industry networks, sustainable economic growth, supply chains, supply chain management, sustainable economy, green logistics, green growth, sustainable firms, collaboration for sustainability, corporate sustainability, corporate social responsibility, green banking, operations management, engineering management, sustainable value chains, sustainable business, circular economy, sustainable production, sustainable consumption, waste, resource management, productivity strategies, sustainable management, resource conservation, sustainability strategies, innovation processes, sustainable innovation, renewable resources, among others. Contributors to the series are researchers, business strategists, civil society thought leaders, and government policy makers.

The series is associated with ***The Greening of Industry Network***, an international network of professionals from research, education, business, government and civil society organizations located in more than 50 countries, with the goal of aligning industrial development strategies with sustainable development principles. The network stimulates, coordinates and connects high quality research to policies, strategies and actions in ways that contribute to a more sustainable society. Through a variety of arenas, it provides an open forum for creative debate to engage researchers, business, workers, government, consumers and other actors in developing a shared understanding of the changes required for creating a more sustainable future.

Now indexed by Scopus!

Ani Melkonyan-Gottschalk • Yanying Li
Editors

Urban Logistics Transformation

E-Commerce and Sustainable Circular Economy Approaches

Springer

Editors
Ani Melkonyan-Gottschalk
Clausthal University of Technology
Clausthal-Zellerfeld, Germany

Yanying Li
Alliance for Logistics Innovation through
Collaboration in Europe (ALICE)
Brussels, Belgium

ISSN 2543-0246 ISSN 2543-0254 (electronic)
Greening of Industry Networks Studies
ISBN 978-3-032-08386-9 ISBN 978-3-032-08387-6 (eBook)
https://doi.org/10.1007/978-3-032-08387-6

This Springer imprint is published by the registered company Springer Nature Switzerland AG
The registered company address is: Gewerbestrasse 11, 6330 Cham, Switzerland

Foreword

I am delighted to present this book showcasing urban logistics projects mainly supported by CINEA, the European Climate, Infrastructure and Environment Executive Agency. As a Project Adviser, I have had the privilege of managing several pioneering research and innovation projects in the field of urban logistics, funded under the Horizon 2020 and Horizon Europe programmes. These projects have played a key role in shaping the evolving landscape of sustainable last-mile delivery in European cities.

Their journey began in 2020 with three foundational projects: LEAD (2020–2023), SENATOR (2020–2025), and ULaaDS. These initiatives explored the potential of digital twins for urban logistics, offering valuable tools for experimentation and informed decision-making in public–private urban environments. Building on this momentum, the URBANE project (2022–2026) introduced a significant breakthrough by embedding the concept of the Physical Internet (PI) into the urban logistics sector. Its focus on developing digital tools for shared logistics assets marked a key step towards maximizing efficiency in last-mile deliveries.

The exploration of the Physical Internet concept continued with the DISCO project (2023–2027). This project takes an integrated, data-driven approach to urban freight. DISCO tackles the growing spatial pressures caused by last-mile delivery needs through solutions like flexible curbside use, temporary micro-hubs, and shared logistics centres in dense urban areas, supported by cutting-edge technologies, such as blockchain, digital twins, and artificial intelligence. Complementing these efforts are several sister projects—DECARBOMILE, GREEN-LOG, and UNCHAIN—all of which have advanced the digitalization and decarbonization of last-mile logistics.

More recently, in 2024, CODEZero and GreenTurn were launched to deepen engagement with consumers and city authorities, supporting the transition to sustainable e-commerce. By partnering with major stakeholders such as IKEA and DPD, these projects aim to develop a common framework to mainstream sustainable practices in e-commerce and delivery services.

These projects directly support the objectives of the European Commission's Green Deal and its target to achieve a 90% reduction in transport-related greenhouse gas emissions by 2050, together with the New European Urban Mobility Framework, which sets out the ambition to achieve zero-emission city logistics in major urban centres by 2030. By leveraging digitalisation, fostering public–private collaboration, and promoting multimodal and shared delivery solutions, they offer concrete pathways towards cleaner, more efficient, and citizen-centric last-mile systems. The innovations captured in this book contribute meaningfully to advancing this EU vision, helping cities transition towards sustainable, smart, and resilient urban freight systems.

The contributions in this book offer a timely overview of the evolving field of urban logistics and highlight a range of innovative approaches under exploration across Europe. Researchers, practitioners, engineers, and decision-makers will find the insights presented here useful in informing their own thinking and activities. May this work inspire new collaborations, innovations, and bold steps towards making last-mile delivery more sustainable for the benefit of our cities, citizens, and environment.

Yannick Bousse
Project Adviser
European Climate, Infrastructure and Environment Executive Agency (CINEA), Established by the European Commission, Brussels, Belgium

Recommendations

Logistics plays a vital role in everyday life, ensuring that people receive the goods they need, when and where they need them. As urban populations grow and lifestyles evolve, the surge in e-commerce has intensified consumer demand and expectations around speed, cost, and convenience of deliveries. Circularity adds yet another important dimension to this evolving landscape. By focusing on resource efficiency and waste reduction, collection points, extended producer responsibility schemes, and reusable packaging use emerge. Additionally and inevitably, logistics contributes to congestion, safety risks, environmental impacts, and competes for limited urban space.

These dynamics highlight the need for innovation and call for a profound transformation of urban logistics systems. It is in this context that the edited volume by Professor Ani Melkonyan-Gottschalk and Dr. Yanying Li offers both a road map of innovation and a manifesto for shaping sustainable, competitive, and joyful urban environments.

This book contributes to current literature in urban mobility, planning, and logistics innovation by presenting state-of-the-art research and practices across technological, business, and policy dimensions. In this ever-evolving landscape, the need for sustainable and resilient practices has become more urgent than ever. This book aims to contribute to more efficient and more sustainable last-mile delivery that is better aligned with cities' broader climate actions and embrace circular economy. This book is not just a collection of series publications but presents a journey through the evolving world of urban logistics.

The journey from a warehouse to a buyer has undergone a profound transformation. Digitalization is unlocking new possibilities, with technologies such as data spaces, blockchain, artificial intelligence (AI), digital twins, and the Internet of Things (IoT) offering unprecedented potential for efficiency, visibility, and optimization, enhancing the management of logistics flows, optimizing the use of infrastructure, and minimizing disruptions impacts.

Cities and companies are taking bold steps in addressing challenges and realizing opportunities. Traffic and routing management systems, access regulations, micro-hubs, electric trucks, cargo bikes, parcel lockers, and delivery stores are reshaping the urban logistics landscape. Together, these innovations reduce emissions, cut congestion, and improve delivery reliability. Looking ahead, the future of urban logistics hinges on interconnecting assets, resources, technologies, regulatory frameworks, and urban space. Strong collaboration between public authorities and private sector operators will be essential to succeed and create competitiveness advantages.

As we learn from the knowledge shared in this book, we can envision a future where urban logistics is seamlessly integrated into daily life, making our cities not just more efficient, but also more vibrant and resilient for generations to come.

Secretary General
ALICE, Brussels, Belgium

Fernando Liesa

Contents

Conclusions and Recommendations

Introduction to Urban Logistics, e-Commerce, and Circular Transformation

Ani Melkonyan-Gottschalk and Yanying Li

1 Overview

Urban freight transport is a critical component of urban life, ensuring the flow of goods that sustain city economies and daily activities. It accounts for approximately 10% to 20% of vehicle kilometres travelled (VKT) in cities and contributes to 2% to 5% of urban employment. According to Eurostat, urban freight was responsible for nearly 15% of greenhouse gas (GHG) emissions and 30% of air pollutants in urban areas in 2017 (Eurostat, 2020).

Urban freight volumes continue to increase, driven by urbanization and, more recently, by the rapid rise of e-commerce. Since the COVID-19 pandemic, on-demand and express deliveries have surged, fundamentally reshaping urban logistics patterns. Online shopping is steadily increasing across the EU, according to the latest annual survey on the use of ICT in households and by individuals. In 2022, 91% of EU residents aged 16 to 74 used the internet, and three out of four (75%) of them purchased goods or services online for personal use. Over the past decade, the share of online shoppers has risen significantly, growing from 55% in 2012 to 75% in 2022—a 20 percentage point increase (Eurostat, 2023).

This surge in online shopping has driven significant growth in home delivery services and employment through gig economy platforms. Consequently, the use of light commercial vehicles and cargo bikes for deliveries has expanded sharply in

A. Melkonyan-Gottschalk (✉)
Institute of Geotechnology and Mineral Resources, Clausthal University of Technology, Clausthal-Zellerfeld, Germany
e-mail: ani.melkonyan-gottschalk@tu-clausthal.de

Y. Li
Alliance for Logistics Innovation Through Collaboration in Europe – ALICE, Brussels, Belgium

A. Melkonyan-Gottschalk, Y. Li (eds.), *Urban Logistics Transformation*, Greening of Industry Networks Studies 15,
https://doi.org/10.1007/978-3-032-08387-6_1

many European cities, creating new challenges related to infrastructure needs and space management, e.g. urban logistics hubs in central areas, also called micro-hubs, cycling lanes, and parking spaces. Many cities have implemented low-emission zones or other measures to reduce air pollution in central areas, which has motivated logistics service providers to purchase electric vehicles. The adoption of electric vehicles in last-mile delivery depends on the availability of charging infrastructure and sufficient electricity supply. At the same time, the growth of platform-based employment in this sector has introduced new challenges related to workers' rights.

As the range of products purchased online has expanded—particularly in categories like fashion—the volume of returned goods has increased significantly. The growing volume of returned goods poses significant challenges for urban logistics, particularly in terms of sustainability. Inefficient return flows and poorly optimized reverse logistics processes not only strain operational capacity but also contribute to increased environmental impacts. Tackling these challenges calls for integrated approaches that combine advanced digital solutions, coordinated logistics practices, and circular economy (CE) principles. The circular practices emphasize designing out waste and pollution, keeping products and materials in use, and regenerating natural systems, aligning logistics operations with broader ecological and social objectives of a circular society. While concepts such as reuse, recycling, and remanufacturing are increasingly recognized, their practical integration into urban logistics systems remains limited. Strengthening this connection is essential to reduce waste, improve resource efficiency, and support more sustainable delivery and return operations.

2 Transforming Urban Logistics Through Research and Innovation

As e-commerce continues to expand, its impact on urban mobility has become a critical challenge for cities across Europe. To help tackle these pressures, the European Commission has supported several research and innovation projects focused on sustainable urban logistics. These efforts focus not only on improving operational efficiency but also on aligning logistics with Europe's broader climate and sustainability goals.

A central theme across these projects has been the use of advanced digital technologies to tackle the complexity of last-mile delivery. Digital platforms, real-time data analytics, blockchain, AI, and digital twins are being harnessed to optimize freight flows, reduce congestion, and support better planning. By improving the integration of logistics with urban planning, these technologies are helping cities move towards smarter, greener delivery systems. When combined with CE principles, these tools enable new forms of distributed production and logistics that support closed-loop supply chains and resilient urban ecosystems.

Crucially, these projects recognize that technical innovation alone is not enough. Many have focused on fostering collaboration between city authorities, logistics operators, technology providers, and even citizens. Co-creation processes have ensured that solutions respond to real urban needs, while testing in living labs across Europe has provided valuable practical insights into what works on the ground.

A common challenge addressed by these initiatives is the need to decarbonise last-mile delivery. From piloting zero-emission vehicles and cargo bikes to developing multimodal solutions that combine public transport and micro-consolidation hubs, these projects have explored multiple pathways to reduce emissions. At the same time, they have laid the groundwork for data-sharing frameworks that enable better coordination and decision-making across stakeholders. Some projects, particularly those aligned with regional circular economy strategies, demonstrate integrated land use, mobility, and resource flows to reduce emissions and support circular urban logistics strategies.

European research and innovation efforts have generated a diverse range of outcomes that are reshaping how urban logistics is planned and managed. One of the key achievements has been the development of advanced digital platforms that integrate real-time data from multiple stakeholders—logistics operators, city planners, and service providers—to optimize delivery routes, reduce congestion, and improve overall freight management in cities. These platforms are already demonstrating their value in real-world scenarios, including the design and implementation of urban freight consolidation strategies.

Another major outcome has been the advancement of decentralized frameworks using blockchain technology to enhance transparency, efficiency, and trust in urban freight operations to enable shared networks and assets. Innovative planning tools have also been introduced, including data-driven meta-models and early prototypes of urban freight data spaces and digital twins, enabling cities to analyse their logistics systems more systematically. These tools provide a structured approach for municipalities to assess their current freight landscape, explore alternative scenarios, and select tailored solutions for improving efficiency and sustainability. In the context of a circular economy, these tools also enable the monitoring of material and energy flows, helping align logistics operations with circularity goals.

Through active engagement with citizens and local communities, these projects have helped co-create logistics solutions that are better adapted to local needs while streamlining administrative processes. Citizen participation has played a crucial role in shaping delivery models, identifying preferred delivery points such as parcel lockers or pick-up locations, and supporting behaviour change towards more sustainable delivery choices. By involving residents in the design and testing of solutions, these projects have not only increased public acceptance but also uncovered valuable local insights that help align logistics systems with everyday urban life. This collaborative approach ensures that urban logistics evolves in a way that serves both the efficiency of the system and the well-being of the communities it operates within.

Beyond improving deliveries, some projects have also started to tackle the often-overlooked challenge of return logistics—a key element in building a truly circular economy. Integrating reverse logistics with circular economy principles—such as reuse, repair, and recycling—offers significant untapped potential. These reverse logistics systems are vital to keeping products and materials in use, acting as a cornerstone for the circular economy vision and strategy. Raising awareness, promoting the use of zero-emission delivery options, and encouraging alternatives to home delivery, like parcel lockers and pick-up points, are key strategies in this transition. The past and ongoing research and innovation projects are actively contributing to building the digital and operational foundations needed to turn this potential into practice.

Collectively, these research and innovation efforts have strengthened Europe's capacity to modernize urban logistics in ways that are both economically viable and environmentally responsible. By testing new models, scaling up successful solutions, and promoting integrated digital systems, they are helping to shape the future of sustainable urban logistics in Europe.

3 Integrating e-Commerce with the Circular Economy

Technological innovation alone is not sufficient to drive the transformation of e-commerce and urban logistics—societal change is equally essential. Consumer behaviour, marketing strategies, and coordinated action across the entire system play a critical role in shifting towards more sustainable practices. Incentive-based approaches—such as discounts for choosing sustainable delivery options or rewards for reducing return behaviour—can help encourage environmentally responsible choices. However, these strategies must be designed carefully to avoid inadvertently promoting overconsumption. Tools like persona-based segmentation offer promising ways to tailor communication and incentives to specific customer groups, making interventions more effective and targeted.

At the same time, integrating circular economy (CE) principles and extended producer responsibility (EPR) frameworks into e-commerce and logistics is vital for creating sustainable consumption and production systems. These approaches focus on extending product lifecycles, promoting reuse and recycling, improving packaging design, and establishing efficient take-back and reverse logistics systems. Within a circular society, these responsibilities extend to the social and ecological consequences of consumption, demanding inclusive and democratic participation in shaping sustainable futures of societies. As online retail continues to grow, the volume of packaging waste, product returns, and discarded goods is increasing, putting significant pressure on waste management systems and natural resources.

By embedding CE principles into supply chain operations—such as product design for durability and reuse, reverse logistics, and systems for product take-back—businesses can shift from linear models of production and disposal to more regenerative, circular approaches. Similarly, EPR frameworks require producers to take responsibility for the environmental impacts of their products throughout the entire lifecycle, incentivizing better product design, facilitating reuse and recycling schemes, and promoting closed-loop logistics solutions. Together, these strategies help reduce resource extraction, minimize waste, and build more resilient and responsible e-commerce ecosystems. By embedding environmental costs into product pricing and supply chain processes, CE and EPR can help shift responsibility for sustainability across the entire value chain.

Despite growing interest and progress in integrating circular economy practices into urban logistics, significant challenges remain. Persistent gaps in governance, infrastructure, and stakeholder coordination often hinder effective implementation. Overcoming these obstacles requires stronger alignment of logistics planning with circular economy principles and embedding sustainability goals into policy frameworks at local, regional, and global levels.

A holistic, systems-level approach is essential to drive the transition to circular urban logistics. This approach must connect logistics operations, consumer behaviour, product design, lifecycle management, and regulatory frameworks, supported by coordinated political leadership. Only through such integrated efforts can the full potential of circular urban logistics be realized, delivering meaningful environmental and societal benefits.

4 Outlines of the Book

This book aims to bridge the gaps between e-commerce, urban logistics, and circular economy strategies by offering real-world case studies, policy insights, and scalable technological solutions that promote sustainability. It highlights emerging topics like the role of micro-hubs in last-mile delivery, addressing challenges around their location, multifunctionality, and public-private collaboration. The book also tackles the need for modern urban logistics infrastructure and cohesive policies, examining how urban vehicle access regulations (UVAR) and digital tools like digital twins can optimize delivery and return flows while reducing emissions and congestion.

Recognizing that technology alone cannot drive transformation, the book emphasizes the importance of consumer behaviour, marketing strategies, and system-wide coordination. It explores incentive-based approaches to encourage sustainable choices without promoting overconsumption, using frameworks like persona-based segmentation for more effective engagement. Drawing on European research, pilot projects, and city experiences, it offers a roadmap for reimagining urban space, fostering collaboration, and embedding circular principles within urban logistics.

Ultimately, this work presents a comprehensive, action-oriented perspective, arguing that the future of sustainable urban logistics depends on the intelligent integration of technology, policy, consumer engagement, and shared societal goals.

The book begins with an introductory chapter that sets the stage by exploring the revolution in urban logistics. It outlines the key challenges cities face today and highlights the importance of innovation and sustainability in transforming urban freight systems.

Part I focuses on innovation in last-mile delivery, starting with how blockchain technology and smart contracts can enable trust and transparency in collaborative logistics. It then explores the role of digital control towers in managing last-mile operations through real-time data integration. The part also covers the use of deep learning and AI for demand prediction to support sustainable logistics-as-a-service models. Finally, it examines emerging delivery technologies such as drones and delivery robots, presenting public perceptions and potential impacts on urban deliveries.

The second part turns to sustainable e-commerce, presenting frameworks for evaluating the environmental performance of last-mile delivery strategies with applications in European cities. It highlights living labs as experimental settings where innovative e-commerce solutions are tested alongside stakeholder collaboration. Effective communication strategies aimed at increasing consumer adoption of sustainable delivery options are discussed, along with case studies on circular economy principles applied to returns, packaging, and the broader e-commerce ecosystem. Consumer behaviour is analysed through persona mapping to develop tailored incentives promoting greener delivery and return choices.

Part III addresses sustainable cities and the circular economy by analysing freight demand suitability and urban planning implications, with examples from the Paris region. The use and optimization of urban space for logistics are explored, emphasizing the importance of finding appropriate locations for logistics hubs and facilities. Practical approaches to enhancing out-of-home delivery systems are reviewed, including local authority strategies in France. The section also discusses the implementation of urban collection points designed to be people-friendly and environmentally responsible, alongside frameworks for extended producer responsibility and its impact on urban logistics. It concludes with a case study on intermodal transport for commuters as part of broader climate neutrality efforts.

The final part presents conclusions and recommendations, emphasizing the critical role of public authorities in fostering the transition towards sustainable urban logistics through effective policy and governance. It discusses the integration of the twin transition—digitalization and sustainability—in urban logistics planning. The book closes with a forward-looking view on the transformative potential of collaborative urban logistics powered by AI, data spaces, and living lab ecosystems, outlining pathways for future innovation and sustainability.

References

Eurostat (2020) Energy, transport and environment statistics—2020 edition. https://ec.europa.eu/eurostat/web/products-statistical-books/-/ks-dk-20-001

Eurostat (2023) E-commerce continues to grow in the EU. https://ec.europa.eu/eurostat/fr/web/products-eurostat-news/-/ddn-20230228-2

Innovation in Last Mile Delivery Towards Zero Emission

Blockchain and Smart Contracts: Enabling Trust in Collaborative Logistics

Rod Franklin

1 Introduction

Cities are complex entities composed of multiple interacting networks of actors, systems, technologies, and regulations (Batty 2013). As a structure built from the interaction of multiple networks, a city is not a planned construct, but an emergent concept reflecting the multitude of interactions between its underlying components (e.g., Fromm 2004). Historically, the approach cities have taken to managing this complexity has been to organize functionally and, within function, to focus on either technical or geographic components of the function (Minett 1975). Unfortunately, such a structure has encouraged the development of "silo" thinking, which ignores the interconnections of the various networks that operate within a city. As Christopher Alexander noted in the article that John Minett was responding to in his 1975 article, a city is not a tree, i.e., it is not a hierarchical construct that can be neatly laid out and managed as if what happens along branch "x" has no impact on things going on along branch "y" (Alexander 1965). The complex interactions that people, organizations, services, etc. have within a city preclude this tidy tree-like thinking (Jacobs 1961).

Complexity makes the management of a city's infrastructure a difficult task. An urban street, for example, may have retailers located along it who need to be resupplied with goods for sale, parking for people to come by and shop, electricity to operate lights and internal systems, sewage to provide for the comfort of employees and shoppers, telephone lines to make calls, Internet access to order supplies and replenish stocks, garbage collection to pick up waste and recyclables, gas for

R. Franklin (✉)
Kuehne Logistics University (KLU), Hamburg, Germany
e-mail: rod.franklin@klu.org

A. Melkonyan-Gottschalk, Y. Li (eds.), *Urban Logistics Transformation*, Greening of Industry Networks Studies 15,
https://doi.org/10.1007/978-3-032-08387-6_2

heating, street cleaning to entice shoppers to a neat shopping district, police and fire personnel to ensure safety, etc. The myriad of interacting systems operate to allow this retailer to provide services to their customers, rent to the building owner, pay its employees, and pay taxes to the city. Numerous other sets of separate networks interact within the city to generate economic, social, educational, and artistic benefits for the city's citizens.

Overseeing the various services that are accessed by the entities that deliver services and benefits to the city's citizens are the city's public works department, the transportation department, the police and fire departments, the tax department, a building inspection department, a business license department, an education department, a social services department, a parks and recreation department, planning departments, etc. In addition, water companies, electric companies, Internet service providers, gas companies, and telephone companies are also managing a set of services that allow these service entities to operate. And, of course, there is that logistics service provider who is being asked to replenish the stock of the retailer, deliver a package to the consumer, or deliver office supplies to the various city departments that enable the retailer to sell to the citizen who wishes to park their car in front of the retailer's shop and shop at the retailer.

As cities have grown, the services that they are expected to deliver to their citizenry have also grown. As the example of the retailer indicates, these services, provided by numerous city functional departments and external commercial entities, intersect, facilitate, and constrain the various recipients of the services in an emergent manner that no individual department or company can foresee or control. Given the city's inability to systemically control or understand what is happening at the operational level means that interventions are as likely to harm as they are to help in achieving higher level goals being set by citizens or regulatory authorities. Another approach to managing the operations of the city is required.

The remainder of this chapter begins a discussion on how a city might tackle the complex systemic challenges that it faces in delivering services to its citizens. This chapter begins with a short discussion of the smart city concept, a concept that is not new but is becoming the focus of most cities as they implement modern Internet-enabled technologies in an attempt to better address the various issues that each of the city's departments and their constituents face. Following the smart city discussion, this chapter highlights the extremely important activity of urban logistics. This is important because without effective urban logistics operations a city ceases to function. Modern cities neither produce the food necessary for their citizens nor the goods needed for these citizens to live and prosper. The challenges faced in effectively managing urban logistics operations, their negative impacts of congestion, environmental pollution, and societal effects are discussed in an attempt to challenge the current laissez-faire approach employed by most cities to this critical city function. The Physical Internet model for logistics operations is discussed next as a potential approach to addressing the urban logistics problem. The Physical Internet draws an analogy between freight shipments over logistics networks and that of Internet data packages shipped over communications networks. Employing this analogy, the Physical Internet is built on the idea of asset sharing and collaboration amongst logistics service providers, an approach that reduces the negative impacts of urban logistics while increasing its efficiency.

While the Physical Internet provides an interesting framework within which to explore a more efficient approach to urban logistics, any system built around the twin concepts of collaboration and asset sharing faces the inevitable problem of user trust. The failure of many attempts by cities to engage their logistics service providers in some form of collaborative behavior means that implementation of a framework like the Physical Internet requires significant planning and stakeholder involvement (e.g., Sternberg et al. 2022; Ostrum 1990; Hardin 1968). A key missing component in most of the failures is the inability of the participants to build trust in the system or systems employed for collaboration and sharing (Giampoldaki et al. 2021). Blockchain, smart contracts, and blockchain-mediated security are proposed as a potential technical solution for enabling trust in the technical systems used to manage the collaborative sharing of assets and services. While systems like blockchain, automated contractual processes, and secure system access provide a means to enable technological trust, this is only one element of the complex problem of generating trust in a socio-technical system (Sztompka 2000). The issue of governance, not addressed in this chapter, is an even more challenging barrier and one that must be addressed at the same time that technical system trust issues are addressed.

This chapter ends with a discussion of the implications of these ideas for the enablement of a real smart city in which the city addresses the issue of the systems nature of its infrastructure and the management of this complex system. Returning to the concept of the city as a large data center introduced in the second section of this chapter, a set of policy recommendations for city management are made. These recommendations summarize and integrate the prior discussion proposing that a city think of itself not as a set of separate services delivered to its citizenry as independent silos of action, but as an integrated system in which each service is linked to every other service and managed in a wholistic manner to ensure that city assets are, like the urban logistics discussion that is the focus of this chapter, managed in a collaborative and shared manner leading to the most efficient and effective delivery of services to the city's citizens.

2 The Smart City

Several proposals have been put forward to try to address the problematic nature of managing a modern city (e.g., Batty 2018). No single approach has gained more attention in the last 50 years than that encapsulated in the term "smart city."[1] Smart

[1] Note that the actual smart city concept, although not called a smart city, was probably in the minds of individuals in the early 1950s as can be seen in an article by Norbert Wiener in a December 1950 *Life* magazine article where he spoke of the city as a "communications net" (Kargon and Molella 2004).

that urban freight data, if collected at all, is collected in multiple different ways by different technologies and for different reasons by different cities, resulting in difficulty in comparing freight operations between cities and limiting the usefulness of the data for city operational planning purposes. As noted by Ballantyne and Lindholm (2014), cities view urban freight as a disturbing factor and the awareness and knowledge of freight movements are low. In this same article, Ballantyne and Lindholm conclude that there is limited involvement of impacted stakeholders in policy setting for freight activities within cities and this leads to less than optimal outcomes and numerous unintended consequences.

The lack of historical interest in urban freight operations, and the general view in city operations that freight logistics is a "disturbing factor," is unfortunate. Urban freight operations, like other flows within a city, are essential to the ability of a city to provide its citizens with the goods and services that they expect and require. Simply bracketing urban freight as a disturbance ignores its criticality to efficient city operations and backs city administration into a corner when the volumes of freight activities increase, a situation that arises both from economic and population growth as well as advances in the "convenience economy" (online shopping and home delivery).

Thinking of the city as a large data center in which flows are tracked and managed to ensure that city resources are utilized in the most efficient and effective manner could help city managers in understanding the importance of urban logistics operations to the city and their impacts on the complex flows of other city services (e.g., personal mobility activities, business and economic activities, etc.). This would allow cities to better understand how urban logistics operations impact the city, but offer the city an opportunity to develop integrated and coordinated plans for ensuring that urban logistics provides the greatest benefits to the citizens of the city, enhancing their environment and not burdening it. To do this, city managers and the stakeholders in the urban freight sector require a new paradigm of operation if future growth in urban logistics operations and their negative consequences are to be avoided.

4 The Physical Internet

While cities have been following various paths in attempting to become "smart," an effort has been underway in the logistics and supply chain world to tackle the increasing complexities of managing freight moving across global, regional, and local logistics networks. Beginning in the early 1990s with the publication of *The Machine that Changed the World* (Womack et al. 1990), discussion of supply chains not as chains of interactions but as networks of interactions gained interest in both the academic and business worlds. These ideas coalesced in 2011 with a paper written by Benoit Montreuil in which the Physical Internet (PI), as an approach to supply chain operations using the digital Internet as a metaphor, was introduced (Montreuil 2011). The argument presented was that if every physical entity is instrumented with a real-time trackable device, why can't these physical entities be managed on their trips from origin to destination like packets over the Internet? The Physical Internet concept is a particularly useful model when the entities being tracked are freight packages, which are very close conceptually to packets being sent over the Internet.

Using a PI model for addressing urban transportation issues, particularly urban logistics, has been shown to provide considerable benefits (Kim et al. 2021). The basis of the PI model is the collaborative sharing of logistics assets to improve logistics efficiency, effectiveness, and mitigate environmental negative impact. The PI has been demonstrated to lower overall costs in a network through efficiency improvements while also lowering emissions (Pan et al. 2013). In the urban context, a PI structured logistics model would incorporate shared consolidation centers, micro hubs, parcel lockers, and integrated planning and operational tools to ensure that vehicle loads were optimized and that vehicle journey lengths were minimized (Kim et al. 2021). The use of low or no-emission vehicles (electric vans, cargo bikes, etc.) would further lower the emissions from logistics operations.

While the PI offers many advantages by increasing logistics efficiency, lowering costs and emissions, and potentially improving the quality of service, it requires logistics service providers to collaborate in delivering their loads. In a highly competitive and low-margin business like logistics, this is a difficult hurdle for companies to overcome (e.g., Basso et al. 2019). In an interesting article by Fawcett et al. (2015) looking at why supply chain collaboration fails, the authors found that a series of relational resistors creates friction that impedes and ultimately causes failure of collaborative efforts between supply chain participants. They created a model of the various resistors that cause failure. Resistance arises within two different groups of potential collaborators. The entrenched individuals see no reason to collaborate, while the emerging resistors fail to get sufficient motivation from their leadership teams to overcome the friction caused by the entrenched resistors and the operational challenges that any new relationship brings with it. This study provides a very good overview of why, even when the benefits of collaboration are known, it is very difficult to get parties to collaborate. Resistance to collaborating is important as this resistance acts to slow the implementation and adoption of collaborative logistics models like the PI, and similar issues create problems for city organizations to work in an integrated manner to deliver services to citizens and businesses (e.g., Pereira et al. 2017).

Numerous studies of collaborative logistics activities within cities have demonstrated that most of these initiatives fail. Many reasons are provided for these failures, poor governance structure, high costs, limited services, poor execution, failed service commitments, lack of control, etc. (Badraoui et al. 2024; Sternberg et al. 2022). While numerous elements lead to failure, a strong predictor of success or failure is whether the parties in the collaborative activity trust one another. Studies of horizontal collaboration by organizations show that trust is a major factor in whether the collaborative relationship succeeds or fails (Rindfleisch 2000). Factors that influence trust have both a human and a systems component (Rompf 2014). Human trust lies at the foundation of all trust relationships. Whether the trust is of another person, an organization, or a technology, that trust is at its heart a trust in the people who are the receivers or who stand behind the entity itself (Sztompka 1999).

The PI provides cities with an opportunity to develop mechanisms to work with logistics service providers to leverage the utilization of existing city infrastructures while lowering the impacts of ever increasing demands for urban logistics services. However, to successfully implement a PI-like service, the service must be perceived as being trustworthy by all stakeholders. For such a service to receive the trust of the user community, it must operate transparently, providing information needed for service providers to see that the quality and level of service being delivered meets their contractual expectations at a cost that has been agreed to. It must provide shippers and receivers of goods with visibility to the status of their orders. It must provide city administrators and consolidation services with both pre-planning and execution data that allow each entity to properly plan their daily activities. It must perform all of these tasks while remaining secure and allowing only authorized personnel to access information for which authorizations have been provided, and it must comply with GDPR and other privacy regulations.

Using the analogy once more of the city as a large data center, a trust enabling layer providing visibility of logistics events to authorized personnel and non-repudiation services to participants enabling service contracts to be enforced would be a critical support layer for the operational management of urban logistics activities. One technical tool that provides such a foundation for some, but not all, of these requirements is the secure database concept called blockchain, with functionalities enabled through the use of smart contracts.

5 Blockchain

Blockchain technology has the potential to play a significant role in the materialization of the PI paradigm by enabling secure and transparent data sharing across the entire logistics network while also supporting identity management enabling a secure and more user-centric approach to logistics management. A blockchain-enabled urban logistics system would employ smart contracts to increase the transparency of interactions between logistics service stakeholders and automate processes such as the movement of goods, payments, inspections, quality control, and last-mile infrastructure/resource sharing. The employment of automated smart contracts facilitates the real-time collection of data from IoT sensors, providing event management capabilities and triggering actions based on the successful completion of events or their failure.

Blockchains provide cryptographic proof of transactions without requiring a trusted third party to maintain the integrity of each transaction. Using a blockchain-based service, authorized nodes in a network have a fully transparent view of all historically accepted transactions. Agreements on whether to post blocks to the blockchain can be managed through various approval processes, generally called "proof of x" processes. Proof of work processes that are used in managing block submittals for the Bitcoin cryptocurrency, a public network of peers, are generally not used in what are termed private peer-to-peer networks. The overhead, cost, and speed of such processes are unsuitable for a network whose membership is controlled by a governing entity that invites entities to participate.

The original description of the blockchain concept was published in 2008 in a paper written by an individual or individuals using the pseudonym Satoshi Nakamoto (2008). In this paper, the blockchain's operation is described in the following manner. All new transactions are broadcast to all nodes participating in the network. These transactions are collected by each node into a block. When the size of the block reaches a network-defined maximum, each node tries to solve the difficult proof-of-work puzzle for its block. When the node solves the proof-of-work puzzle, it broadcasts its block to all of the nodes in the network. The other nodes in the network accept the block if all of the transactions contained in it are valid. The network expresses its acceptance of the block by adding the block to the current blockchain and starting work on the next block using the hash of the newly accepted block as the previous hash (Nakamoto 2008).

There are four main types of blockchain networks: public, private, consortium, and hybrid (Dhillon and Sivabalan 2025). Public blockchains, like Bitcoin mentioned previously, are open to anyone to join. Private blockchain networks are restricted networks in which only invited participants are allowed. They are managed by a single entity or a formal governing body that provides the invitations and oversees their operations. Consortium networks operate similarly to private blockchains, but utilize a governance structure defined by the organizations making up the particular consortium. Hybrid structures integrate elements of public and private or consortium blockchains to realize various structures related to the participants' needs.

The PI paradigm, if implemented within an urban context to manage logistics operations, would logically operate using a private or consortium structure. Cities today regulate which logistics service providers can operate within their boundaries through licensing arrangements. It would be expected that similar types of regulatory tools would be utilized should a blockchain structure be used to facilitate trust in the urban logistics model of the city. Such a structure would allow the city to approve who utilizes the blockchain infrastructure and control the process for posting blocks to the chain.

The cryptographic nature of the blockchain database ensures that once a block has been approved for attachment to the chain its contents cannot be changed. This non-repudiation function is at the core of blockchain's usefulness for enabling trust (Lemieux and Bravo 2021). If the network of peers consents to attach the block to the chain, it has voted on the veracity of what the block contains and no one can change what is in the block. While this provides a good reason to trust the data that is on the blockchain, it should be noted that the consensus process needs to provide trustworthy approvals for the blocks. Consensus processes are different from the "proof of…" process. The "proof of…" process simply provides a block proposer with the opportunity to propose their block to the network for addition to the chain. The consensus process is the process by which the voting peers in the network agree to accept the block. Processes such as Practical Byzantine Fault Tolerance (PBFT) or Raft are often used within private or consortium blockchain structures to approve blocks for addition to the chain. These approaches vary in that PBFT can be used where it is assumed that some of the participants in the consortium may be malicious. Raft, on the other hand, is used where node failures (crashes) are expected to arise (Lashkari and Musilek 2021).

While the concept of a blockchain is an interesting technical artifact, blockchains are relatively uninteresting when it comes to managing an urban logistics operation. This is because blockchains are simply flat linked data files without external relevance. They are secure storage systems; however, what is being stored is the interesting component of the blockchain concept. What is being stored requires a collection mechanism, and this is the role of smart contracts.

6 Smart Contracts

Smart contracts are code that interacts with the blockchain based on rules instantiated in the code and events monitored by the blockchain infrastructure (Alqarni et al. 2023). Contracts could cover service level agreements in which certain requirements, if performed, automatically generate a payment. They can also be developed such that if problems arise notifications can be sent to responsible parties in an automated manner to take corrective action before failures occur. Contract execution requires connections to external sources of events. These connections can be made through appropriately configured Application Programming Interfaces (APIs) that connect to external IoT devices or sensors. Communications with users can be enabled through the contract by user interface updates and/or direct contact through email, SMS, or other real-time communications mechanism.

Depending on the type of contract that is required for the parties and the processes that can be monitored through IoT and sensor technologies, either the smart contract system can be manually built or an automated system can be constructed to create smart contracts based on the event structures to be monitored. Automated systems reduce the amount of coding that must be performed when new partners are added to the blockchain, but they also restrict the events that can be monitored to the ones enabled in the code-generating software.

As an example of a smart contract generator, Fig. 1 shows a diagram of an automated contract generator that was developed for the URBANE project, funded by the EU Horizon programme. Inputs to the configurator are made through a user interface that requests certain information concerning the logistics transport and storage events to be monitored, their sequence, their description/classification, and any temporal requirements for their execution.

The smart contract generator is code that sits within the platform and above the blockchain layer. It instantiates an event monitoring "contract" that receives information concerning events subscribed to be based on the contract and posts them to

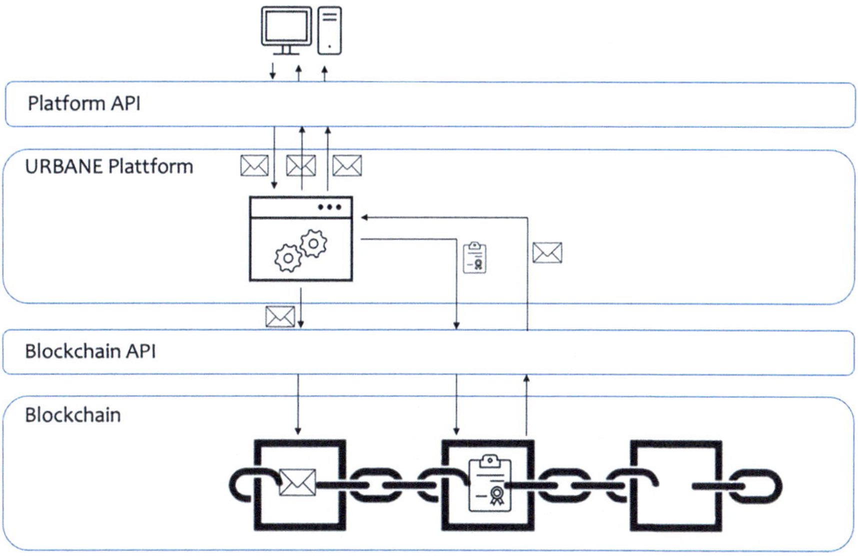

Fig. 1 Smart contract generator diagram from the URBANE project

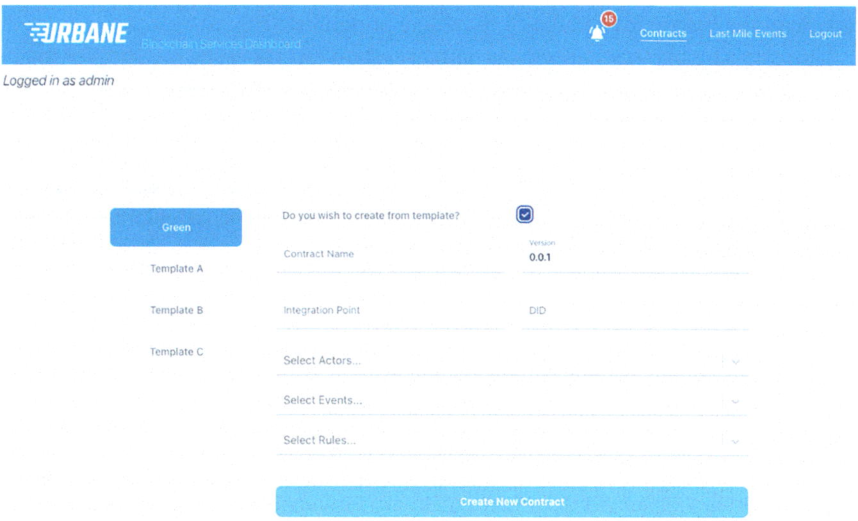

Fig. 2 URBANE smart contract event setup interface

the blockchain. Setting up the contracts is performed through a user interface, which is shown in Fig. 2.

Using smart contracts reduces monitoring requirements and the potential for third parties to shirk when performing contracted services. Both of these factors assist in creating trust in a system that requires participants to collaborate with third parties that they may never have worked with before or that they have limited knowledge about. Because authorized peers are able to monitor the actions of those who are executing the smart contracts, this visibility provides insights into the performance of the various service providers and also builds knowledge about their trustworthiness within the urban logistics environment.

7 Blockchain and Security

While visibility through the use of smart contracts and non-repudiation through blockchain's encryption and consensus processes facilitate trust, security of operations is also a requirement of a trustworthy system. Blockchains can be used to manage the identity of system users, devices accessing the system, and external systems that may be employed by organizations to post or retrieve information from the system. These interactions can all be tracked and posted to the blockchain. As with all events posted to the blockchain, this information is immutable and can be used to determine if a malicious or unauthorized user has accessed the system. The URBANE platform's identity management approach follows, for example, the SSI paradigm (Preukschat and Reed 2021) that gives full control of the data to the user and enables GDPR[3] to the greatest extent possible, by allowing the right of data erasure (to be forgotten) and the right of data portability (Fig. 3).

Utilizing a blockchain and smart contract system incorporating security for managing urban logistics operations addresses many of the concerns organizations have with the trustworthiness of an external system for collaborative operations. Such a system provides the ability to generate, manage, and execute smart contracts providing transparency to key delivery operations, security in tracking and system access (e.g., DIDs[4] and VCs[5]), and non-repudiation through the immutable nature of the blockchain data structure.

8 Limitations, Policy, and the Way Forward

While this chapter has generally provided a positive spin to the use of blockchain, smart contracts, and blockchain-enabled security systems for overcoming the hesitancy of logistics stakeholders to participate in a city-enabled collaborative logistics model, the reader should understand that this is only a necessary step to move to this type of logistics model. It is insufficient to create movement unless there is

[3] Regulation (EU) 2016/679, European Parliament, Council of 27 April 2016, http://data.europa.eu/eli/reg/2016/679/oj

[4] Decentralized Identifiers (DIDs) v1.0, World Wide Web Consortium (W3C), thtps://www.w3.org/TR/did-core/

[5] Verifiable Credentials Data Model v1.1, W3C, https://www.w3.org/TR/vc-data-model/

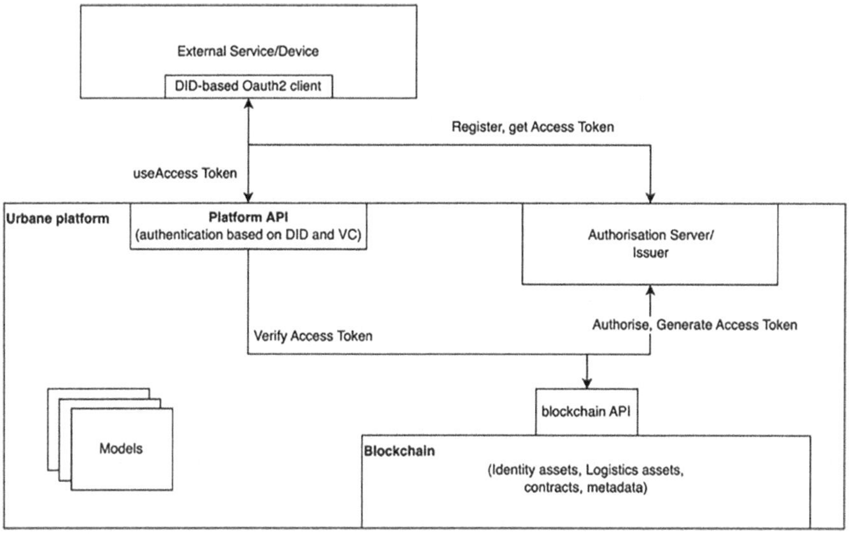

Fig. 3 URBANE platform digital ID security process

concerted involvement by the city to involve all logistics stakeholders in the development process. Each stakeholder will have their own concerns, and these concerns will need to be addressed to ensure that sufficient participation in the effort occurs. In addition, the city will need to ensure that a proper governance model, along with appropriate regulations and incentives, is developed, so that the collaborative model becomes self-sustaining. City personnel will need to become much more engaged with the logistics operations occurring within their borders and begin to be part of the process rather than entities that view logistics as problematic without a solution.

The problem of urban logistics falls into the emerging concept of a city as a commons (e.g., Sassen 2013, Jacobs 1961, Ostrum 1990). How should a city manage a common service for its citizens? Current practices would seem to indicate that cities would generally look at this question as one that involves the outsourcing of the collaborative urban logistics activity to an external third party. This entity, in partnership with city regulators, planners, and regulations, would enforce collaboration amongst the, as previously noted, "disturbing factor" urban logistics operators, much as city police enforce the rules of the city. This approach has unfortunately failed for most urban collaboration projects (Giampoldaki et al. 2021).

Legal and political science scholars look at the issue of collaboration as a game and apply game theoretic concepts to the issue of governance and enforcement. In both the legal and political science literature, regulation and enforcement strategies based on punishments for defecting are seen to be effective only if they are part of a portfolio of regulatory and enforcement actions that escalate penalties based on the activity being enforced (Downs 1998). These views encourage examining governance policies in light of frameworks that are based on punishments that are proportional to the violation. This concept is central to Ostrum's framework for collective management of common resources (Ostrum 1990). In Ostrum's framework, eight constructs are used to frame the regulatory policy for collaborators. These eight constructs are as follows (Herzberg 2020):

1. Clearly defined boundaries: what stakeholders can do and what they cannot do.
2. Congruence between usage and costs based on local conditions.
3. Collective choice arrangements allowing stakeholders to make and change the rules.
4. Active monitoring to ensure compliance.
5. Graduated sanctions starting low but increasing as violations recur.
6. Conflict resolution mechanisms that are rapid, low cost, and local.
7. Recognition by authorities of the rights of users to make their own rules.
8. Nested enterprises and associated governance rules.

The policy implications for cities in attempting to positively address the issue of urban logistics within their boundaries require creativity, stakeholder involvement, and a recognition that collaboration and sharing are social relationships, not regulatory outcomes. Thinking of urban logistics as a common-pool resource and structuring policy thinking along the lines of Ostrum's eight constructs has the potential to achieve results that other efforts have not been able to achieve (Sternberg et al. 2022).

This chapter began with a discussion of how cities today are failing to integrate the multitude of data flows that are being fed into the various city departments through independently deployed sensor technologies. Integrating these services into a structure similar to how large data centers are organized and managed provides the city with the capability to normalize and share data across functional boundaries. Integrating a blockchain infrastructure that supports logistics collaboration into such a data center-like approach to the management of city infrastructures would enable users to develop trust in the city's data management processes and facilitate the beginning of a collaborative logistics model. Like the Internet where networks of networks are linked together through common protocols, a Physical Internet-like conception of the data flows moving through a city's infrastructure will provide a solid foundation for building leverageable services for collaboration not only for logistics stakeholders, but citizens as well. It will also enable the city to better manage its infrastructure and ensure that it can accommodate the effects of environmental, economic, and social change. However, cities will still need to address the tension between traditionally adversarial organizations and develop not only a trustworthy infrastructure management system, but also a governance system that addresses the human side of the socio-technical system known as urban logistics.

9 Conclusion

Building a city-based urban logistics structure around the Physical Internet model and integrating this structure into a city's integrated "smart city" organization has the potential to create a controlled Logistics as a Service approach to city logistics operations. Such an approach, delivered through a set of trustworthy services enabled by blockchain and smart contract technologies, can lead to the reduction of logistics congestion and environmental problems that currently cause problems for city administrators and their constituencies.

This chapter explores the potential for organizing city logistics activities according to the collaborative and sharing paradigm that is the foundation of the Physical Internet. The concept envisions employing a trustworthy system model based on blockchain, smart contract systems, and blockchain-enabled security to achieve the smart city vision of optimizing the use of shared common resources (city infrastructures). It examines how a blockchain service could be employed to create trust in a system requiring collaboration amongst traditionally adversarial service providers. Using the inefficient use of city infrastructure as a starting point, the paper examines the development of a PI-enabled urban logistics model built using blockchain, smart contract services, and blockchain-based security enabled by edge-based sensor technologies to control within city boundaries the delivery of freight, accounting for interoperability, security, resilience, and sustainability. The benefits of such an approach are discussed, and the limitations of the model are identified. Finally, recommendations for how a city might move forward in the development of its own Logistics as a Service platform are discussed, focusing particularly on the issue of governing a common-pool resource.

References

Albino V, Berardi U, Dangelico RM (2015) Smart cities: definitions, dimensions, performance, and initiatives. J Urban Technol 22(1):3–21

Alexander C (1965) A city is not a tree. Ekistics 139:344–348

Alqarni M, Alkatheiri M, Chauhdary S, Saleem S (2023) Use of Blockchain-based smart contracts in logistics and supply chains. Electronics 12:1340. https://doi.org/10.3390/electronics12061340

Badraoui I, van der Lans IA, Boulaksil Y, van der Vorst JG (2024) Horizontal logistics collaboration success factors: expectations versus reality. BIJ 31(1):29–52

Ballantyne EEF, Lindholm M (2014) Identifying the need for freight to be included in local authority transport planning. In: Gonzalez-Feliu J, Semet F, Routhier JL (eds) Sustainable urban logistics: concepts, methods and information systems, ecoproduction. Springer, Berlin, Heidelberg. https://doi.org/10.1007/978-3-642-31788-0_3

Barroso LA, Hölzle U, Ranganathan P (2019) The datacenter as a computer: designing warehouse-scale machines, 3rd edn. Springer Nature Switzerland AG

Basso R, Kulcsár B, Egardt B, Lindroth P, and Sanchez-Diaz I (2019) Energy consumption estimation integrated into the electric vehicle routing problem. Transportation Research Part D: Transport and Environment 69:141–167

Batty M (2013) The new science of cities. MIT Press

Batty M (2018) Inventing future cities. MIT Press

Bibri SE, Krogstie J (2020) The emerging data–driven Smart City and its innovative applied solutions for sustainability: the cases of London and Barcelona. Energy Inform 3:1–42

Dablanc L, Francois A (2025) New ways of collecting urban freight traffic data and applications for urban freight policies and research. Case Stud Transp Policy 19:101315, ISSN 2213-624X. https://doi.org/10.1016/j.cstp.2024.101315

Dameri RP (2013) Searching for smart city definition: a comprehensive proposal. International Journal of Computers & Technology 11(5):2544–2551
Dhillon R, Sivabalan P (2025) Exploring dimensions of governance for different types of blockchain systems. Br Account Rev 101588, ISSN 0890-8389. https://doi.org/10.1016/j.bar.2025.101588
Downs GW (1998) Enforcement and the evolution of cooperation. Michigan J Int Law 19(2):319–344
Fawcett SE, McCarter MW, Fawcett AM, Webb GS, Magnan GM (2015) Why supply chain collaboration fails: the socio-structural view of resistance to relational strategies. Supply Chain Manag Int J 20(6):648–663
Franklin R, Fergadiotou I, Kampa M, Rizopoulos D, Tavasszy L (2023) The urban cloud: Linking city services, cloud computing, and the Physical Internet to achieve smart city objectives. In: IPIC 2023 conference papers and posters contributions proceedings (Version 8th of June 2023), Athens
Fromm J (2004) The emergence of complexity. Kassel University Press GmbH
Giampoldaki E, Madas M, Zeimpekis V, Vlachopoulou M (2021) A state-of-practice review of urban consolidation centres: practical insights and future challenges. Int J Log Res Appl 26(6):732–763. https://doi.org/10.1080/13675567.2021.1972950
Gonzalez-Feliu J, Semet F, Routhier JL (eds) (2014) Sustainable urban logistics: concepts, methods and information systems. Springer, Berlin, Heidelberg, pp 113–143
Haque AKMB, Bhushan B, Dhiman G (2021) Conceptualizing smart city applications: requirements, architecture, security issues, and emerging trends. Expert Syst 39, no5
Hardin G (1968) The tragedy of the commons. Science 162:1243–1248
Herzberg RQ (2020) Elinor Ostrum's governing the commons. Independ Rev 24(4):627–636
Jacobs J (1961) The death and life of great American cities. Random House, New York
Kargon R, Molella AP (2004) The city as communications net: Norbert wiener, the atomic bomb, and urban dispersal. Technol Cult 45(4):764–777
Kim N, Montreuil B, Klibi W, Kholgade N (2021) Hyperconnected urban fulfillment and delivery. Transport Res Part E Logist Transport Rev 145:102104
Lashkari B, Musilek P (2021) A comprehensive review of blockchain consensus mechanisms. IEEE Access. https://doi.org/10.1109/ACCESS.2021.3065880
Law KH, Lynch JP (2019) Smart city: technologies and challenges. IT Prof 21(6):46–51
Lemieux VL, Bravo M (2021) Introduction: Theorizing from Multidisciplinary Perspectives on the Design of Blockchain and Distributed Ledger Systems (Part I). In: Lemieux VL, Feng C (eds) Building decentralized trust. Springer Nature Switzerland AG
Lim Y, Edelenbos J, Gianoli A (2019) Identifying the results of smart city development: findings from systematic literature review. Cities 95:102397
Miladinovic I, Schefer-Wenzl S (2018) Dynamic allocation of Smart City applications. Open J Inter Thing 4(1):144–149
Minett J (1975) If the City is not a tree, nor is it a system. Plan Outlook 16(1–2):4–18
Montreuil B (2011) Towards a physical internet: meeting the global logistics sustainability grand challenge. Logist Res 3(2–3):71–87
Nakamoto S (2008) Bitcoin: a peer-to-peer electronic cash system
Ostrum E (1990) Governing the commons: the evolution of institutions for collective action. Cambridge University Press, Cambridge
Pan S, Ballot E, Fontane F (2013) The reduction of greenhouse gas emissions from freight transport by pooling supply chains. Int J Prod Econ 143(1):86–94
Pereira GV, Cunha MA, Lampoltshammer TJ, Parycek P, Testa MG (2017) Increasing collaboration and participation in smart city governance: a cross-caseanalysis of smart city initiatives. Inf Technol Dev 23(3):526–553
Preukschat A, Reed D (2021) Self-sovereign identity. Manning Publications
Rindfleisch A (2000) Organizational trust and interfirm cooperation: an examination of horizontal versus vertical alliances. Mark Lett 11:81–95
Rompf SA (2014) Trust and rationality: an integrative framework for trust research. Springer

Sassen S (2013) The global city, New York, London, Tokyo

Sinaeepourfard AJ, Garcia XM-B, Marín-Tordera E, Cirera J, Grau G, Casaus F (2016) Estimating smart city sensors data generation current and future data in the City of Barcelona. In: Proceedings of Conference: The 15th IFIP Annual Mediterranean Ad Hoc Networking Workshop

Sternberg H, Linan I, Prockl G, Norrman A (2022) Tragedy of the facilitated commons: a multiple-case study of failure in systematic horizontal logistics collaboration. J Supply Chain Manag 58(4):30–37. https://doi.org/10.1111/jscm.12278

Sztompka P (1999) Trust: A sociological theory. Cambridge University Press

Sztompka P (2000) Trust: a sociological theory. Cambridge University Press, Cambridge

Vallianatos M (2015) Uncovering the early history of "Big Data" and the "Smart City" in Los Angeles, Boom California, accessed 25 May 2025

Womack JP, Jones DT, Roos D, Carpenter DS (1990) The machine that changed the world. Macmillan Publishing Company, New York

Last-Mile Digital Control Tower

Amalia Ntemou, Alexandros Karalis, Dariya Rublova, Babis Magoutas, Ioannis Alexandrakis, Amalia Bozinaki, and Panagiotis Georgakis

Towards efficient, transparent and sustainable urban last mile logistics.

1 Problem Formulation, Challenges, and Objectives

Urban logistics have been significantly transformed due to the rapid growth of e-commerce. This has subsequently led to the introduction of significant complexities and the creation of multiple challenges in cities, which traditional systems are not always able to address. Looking to the future, there are forecasts that estimate that, in less than 10 years, the global market size will exceed $75 trillion (https://www.precedenceresearch.com/e-commerce-market). This fact implies a significant increase in deliveries and consequently in the number of vehicles, traffic congestion, and pollution. Notably, statistics show alarming evidence that, without effective interventions, these impacts could increase by 60% in 5 years (https://www.weforum.org/stories/2025/01/companies-creating-more-sustainable-future-urban-deliveries/#:~:text=Since%202020%2C%20last%2Dmile%20deliveries,much%20as%2060%25%20by%202030).

A. Ntemou (✉) · A. Karalis · D. Rublova
Netcompany S.A., Luxembourg, Luxembourg
e-mail: Amalia.NTEMOU@netcompany.com; Alexandros.KARALIS@netcompany.com; Dariya.RUBLOVA@netcompany.com

B. Magoutas · I. Alexandrakis · A. Bozinaki
Frontier Innovations I.K.E, Athens, Greece
e-mail: b.magoutas@frontier-innovations.com; ioannis.alexandrakis@frontier-innovations.com; amalia.bozinaki@frontier-innovations.com

P. Georgakis
School of Architecture & Built Environment, University of Wolverhampton, Wolverhampton, UK
e-mail: P.Georgakis@wlv.ac.uk

A. Melkonyan-Gottschalk, Y. Li (eds.), *Urban Logistics Transformation*, Greening of Industry Networks Studies 15,
https://doi.org/10.1007/978-3-032-08387-6_3

Last-mile logistics encounter challenges such as a lack of real-time situational awareness, fragmentation in stakeholders' coordination and collaboration, the existence of operational complexity, and the limited adoption of robust and proactive decision-making capabilities (Demir et al. 2022; Bosona 2020).

First of all, limited access to comprehensive and real-time data across the entire supply chain is one of the primary challenges in the logistics sector. Competition among stakeholders, lack of transparency or existence of multiple systems handling data prevents companies from having a unified and coordinated real-time view of delivery and fleet operations, which as a result can lead to delays, increased costs, and poor customer experience.

Another important matter is related to the involvement of multiple stakeholders in the logistics sector. Logistics service providers (LSPs), suppliers, retailers, and other involved entities manage their own systems and follow their own priorities, which can cause miscommunications, delayed responses, and dissatisfaction from service recipients. This fragmentation disrupts responsiveness and the seamless and fully coordinated delivery of last-mile logistics services.

Focusing on the operational aspects of e-commerce, one inevitably encounters complexities arising from fluctuating and dynamic demand. Traditional systems often lack the scalability and flexibility required to quickly adapt to rapid changes or unexpected disruptions (such as demand spikes, fleet unavailability, or road closures).

On the other hand, the wider adoption of AI and its integration into daily last-mile logistics operations has recently become increasingly important due to its groundbreaking potential. Demand fluctuations and potential disruptions can be predicted, while also missed delivery windows and costs can be prevented through more proactive and dynamic service management that can only be achieved through extensive use of decision-making and predictive analytics systems.

The aforementioned challenges and limitations have driven the adoption of digital control towers as a strategic solution to enhance situational awareness, operational coordination, and decision-making across the entire last-mile delivery ecosystem. The last-mile digital control tower (DCT) developed within the GREEN-LOG project (GREEN-LOG), which is presented in detail in this chapter, represents an innovative way AI technologies and urban freight network simulations are interlinked. It enables real-time monitoring and optimization of delivery flows and supports heterogeneous transport data fusion. Moreover, it gives the means the users monitor every aspect of the delivery network and generate real-time decisions to optimize delivery flows while considering the specificities of zero-emission vehicles, cargo-bikes, policy constraints, etc.

2 Literature, Current Practices, and Knowledge

The increasing complexity of urban logistics is driven by the rapid e-commerce growth, creating expectations for fast delivery and environmental impacts of freight transport in cities. According to recent studies and strategies applied across major

European metropolitan areas, urban freight contributes up to 25% of traffic emissions and occupies more than 30% of urban road space during peak hours (https://www.lvmt.fr/wp-content/uploads/2024/05/Harris-and-Dablanc-European-initiatives-of-urban-logistics-report-20231.pdf; https://www.etp-logistics.eu/wp-content/uploads/2022/08/Urban-Freight-Roadmap.pdf), which has created an urgent need for integrated, powerful solutions that offer better coordination, optimization, and resource allocation.

Innovative digital solutions such as digital control tower (DCT) are the means to address the increased challenges related to complexity and sustainability of urban last-mile logistics. Collaborative strategies, along with data-driven digital solutions, offer a foundational way for shaping sustainable urban landscapes, with a view to effectively scale up this digital solution, coordination between operators and authorities is essential. Within this context, the GREEN-LOG DCT is a structured, data-driven solution that empowers local authorities and LSPs, resulting in enhanced operational performance, reduced operational cost, and formulation of regulations incorporating elements that will ultimately improve service quality and urban sustainability. These platforms integrate real-time and historical data, advanced analytics, and algorithms while also allow multi-stakeholder collaboration, therefore enabling enhanced decision-making, and operational optimization.

Moreover, this approach not only enhances transparency and accountability but also fosters co-designing business models aligned with sustainable goals. Under this principle, the increasing effectiveness and support for environmental sustainability can be done by combining technologies used both by the stakeholders, such as GPS and IoT, and part of operators' production line, such as warehouse management systems (WMS), to optimize routes, consolidate shipments, and provide real-time delivery updates. In areas like the United Arab Emirates, where the number of addresses is limited or erroneous and antiquated tracking systems impede logistics (Wanganoo and Patil 2020), an intelligent last-mile solution might significantly increase consumer satisfaction, combined with a successful delivery rate. The DCT promotes creative, cross-sector alliances required for creating smarter and more resilient cities by allowing urban actors to break out from conventional silos.

An overview of how the GREEN-LOG DCT concept addresses the current challenges in last-mile logistics and the identification of current market gaps are presented in the following part of this chapter. It also provides insights about contemporary practices that foster collaboration between city authorities and LSPs to cultivate more sustainable urban ecosystems.

The concept of the DCT has been applied across various contexts and is particularly prevalent in the EU market, but its full potential and adoption have yet to unfold. Indicative examples include DB Cargo's Control Tower for customer communication and network control (https://www.dbcargo.com/rail-de-en/logistics-news/db-cargo-control-tower-13357640), DHL's transport-connected control towers to optimize transport flows (https://dhlsupplychain.dhl.com/EMEA-CCT) and Netcompany's PULSE, a control tower for a real-time data ecosystem that allows businesses to share and act on a single source, implemented in P&O Ferries and Copenhagen Airports (https://netcompany.com/platforms/pulse/).

Real-world examples of such change are becoming more widespread across Europe. Pilot initiatives in cities such as Barcelona and Amsterdam have shown how multi-stakeholder logistics platforms can reduce delivery mileage while increasing vehicle load efficiency by coordinating deliveries and sharing infrastructure, such as micro-hubs and pedestrian delivery corridors.

Barcelona's city council has implemented micro-hubs to facilitate and promote last-mile deliveries with the use of cargo bikes, aiming to reduce traffic congestion and emissions. The initiative, which used micro-hubs as consolidation spots, improved shipments' distribution inside the city center. The completion of 62 deliveries via a single route of cargo-bike in one day highlights the possibility of upgrading such systems with the goal of scaling up micro-hubs and, eventually, urban logistics situations (Sochor et al. 2023). Similarly, Amsterdam, a more mature and complicated city, investigated the use of pedestrian delivery routes and rivers to improve urban freight transportation. Experiments by extending a routing optimization method, a key feature of DCT, for use in a multi-modal two-echelon location and routing problem with increasing time windows in urban logistics. The city's endeavor to shift deliveries from roadways to canals has demonstrated significant reductions in driven kilometers and emissions. This program demonstrated both the impact of decreasing road congestion and the use of canal networks to give a more sustainable logistical solution (Pourmohammad-Zia and Mark 2023).

Despite this progress, industry comments and studies show ongoing gaps in current developments resulting from most logistics companies' reluctance to share data owing to either financial or competitive concerns, underscoring areas where innovation is still needed. This hesitation has a domino effect on the last-mile system performance generally. DCT's capacity to produce accurate, system-wide insights is restricted without data exchange, thereby leading to fragmented operations and inefficiencies (Ippolito et al. 2023). Key technical characteristics of the DCT—optimization engine as part of dynamic routing and delivery bundling—are hampered by this limitation, therefore producing longer delivery times, more fuel consumption, and low-quality scalability. Further compromising the DCT's ability to execute adaptive, multi-modal routing schemes in real time is rigidity in reacting to fast urban changes, such as congestion, weather variations, or consumer behavior (Nathanail and Papoutsis 2013).

Likewise, the absence of real-time data not only limits visibility but also directly reduces the usage of AI-driven analytics as predictive models rely on continuous, high-quality data sources to project demand patterns, traffic conditions, or service interruptions (Bruni et al. 2023). In addition to that, data limitations not only hinder operational flexibility but also hamper strategic planning for urban sustainability. Local authorities, such as city planners, cannot accurately assess the environmental impact of freight activity or make informed decisions about policy interventions without shared, high-resolution logistics data (https://sustainablemobility.iclei.org/the-data-gap-what-we-still-dont-know-about-urban-freight/). Furthermore, the constant environmental load of disorganized logistics opposes one of the main objectives of DCT, facilitating sustainable urban development (Nathanail and Papoutsis 2013). These difficulties together compromise the main functionalities of the DCT:

real-time monitoring, predictive analytics, dynamic optimization, and stakeholder coordination; hence, they highlight the need for next-generation solutions that remove these systematic obstacles by means of improved data governance, interoperability standards, and incentive-driven cooperation (Adetiloye 2012).

In essence, although the concept of the DCT has great potential to revolutionize urban landscapes, there are still some persistent factors including data sharing that still limits its whole potential. Difficulties, as those detailed in this section, directly affect DCT's fundamental capabilities: advanced analytics and logistics operations' optimization, including routing and real-time visibility. The GREEN-LOG DCT is a web application designed and developed to try to tackle these obstacles, combining advanced analytics, optimized dynamic (re-)routing, and policy recommendations through a user-centric and user-friendly user interface (UI). These, along with design rationale and pilot deployment, are presented in the next section to showcase how it has been adapted to fit particular urban environment's needs.

The subsequent section will delineate the functional architecture and core functionalities of the GREEN-LOG DCT, demonstrating the platform's implementation and adaptation within the context of several European urban living labs (LLs). These case studies will offer insights into the system's functional realization, the behavioral changes it tried to impose, and the wider implications for the future of sustainable urban freight.

3 Case Studies, Data Collection, and Model Development

This chapter focuses on the development and implementation of the essential functionalities of the GREEN-LOG DCT and its demonstration in real-world urban environments. The implementation has been driven by the end users to address the specific needs of several European urban LLs. The GREEN-LOG DCT has been developed as a flexible and modular, policy-aligned platform that combines advanced analytics, stakeholder collaboration means, and sustainability goals in a single operational environment through a succession of co-creation LLs deployments and iterative design processes. It was designed to allow progressive integration and adaptation, ensuring compatibility with varying levels of digital maturity across participating urban ecosystems.

The following subsection describes the system's design, key functionalities and components (real-time monitoring through the LaaS marketplace, the nowcasting and forecasting engine, and multimodal and cross-service dynamic (re-)routing) while also its validation in cities (LLs), tackling their specific urban logistics challenges.

The functionalities have been developed, enhanced and validated through operational pilots, in collaboration of LSPs, local authorities, research institutions and technology providers. Together, they enabled DCT to function as a mediator layer between policy enforcement and logistics operations, dynamically incorporating urban access regulations, congestion conditions, and delivery preferences into

offer forward-looking analysis of traffic congestion, locations of peak demand, or areas of compliance concern, supporting the policy assessment goals stated.

Across all these interfaces, user interaction is captured not merely as front-end activity but as a vital component of the system's data ecosystem. Each interface contributes structured inputs into the processing chain, subscribes to system outputs in real time, and participates in the broader orchestration logic that defines the DCT's adaptive capacity. The interface layer, therefore, acts as the human-system boundary where decisions are made, validated, and refined, ensuring that the platform remains responsive, transparent, and aligned with stakeholder needs at every stage of deployment.

3.1.1 Real-Time Monitoring Capabilities

The DCT's real-time monitoring system is a fundamental capability as it offers thorough situational awareness over last-mile distribution activities. Two complementary systems give this capability. Operatively, the platform uses a BYOD concept. Embedded inside a specific mobile app created under the GREEN-LOG project are lightweight software development kits (SDKs). These SDKs let carriers track delivery-related events, register goods, access package data, create visualized best paths, and set rerouting in response to interruptions. Standard real-time protocols continuously stream generated data, such as track and trace events and disruption events, into the dataspace, where the Orchestrator consumes it for additional analysis and integration into downstream systems, including routing and compliance monitoring. Concurrently, the DCT is closely linked with the LaaS marketplace, which serves as a front-end user interface as well as a data source for system components built under the GREEN-LOG architecture. This information also facilitates the creation of operational reports available to public bodies via DCT simple dashboard.

3.1.2 Predictive Analytics and Demand Forecasting

By providing accurate, near-real-time predictions for both road network traffic and logistics service needs, the NFE addresses traffic congestion and shifting logistical needs in metropolitan contexts. The NFE increases the operational efficacy and decision-making capacity of municipal authorities and LSPs by using historical data together with real-time inputs from the GREEN-LOG dataspace. Through short-term traffic forecasts, the engine improves traffic management by means of dynamic rerouting solutions and safety improvements, especially help metropolitan regions like Athens and Barcelona. Based on parcel demand projections, the NFE helps LSPs to maximize resource distribution and scheduling, hence supporting more sustainable, responsive urban logistics operations. Forecasting tools included in the GREEN-LOG platform also enable scalable future improvements, thereby assuring

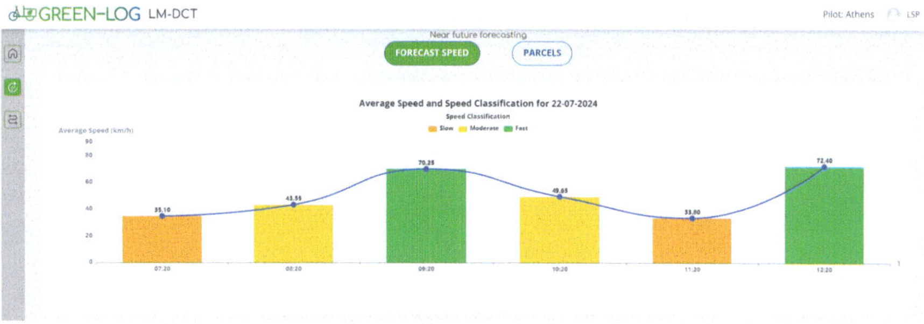

Fig. 2 Forecast of average speed of all vehicles for the next hours

that both long-term strategies and urgent operational adjustments are improved for sustainable urban development (Fig. 2).

3.1.3 Multi-Modal Dynamic Routing and Optimization

The routing engine, an API-driven tool in charge of computing the best and sustainable delivery pathways, controls routing activities inside the DCT. While considering several limitations like delivery time frames, geofenced access zones, and vehicle characteristics, the engine allows routing across several vehicle types. Real-time and predicted data allow the engine to create dynamically changing route plans in response to unforeseen events such as traffic congestion or failed delivery attempts. Most of the LLs in GREEN-LOG utilize routing outputs. LSPs have comprehensive daily operations schedules; local authorities have visibility into amenities provided, including consolidation hubs and depot stops—and other services, like the BYOD application, which is updated with the route information for smooth alignment (Fig. 3).

3.1.4 Sustainable Decisions and Incentives Imposed Through Policy Constraints

The DCT's capacity to translate environmental policy into practical logistical choices via an integrated sustainability incentive system is among its most original features. Through this feature, local authorities specify geolocated zones, desired time frames, allowed delivery options, and incentive structures, thereby interacting with the system. These setups are turned into a structured policy output that shapes two main components—the LaaS marketplace and the routing engine.

Within the routing engine, the policy output acts as a set of restrictions, directing route optimization to suit policy-compliant deliveries. The optimization approach gives deliveries in line with local sustainability targets, such as restricted deliveries

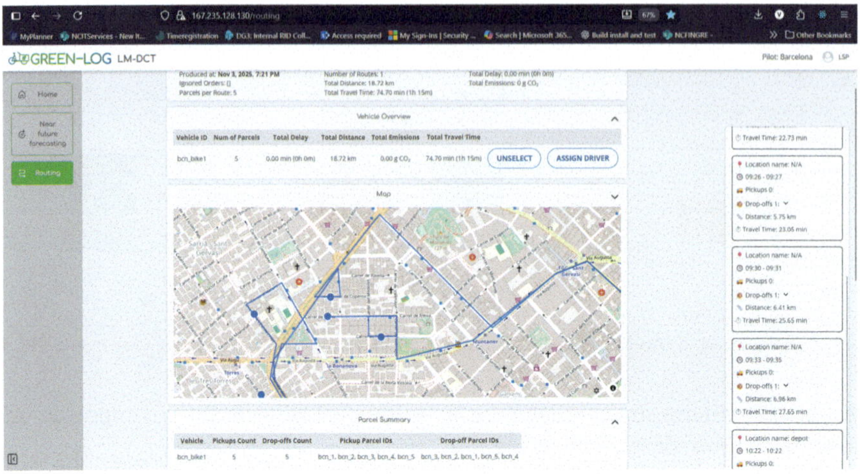

Fig. 3 Routing page in DCT

in a specified area (school zones). Concurrently, the identical output is transferred to the LaaS platform, where customers are given delivery choices denominated with matching incentives (e.g., "+10 Green Credits"). Operating two main subcomponents- (the Credits Engine which links delivery performance to credit calculation and the Incentive Engine which measures management API, to convert regulatory rules into quantifiable conditions (e.g., off-peak compliance or vehicle eligibility) -the Nudging Engine calculates the corresponding incentives (Fig. 4).

3.2 *Pilot Deployment in European Urban Living Labs*

Five urban LLs have been established under the GREEN-LOG initiative in five European regions: Athens, Barcelona, Flanders (including Ghent, Mechelen, and Leuven), Oxfordshire, and Ispra. The cities have been selected for their unique urban logistics challenges and governance structures, each aiming to solve various but concurrently related goals with reference to logistics. These LLs served as real-world testing grounds where the GREEN-LOG DCT could be tested, assessed, and modified collaboratively with local stakeholders, hence supporting this iterative and co-creative development process.

The LLs ensured the platform responds to both technological boundaries and users' requirements by allowing either context-specific customization or a combination of fundamental components and their functionalities, such as dynamic routing and real-time monitoring, instead of imposing a one-size-fits-all approach due to the technological approach followed. Additionally, the diversity of the LLs provides an interesting examination of how the DCT's components can be applied across several urban situations and regulatory contexts.

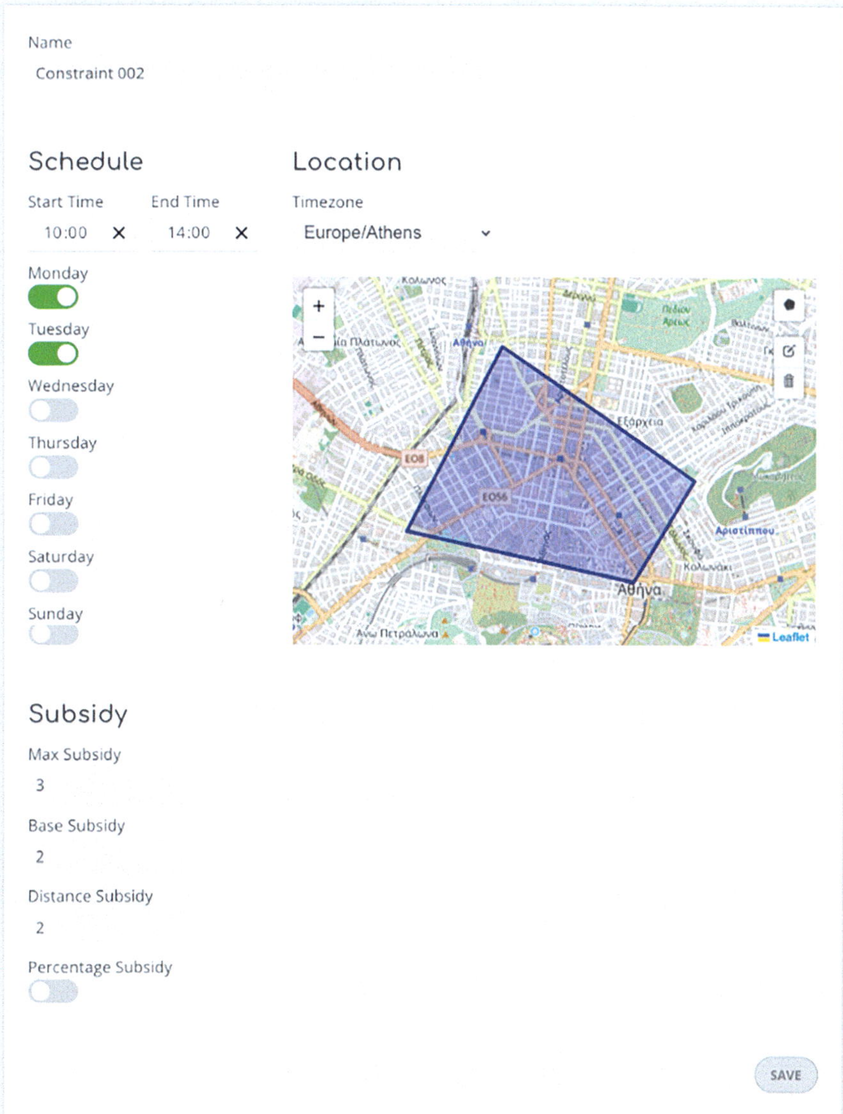

Fig. 4 Dynamic maps and creation of policy constrains efficient and compliant transport operations

Each LL represented a unique combination of local concerns and strategic goals. During the GREEN-LOG lifetime, three LLs were identified as having benefited more significantly from the solution. The following table provides an overview of the principal characteristics of each LL, including the issues they aim to resolve, the stakeholders engaged, and the primary areas of innovation (Table 1).

Table 1 Application of DCT in GREEN-LOG living labs (Athens, Flanders and Barcelona)

LL/city	Stakeholders involved	Urban logistics challenge	DCT feature used	Innovation focus
Flanders LL/Leuven	Stad of Leuven, local shop owners, urban LSP, social organizations	Reduce empty routes of LSPs, impose LEZ, reduce Van routes	Policy constraints and incentives, real-time monitoring, BYOD	Incentives for free-emission routes, bundling effect and off-peak hours
Flanders LL/Ghent	Stad of Ghent, local producers, urban LSPs	Reduce transportation, LEZ enforcement	Policy constraints and incentives, real-time monitoring	
Flanders LL/ Mechelen	Stad of Mechelen, citizens, city hub, urban LSP	Reduce transportation in Mechelen's neighborhoods, LEZ enforcement	Policy constraints and incentives, real-time monitoring	
Athens LL/ Athens	Municipality of Athens (DAEM), 2 urban LSPs	Reduce traffic in Athens historical city center, optimize operations	Policy constraints, dynamic rerouting, BYOD	Consolidation hub, restricted school zones
Barcelona LL/ Barcelona	Urban LSP, train authority, municipality of Barcelona	Traffic and inaccessible zones, multimodality for reduction of operational cost	Policy constraints, multi-modal routing, real-time monitoring, BYOD	Robot delivery, cargo bike hybrid models

The first case study to deploy DCT was Flanders Region, where the cities of Leuven, Ghent, and Mechelen played a pivotal role in the tool's development and contextual adaptation. These three urban landscapes were linked by a shared goal: to improve city center sustainability through the implementation of Low Emission Zones (LEZs). Local authorities aimed not just to reduce urban transportation congestion in their city centers, but also to encourage ecologically responsible behavior by local LSPs, shop owners/producers, and residents. In addition, efforts were made to reduce operating costs through bundling and consolidation.

Each of the cities involved in Flanders had different yet complementary goals within the respective LL, reflecting the region's diversity of urban difficulties and strategic agendas as depicted in Fig. 5. Leuven, for example, placed strong emphasis on the socio-environmental aspect of logistics reform. To do so, Leuven's primary initiative focused on reducing food waste generated by supermarkets and restaurants by facilitating the redistribution of surplus food to local social organizations. This targeted initiative aimed to both waste reduction and raise of social equity within the community. Ghent, on the other hand, concentrated on fostering collaboration between local producers and LSPs. Through the use of the DCT, Ghent aimed to create more integrated, cost-effective delivery models that would

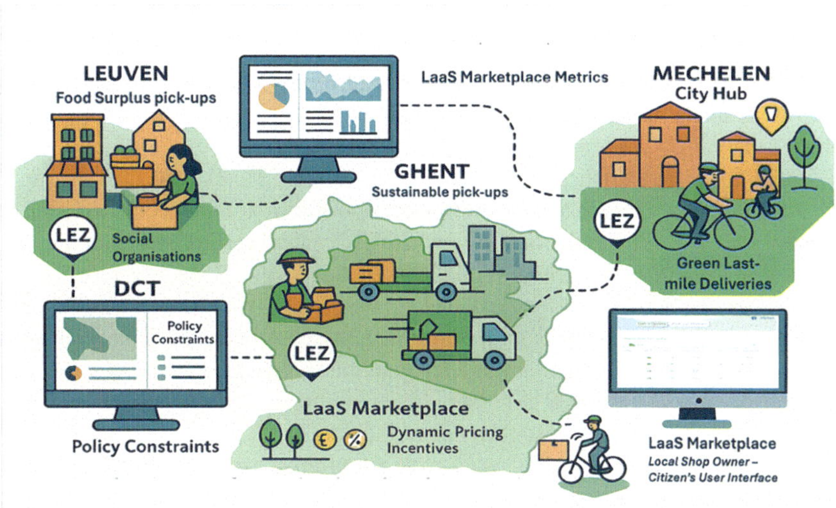

Fig. 5 Flanders LL case study

help reduce redundant routes and encourage cooperation in the supply chain. This collaborative approach was expected to improve service quality, minimize environmental impact, and promote efficiency in urban deliveries.

In contrast, the city of Mechelen pursued a more citizen-centered strategy by attempting to raise public awareness regarding the environmental impacts of the growing e-commerce sector. To this end, the city introduced the concept of City Hub, serving as a consolidation center where the e-commerce parcels gathered before their distribution via last-mile delivery. This hub was envisioned as a way to reduce delivery traffic in residential areas while also teaching residents about the necessity of sustainable logistics operations and minimizing the kilometers traveled by LSPs through optimized bike routes.

To achieve their urban objectives, local authorities in all three cities in Flanders combined the capabilities of the LaaS marketplace with the Policy Constraints feature of the DCT. This functionality allowed them to define specific geolocated LEZ boundaries, set time frames (days and time windows) for access restrictions, and specify the types of vehicles permitted. These combined parameters are the criteria that need to be fulfilled, to make the local shop owners and citizens gain credits when requesting transportation service. The credits appeared in the LaaS as part of the results whenever either a local shop owner or recipients (demand side) asked for urban transportation (supply side), accompanied by dynamic pricing results to promote the bundling effect. From the perspective of local authorities, they also utilized the real-time monitoring capabilities of DCT. Through its intuitive dashboard in the marketplace metrics section, they accessed detailed reports and analytics on key LaaS performance indicators and urban logistics flows. The dashboard provided both quantitative and spatial data on pick-up and delivery activity, and overall system performance. These insights helped shape data-driven policy improvements and informed stakeholder conversations about future interventions. Additionally, by visualizing tailored to their city's reports, municipalities were able to identify bottlenecks, evaluate behavioral responses to LEZ enforcement, and optimize logistics network planning based on current knowledge.

Athens LL characteristics contributed to the enhancement of the DCT features and functionalities. This case study showed a very different urban context than the Flanders LL, providing insights into how the DCT may be tailored to suit a more complex metropolitan area with a strong historical and cultural identity, resulting in seasonal sustainability challenges. Athens LL was established through a collaboration between representatives from the Municipality of Athens and multiple LSPs, who worked together with the aim of testing an innovative and collaboratively last-mile delivery model, designed to reduce driven kilometers and congestion, particularly in the area near the historical city center of Athens.

The core objective of the Athens LL was to optimize urban freight distribution while aligning logistics operations with broader urban mobility and environmental policy goals. A significant challenge in the Athens context is the high density of commercial deliveries in the central urban area, which experiences chronic traffic congestion due to limited road space and increased sensitivity stemming from its archaeological and touristic importance, compounded by inadequate urbanization. To that extent, the Municipality of Athens appointed a specific public space serving as an urban consolidation hub, strategically located near to the city center to intercept freight flows prior to their entry into the inner city. This hub served as a central node in the GREEN-LOG experimental framework, enabling the LSPs to reorganize their delivery strategies through shipment consolidation and the reduction in vehicle entries into specific zones by combining their routes into a common one.

Athens LL also advanced the Policy Constraints DCT feature, extending its application to interventions in LSP operations. The municipality established vehicle restrictions in specific areas, particularly around schools, during pre-defined time windows. This intervention was designed not only to reduce emissions and congestion during critical periods of the day but also to improve safety for children and residents. The DCT facilitated this by enabling the municipality to create policy constraints, pertaining to time, location, and vehicle type, directly integrated into the routing algorithms developed under the GREEN-LOG initiative, which was utilized and assessed by the participating LSPs. This allowed LSPs to evaluate the route produced by their planning system against the GREEN-LOG routing component, provide their feedback on route feasibility, and modify their schedule for demonstration purposes in light of policy changes. Moreover, LSPs placed significant emphasis on the measurement of the impact of these new operating procedures on their service levels and customer satisfaction (Fig. 6).

During the Athens demonstration, the BYOD app was used to facilitate mobile access, enabling dynamic routing and operational control, thereby enhancing visibility of shipments through real-time tracking. The driver serving both LSPs utilized a personal smartphone to access information generated by the DCT, receive real-time updates, and confirm the completion of delivery tasks. The interplay of

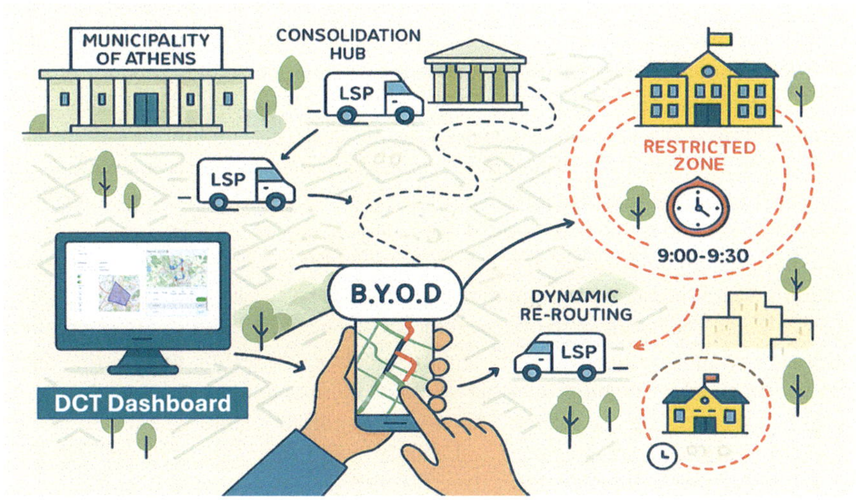

Fig. 6 Athens case study

BYOD with policy constraints and dynamic routing enhanced adaptability and responsiveness in daily operations by enabling the system to continuously recalculate optimal delivery routes. DCT's real-time monitoring view enabled local authorities to supervise freight activities within the city area and allowed LSPs to obtain insights for their collaborative last-mile deliveries. This capability for real-time visibility represented a notable improvement in the municipality's capacity to interact with the goods sector in a proactive and transparent manner.

Similar to Athens, Barcelona is a significant site that provided valuable insights for improving DCT's functionality. The principal stakeholders engaged in this living lab were aligned with Athens living lab objectives. In this case, the demonstration was conducted by a sole LSP focusing solely on last-mile delivery. LL's co-creation and demonstration aimed to tackle the intricacies of Barcelona's densely populated metropolitan area, where the accessibility to certain areas poses considerable logistical obstacles for traditional delivery vehicles, including cargo bikes.

Considering these transport policies, the Municipality of Barcelona, in coordination with the rail authority and the local LSP, commenced a thorough evaluation of multimodal delivery frameworks to be tested as part of the GREEN-LOG project. An essential innovation devised for evaluation in this living lab was the combination of train-based freight transit with autonomous delivery robots, intended to convey cargo in regions where the logistics service provider could not operate. The robots were put onto rail wagons and positioned at strategic transfer locations beyond the metropolitan core, allowing LSP to evaluate the delivery process outside of their operations territory in order to decrease transportation routes and costs ultimately.

Concurrently, the Train Authority examined the execution of a hybrid delivery model in specified zones and train timetables in collaboration with the LSPs to assess the effects of this strategy on both Train and LSP's SLAs, as well as on the residents of Barcelona. In parallel, LSP designed and tested the model, which involved pairing cargo bikes with on-foot pedestrian routes, creating an adaptable and environmentally sustainable delivery framework capable of operating efficiently in compact, high-traffic, and restricted areas. This hybridization of delivery methods allowed the LSP to trial various combinations of mobility tools in real-world conditions, adjusting strategies according to zone-specific requirements.

To facilitate these experimental configurations, the DCT's routing algorithms were enhanced and expanded to meet Barcelona's special delivery requirements. These encompassed geospatial routing customized for restricted zones, integration of multimodal schedules for alignment with train arrivals, and route optimization for bicycle and pedestrian delivery combinations. The DCT platform, in interaction with the previously outlined BYOD features, included real-time monitoring capabilities that enabled ongoing tracking of both the autonomous delivery robots in transit and the LSP's ground fleet. This dual-monitoring feature guaranteed a high degree of operational transparency and enabled swift reactions to disturbances or inefficiencies.

3.3 Analysis/Impact Assessment

The adoption of the GREEN-LOG last-mile DCT has the potential to generate significant impact across multiple dimensions, i.e., operational efficiency, environmental sustainability, economic performance, and customer experience. The integration of such advanced systems into the last-mile logistics creates substantial added value not only for the overall urban freight transport system but also for individual stakeholders within the city logistics ecosystem, including LSPs, city authorities, and end consumers.

At the operational level, the last mile DCT provides comprehensive operational coordination and real-time situational awareness. This is achieved through the consolidation and visualization of all relevant data from diverse sources (such as fleet type, order status, and historical and real-time data) into a unified and user-friendly dashboard. The integration of sophisticated machine learning algorithms within DCT allows the system to support advanced decision-making processes and any kind of risk prevention, avoiding delays and increasing performance. For instance, the system can enable proactive rerouting and dynamic resource reallocation in the event of forecasted demand increases or disruptions such as traffic congestion or vehicle breakdowns. In addition, the automation that is achieved through the digitalization of traditional and manual processes can significantly lead to human error reduction, thereby leading to throughput enhancement.

The primary objective of routing optimization algorithms, as an integral part of DCT, is to minimize the total distance covered by the delivery vehicles, which directly translates into lower fuel consumption and a corresponding reduction in greenhouse gas emissions. Preliminary simulation results from the pilot demonstration in Athens LL indicate, among others, a significant reduction in key performance metrics. More specifically, the simulated scenarios during the first round of demonstrations showed a 35% decrease in kilometers driven, a 26% reduction in travel time, and a 12% decrease in the number of stops, contributing to a substantial reduction in the overall carbon footprint. Furthermore, the enhanced coordination enabled by the DCT significantly reduces failed delivery attempts, thereby avoiding unnecessary repeat trips and associated emissions. In addition, the system empowers LSPs to make informed decisions, such as prioritizing the vehicles they prefer and consolidating deliveries where possible, further minimizing environmental impact. Considering all the above, these advancements contribute to the broader improvements in urban quality of life, as reduced traffic congestion and lower noise pollution make city environments more livable.

It is evident that the aforementioned interventions contribute to additional benefits that increase the value of the DCT. As mentioned earlier, improved fleet utilization and the reduction in total travel distance result in lower fuel consumption and thus bringing economic benefits for LSPs. Optimized scheduling supports greater operational efficiency and higher last-mile delivery service provision. Moreover, the automation and digitalization of various processes across the logistics chain foster the overall improvement of the services provided. These factors enable LSPs to gain competitive advantages in increasingly demanding urban logistics markets.

Last but not least, the availability of accurate, real-time tracking information and proactive communication among stakeholders enhance their experience. The flexibility offered by the system enables LSPs to select more accurate delivery time windows and access adaptable options such as rescheduling or alternative pickup points. This enhances their overall satisfaction as users. In this line, reliability and customers' trust are strengthened which is an important aspect for the competitiveness. It is worth noting that, by the end of the GREEN-LOG project (June 2026), it is expected that the GREEN-LOG DCT will have been positively evaluated in terms of usability and ability to support decisions by 90% of the users.

4 Conclusion and Recommendations

The integration of DCT and associated advanced technologies into urban logistics presents great potential to significantly improve operational efficiency, reliability, and sustainability in the last-mile delivery ecosystems. The GREEN-LOG DCT represents an integrated and scalable solution for enhancing situational awareness, operational coordination, and decision-making among various stakeholders.

The platform integrates advanced technologies such as real-time data analytics, predictive models, and the LaaS marketplace metrics, as well as it enhances operational efficiency while encouraging environmentally responsible practices. These capabilities allow the system to successfully address key challenges in urban logistics. It streamlines complex operations, optimizes routes, and facilitates dynamic resource allocation, resulting in improved fleet utilization, reduced delivery times, and fewer failed attempts. In the end, this leads to enhanced service reliability and customer satisfaction.

Besides the operational advantages, the environmental benefits that arise from the optimizing routes and consolidating deliveries align with broader urban sustainability goals, helping cities reduce congestion, noise pollution, and carbon footprints. The platform fosters enhanced collaboration and transparency among logistics service providers, city authorities, and customers. The integration of the LaaS (Logistics-as-a-Service) marketplace metrics further facilitates data-driven decision-making and incentivizes sustainable practices.

Future research should focus on expanding AI applications across the entire logistics chain. Equally important is the need to strengthen stakeholder engagement through well-designed incentive mechanisms that encourage active participation and long-term collaboration. Seamless data sharing and system integration should be facilitated by promoting open standards and interoperable architecture across platforms and systems. Attention should also be paid to refining the user experience, for both technical operators and end consumers, ensuring intuitive interfaces, real-time transparency, and flexible and responsive service options. The successful adoption of these technologies will depend not only on technical innovation but also on inclusive policies, cross-sector partnerships, and a shared commitment and dedication to creating livable, resilient cities for the future.

Acknowledgement This work has been supported by the GREEN-LOG project that has received funding from the European Union's Horizon Europe research and innovation programme under Grant Agreement No. 101069892 and from the United Kingdom Research and Innovation (UKRI) agency. Views and opinions expressed are, however, those of the author(s) only and do not necessarily reflect those of the European Union, the European Climate, Infrastructure and Environment Executive Agency (CINEA) or the UKRI. Neither the European Union nor the granting authorities can be held responsible for them.

References

Adetiloye T (2012) Collaboration planning of stakeholders for Sustainable City logistics operations. arXiv. Computers and Society

Bosona T (2020) Urban freight last mile logistics—challenges and opportunities to improve sustainability: a literature review. Sustainability 12(21):8769

Bruni ME, Fadda E, Fedorov SA, Perboli G (2023) A machine learning optimization approach for last-mile delivery and third-party logistics. Comput Oper Res. https://doi.org/10.1016/j.cor.2023.106262

Demir E, Syntetos A, Van Woensel T (2022) Last mile logistics: research trends and needs. IMA J Manag Math 33(4):549–561

GREEN-LOG (No:101069892) has received funding from the European Union's Horizon Europe research and innovation programme and from the United Kingdom Research and Innovation (UKRI) agency

Ippolito N, Mistrice LMB, Gentile G (2023) Last-mile-logistic platform for companies within urban areas. In: 2023 8th international conference on models and Technologies for Intelligent Transportation Systems (MT-ITS), pp 1–6. https://doi.org/10.1109/MT-ITS56129.2023.10241672

Nathanail E, Papoutsis K (2013) Towards a sustainable urban freight transport and urban distribution. J Traffic Logist Eng. https://doi.org/10.12720/JTLE.1.1.58-63

Pourmohammad-Zia N, Mark VK (2023) Urban logistics in Amsterdam: a modal shift from roadways to waterway. arXiv. https://doi.org/10.48550/arxiv.2309.00345

Sochor J, Fenton P, De Bode A, De Azpiazu MC, Borraz RL, Gasparin P (2023) HALLO: hubs for last mile delivery solutions in Barcelona and Stockholm. Transp. Res. Proc. 72:3300–3307. https://doi.org/10.1016/j.trpro.2023.11.797

Wanganoo L, Patil A (2020) Preparing for the smart cities: IoT enabled last-mile delivery. In: Advances in Science and Engineering Technology International Conferences (ASET), pp 1–6. https://doi.org/10.1109/ASET48392.2020.9118197

Deep Learning Demand Prediction for Sustainable Logistics as a Service (LaaS)

Panagiotis Georgakis, Matthew Hocking, and Guanbo Jia

1 Challenges and Objectives

The world is experiencing an unparalleled growth in last-mile transport following the unprecedented growth of e-commerce. During the past decade, e-commerce has risen at a faster-than-ever rate, with sales ratios being tripled from 2014 to 2020, while they are expected to grow to $5.4 trillion in 2022. The steep curve is expected to continue in the years to come, fuelled by a multitude of factors related to societal changes and technological advancements.

Urbanization is the new normal (according to the United Nations, by 2030, 60% of the world population will live in urban areas), and Eurostat statistics indicate that a larger share of individuals uses e-commerce in more densely populated and urban areas.

The purchasing power of the middle class is increasing, resulting in constant growth of online purchases. A total of 2.1 billion people are expected to buy goods online in 2021, and the share of online retail is expected to rise to 20% by 2023. A widening range of products can be purchased online, including non-perishable goods such as groceries and food and medicines. E-grocery alone grows 10% per annum worldwide. New technological developments in the delivery segments allow for instant and time-definite delivery, improving the e-commerce user experience and trust. Same-day delivery experiences a growth of 20–40% and instant delivery 10% per annum.

Delivering multiple-drop parcel routes in urban centres is increasing pollution and congestion. The use of sustainable delivery methods for last-mile delivery is

P. Georgakis · M. Hocking (✉) · G. Jia
The School of Architecture and the Built Environment (SOABE), The Faculty of Science and Engineering (FoSE), The University of Wolverhampton, Wolverhampton, UK
e-mail: M.Hocking2@wlv.ac.uk

A. Melkonyan-Gottschalk, Y. Li (eds.), *Urban Logistics Transformation*, Greening of Industry Networks Studies 15,
https://doi.org/10.1007/978-3-032-08387-6_4

increasingly being specified by regional infrastructure owners and urban residents. The work described here is part of the GreenLog project, funded by European Horizon 2020.

GreenLog accelerates systemic changes in last-mile delivery ecosystems for economically, environmentally and socially sustainable city logistics. The project establishes city platforms comprising inclusive stakeholder urban living labs for nurturing social innovation, designing and deploying innovative delivery solutions while allowing the most effective exchange of ideas, the development of robust, harmonized regulatory and policy frameworks, and cooperative business models that build upon effective public/private-sector collaboration and joint investments. The GreenLog approach provides an innovative simulation environment for scenario building, combining different solutions that allow the integration of last-mile delivery interventions with the highest possible impact on environmental sustainability and traffic reduction, while considering their financial viability. On the operational level, GreenLog provides cargo-bike-based innovations for sustainable micro-consolidation design and deployments, multi-modal parcel deliveries integrating public transportation, Logistic as a Service platforms for interconnected city logistics and automated delivery concepts with the use of autonomous vehicles and delivery droids. The solutions are supported by interconnected city logistics dataspaces that feed dynamic services for proactive optimization of the ecosystem, respecting the interests of consumers, businesses and city stakeholders. The approach is deployed and validated in five cities and areas: Athens, Barcelona, Flanders, Oxfordshire and Ispra. Special focus is given on the transferability of the proposed innovation, starting from three follower cities with high interest in tailoring and replicating the GreenLog solutions, and continuing with a pan-European outreach with intensive transferability acceleration actions, setting the ground for the establishment of multiple city logistics urban living labs.

To ensure the most efficient routing of last-mile delivery vehicles, the likely volume of parcels to a given delivery area must be predicted from previous delivery volumes. Most accurate prediction methods for time-series, stochastic data involve deep learning techniques. The analysis of historical parcel delivery data by the logistics companies studied leads to prediction models described here.

The data studied relate to a single logistics company, Pedal & Post[1], which operates in the City of Oxford, UK. The delivery areas modelled are based on UK postcode areas,[2] which individually hold around 10–40 postal addresses. UK postcodes are in a format similar to 'OX1 1AE'. In this study, the postcode areas modelled are restricted to the range 'OX1 xxx' to 'OX4 xxx'. This corresponds to an area across the centre of Oxford, as will be seen in the visualization of results.

[1] www.pedalandpost.co.uk

[2] en.wikipedia.org/wiki/Postcodes_in_the_United_Kingdom

2 Current Practices in Deep Learning Models

There are various AI methods used in demand prediction techniques in the literature. Deep learning shows better performance and results in terms of forecasting compared to other methods. Although the forecasting based only on the historical data of manufacturing demand is achievable, the accuracy of the prediction results is considerably lower than when considering multiple factors. Researchers set up strategies of the exponential forecasting framework for sales forecasting to optimize manufacturing planning and inventory. Therefore, our aim is to forecast future demands based on historical data coming from past manufacturer orders and retailers' demands based on customers' requirements, just enough time in advance to compensate for manufacturer lead times. Several ML and DL methods are used to develop the accuracy of demand forecasting. Abbasimehr et al. used a multi-layer LSTM method to compare with other time series forecasting methods, such as K-nearest neighbours (KNN), exponential smoothing (ETS), auto regressive integrated moving average (ARIMA), support vector machines (SVM), recurrent neural network (RNN), artificial neural network (ANN) and long-short term memory (LSTM). The results of this study show that the LSTM method is more efficient compared to the tested methods with regard to performance measures (Abbasimehr et al. 2020). Jiang and Dong (2024) used an LSTM and a grey model (GM) to predict the future values based on historical data and the total industrial value. The Statistical Yearbook of GD (2005–2020) was selected to represent the demand of the manufacturing industry.

Other indicators are from this statistical yearbook from 2005 to 2020. The experiment shows that LSTM has excellent results in the previous comparative experiments. The GM model is the classic model in the field of Auto-regression. Nevertheless, it does not give accurate results, which is significantly lower than the forecasting results given by LSTM considering multiple factors (Dou et al. 2021). A study conducted by Raizada et al. is based on a comparative analysis of several supervised machine learning algorithms, such as support vector machine (SVM), random forest regression, K-NN algorithm and extra tree regression to build a forecasting model for future sales of 45 retail outlets of Walmart store in India.

The study shows that the extra tree regression technique is the most efficient model to predict the sales for the selected dataset; however, the predictions obtained from the algorithm may vary based on the variance in training data (Abbasimehr et al. 2020; Etemad-Shahidi and Mahjoobi 2009; Malekmohamadi et al. 2011). Wang et al. compared the performance of classical forecasting models and the latest developing forecasting technologies for perishable products and non-perishable items of a large grocery retailer. The authors made a comparison in terms of performance and accuracy between many algorithms: ARIMA, SVM, RNN and LSTM (Wang et al. 2019; Berbiu et al. 2017). The study shows that SVM, RNN and LSTM have a high predictive performance for perishable products, whereas ARIMA is outstanding in the runtime and LSTM is the most efficient method to deal with non-perishable items due to its advanced prediction performance (Wang et al. 2019; Adamowski and Chan 2011).

3 Model Development

3.1 *Methodological Approach*

According to the research study, the convolutional neural network (CNN) and the long short-term memory neural network (LSTM) are considered very efficient deep learning methods for demand forecasting to warrant a high accuracy level in the real-world environment. Therefore, here we combine the convolutional neural network (CNN) with the long short-term memory network (LSTM) to form a deep learning-based prediction model, CNN-LSTM, in order to conduct the logistics demand prediction efficiently and accurately.

3.2 *Convolutional Neural Network (CNN)*

With the development of the deep neural network, the convolutional neural network has been proposed and is currently one of the most famous algorithms. CNN is a kind of deep neural network. It is more complex in structure and typically consists of an input layer, a convolutional layer, a pooling layer, a fully connected layer, and an output layer, etc. It has successfully been applied in various fields such as detection and segmentation. CNN provided outstanding performance than the previous traditional machine learning algorithms in the above-mentioned fields. It mainly includes several convolutional layers, pooling layers, and fully connected layers. A convolutional neural network (CNN) is a regularized type of feed-forward neural network that learns feature engineering by itself via filters (or kernel) optimization. Vanishing gradients and exploding gradients, seen during back-propagation in earlier neural networks, are prevented by using regularized weights over fewer connections. The details of each component in CNN are introduced as follows. The function of the convolution layer is to collect the features of input data, which contains several convolution kernels. Each cell of the convolution kernel corresponds to a bias vector and a weight coefficient, similar to a neuron of the feedforward neural network. Each neuron in the convolutional layer is related to multiple neurons in the area close to the previous layer. The size of the area depends on the size of the convolution kernel, which is called 'receptive field' in the literature. Its meaning can be compared to the receptive field of visual cortical cells. During the convolution kernel is working, the input features will be scanned regularly, and the input features will be summed and multiplied by the matrix elements in the receptive field, and the deviations are superimposed. After feature extraction in the convolutional layer, the output feature map is going to be passed to the pooling layer for information filtering and feature selection. The pooling layer contains preset pooling functions whose function is to replace the result of a single point with the statistics of the feature map of its neighbouring area in the feature map. The selection of the pooling area in the pooling layer is not different from the steps of the convolution kernel scanning feature map, controlled by the pooling size, step size, and filling. In the traditional feedforward neural network, the fully connected layer in CNN is equal to the hidden layer. The fully connected layer is located in the last part of the hidden layer of CNN and only transmits signals to other fully connected layers. The featured graph loses the spatial topology in the fully connected layer and is expanded into a vector and passes the excitation function. The convolutional layer and pooling layer in the CNN can perform feature extraction on the input data. The role of the fully connected layer is to nonlinearly combine the extracted features to obtain an output. Fully connected layers play the role of 'classifier' in the entire convolutional neural network. The convolutional layer, pooling layer, and activation function layer map the original data to the hidden layer feature space. The fully connected layer plays the role of mapping the learned 'distributed feature representation' to the sample label space. The fully connected layer isn't expected to have features extracting ability but trying to use the existing high-order features to complete the learning goal.

3.3 Long-Short-Term Memory (LSTM)

LSTM is devised to solve the long-term dependence problem of a general RNN. It has been successfully used in various fields such as machine translation, speech recognition, image description generation, video tagging and financial time series. All RNN has a chain form of the repetitive neural network module. It mainly includes the forgetting gate, input gate, and output gate. The LSTM network can effectively mine the temporal dependence of information in the time series by adding the forget gate, input gate, and output gate to the hidden layer. At each moment, the LSTM unit receives the current data through three gates input and the last hidden state as well as the memory unit state. Then the calculation process of LSTM is as follows. The forget gate helps LSTM decide what information will be stored in the memory unit. LSTM uses the input gate to decide what data will be stored in the new unit.

3.4 The Hybrid CNN-LSTM Model

The innovation of the CNN-LSTM prediction model is that it is a deep learning model structure commonly used to process sequence data, especially for sequence data with spatiotemporal features. This prediction model combines two different types of artificial neural network layers to fully capture spatial and temporal features in the dataset and has the following structure as follows: (1) The convolutional neural network (CNN) is used to capture spatial correlations in sequential data. It slides over the input data via a convolutional kernel to extract features from local regions, which helps detect local patterns and features in the input sequence. A pooling layer is usually added after the CNN layer to reduce the spatial dimension of the data to reduce the computational complexity. (2) Long short-term memory neural network (LSTM) is a variant of a recurrent neural network (RNN), which is specifically used to process the modelling of sequence data. LSTM uses gating mechanisms (input gate, forget gate and output gate) to control forgetting and remembering of past information, so as to effectively deal with long sequence dependencies. In time series prediction, LSTM can be used to learn long-term dependencies in the sequence data and capture the time evolution pattern in the sequence.

3.5 *The Applications of the Hybrid Model*

The hybrid CNN-LSTM model combines the convolutional neural networks (CNN) and long short-term memory networks (LSTM), in which CNN has excellent performance at extracting spatial features, while LSTM has outstanding performance at capturing temporal features in real-life data. The hybrid model has illustrated an extraordinary performance for the analysis and prediction of the dataset with both spatial and temporal features. The hybrid CNN-LSTM model has various real-world applications, especially in handling the spatio-temporal data analysis and predication, including the supply chain demand prediction, the financial stock prediction, weather prediction, air pollution prediction, urban flood risk prediction, electric energy consumption prediction and traffic flow prediction. The applications of the hybrid CNN-LSTM model are described later.

Since the supply chain demand forecasting (Na 2024) is an important factor driving the entire supply chain, and low error rates in forecasting are a common goal pursued by the industry, in the supply chain system which regards suppliers, producers, and consumers as an organic whole, and coordinates the information flow, logistics, and capital flow of all members in the overall operation of cross organization, the CNN-LSTM model is used for supply chain demand prediction, in which the input data is processed by CNN to obtain relevant features while LSTM is employed to learn and predict the extracted feature data ultimately yielding the final prediction results to predict the supply chain demand in the future. Investment wealth management (Zhao 2024) referring to the use of funds by investors to arrange funds reasonably including savings, bank financial products, bonds, stocks, commodity spots, real estate, gold and art has become a mainstream of the contemporary era in modern society. The CNN-LSTM model is applied to achieve a accurate of the stock price prediction, and the model constructs a sequence array of historical data and its leading indicators (options and futures), and uses the array as the input image of the CNN framework, and extracts certain feature vectors through the convolutional layer and the layer of pooling, and as the input vector of LSTM for the stock prediction. For the global climate change, the accurate weather forecasting (Devapogu et al. 2025) is increasingly crucial for sectors such as agriculture, energy management, and environmental protection, thus the CNN-LSTM model is good at predicting future weather conditions such as temperature, humidity, and wind speed, in which the weather dataset covers several years with features including date, temperature, wind speed, humidity, and precipitation, and in the model the CNN reduces the dimensionality of the time-series data, while LSTM captures the long-term memory of the massive temperature time-series data to predict the accurate weather. In the modern society the air pollution (Liu and Hu 2025) is a serious problem that affects economic development and people health, an efficient and accurate air quality prediction model would help to manage the air pollution problem, and the CNN-LSTM model is used for air quality prediction in air pollutant data based on various inputs like historical data, meteorological conditions, and even images, from the dataset which includes concentration of nitrogen dioxide (NO_2), carbon monoxide (CO), and sulfur dioxide (SO_2) by integrating spatiotemporal features such as PM2.5 data from adjacent stations and meteorological variables achieving superior performance in hourly PM2.5 prediction for the accurate air quality prediction in the future. Nowadays the flooding remains one of the most devastating and costly natural disasters and the rapid prediction of urban flooding is an important measure to reduce the risk of flooding and to protect people property (Chen et al. 2023), and the CNN-LSTM model is applied for the flood risk prediction in order to allow more time and resources to be spent on transporting materials to assist high flood-risk regions before a flood occurs in the global warming scenario. Generally speaking, the model use the CNN model providing promising output in satellite images processing and classification of regions of known high and low flood risk worldwide to predict the flood risk of an area based on its satellite image, and the LSTM model providing promising output in time-series, series, and continuous data with memory and time management for efficient flood prediction for the global satellite dataset. Innovations in technologies that rely on electricity (Abraham et al. 2022) have led to an uncontrollable rise in power usage, and the analysis and forecasting of energy consumption systems are necessary in order to predict future electricity demand and enhance the power distribution system in intelligent power management system and the CNN-LSTM model is used for the electric energy consumption prediction in order to play a significant role in drawing up a national energy development policy, and in the model the CNN in the first module extracts the important information from several variables in the individual household electric power consumption dataset and the LSTM layer uses the above-mentioned information as well as the trends of time series including the forward and backward states to make predictions, and the obtained values in the LSTM module will be passed to the last module that consists of two fully connected layers for predicting the electric energy consumption in the future. Considering that in the intelligent transportation systems (Toncharoen and Piantanakulchai 2018) the urban short-term traffic flow forecasting becomes increasingly important with the continuous advancement of smart city construction, the CNN-LSTM model is used for urban short-term traffic flow prediction, and the model integrates CNN and LSTM into an end-to-end network framework and the convolutional neural network is used to capture the local spatial characteristics of traffic flow while the long short-term memory-cycle neural network is used to capture temporal characteristics of traffic flow data, and then the output results of the two networks are combined by the corresponding weights to obtain the predicted results through the trajectory data and finally the traffic flow prediction values of the urban areas are obtained by fusing with external factors to predict the traffic flow in the future.

4 Case Study Using Deep Learning for Demand Prediction in Oxford, UK

Stakeholders in the GreenLog project are Oxford City Council and the last-mile delivery company Pedal & Post. Pedal & Post receives daily instructions from national logistics partners, including EVRI and DHL. These daily lists are split into geographical areas which correspond closely to the routes taken by Pedal & Posts' individual e-bike couriers. Before this data can be used to train the DL algorithm, it is aggregated so that one dataset covers the whole delivery area, then anonymized to remove any customer information covered by GDPR.

The summarized data (Fig. 1) is reduced to a count of parcels, per day, per postcode. A summary of parcel deliveries in the period 2022–04 to 2024–04 is thus calculated as shown in Fig. 2.

Considering the complex spatio-temporal features in the dataset, a deep learning-based demand prediction model called CNN-LSTM is used to efficiently process the dataset. This prediction model employs two different types of artificial neural network layers to fully capture spatial and temporal features in the dataset. Firstly, it uses a convolutional neural network to capture spatial correlations in the dataset and slides over the input data via the convolutional kernel to extract features from local regions, which helps detect local patterns and features in the dataset. A pooling layer is usually added after the layer to reduce the spatial dimension of the data to reduce the computational complexity.

After that, it uses the long short-term memory neural network to learn long-term dependencies in the sequence data and capture the time evolution pattern in the dataset and the long short-term memory neural network is a variant of recurrent neural network and is specifically used to process modelling of sequence data since it uses gating mechanisms including input gate, forget gate and output gate to control forgetting and remembering of past information to effectively deal with long sequence dependencies in the dataset.

The prediction model is used to process spatial and temporal information in sequence data. The general procedure of the model to predict spatial-temporal series data is as follows. The first stage is to prepare the input data. In general, sequence

```
{
        "date": "2024-08-01",
        "postcode": "E14 9QG",
        "deliveries": 1
}
```

Fig. 1 json format of aggregated parcel data

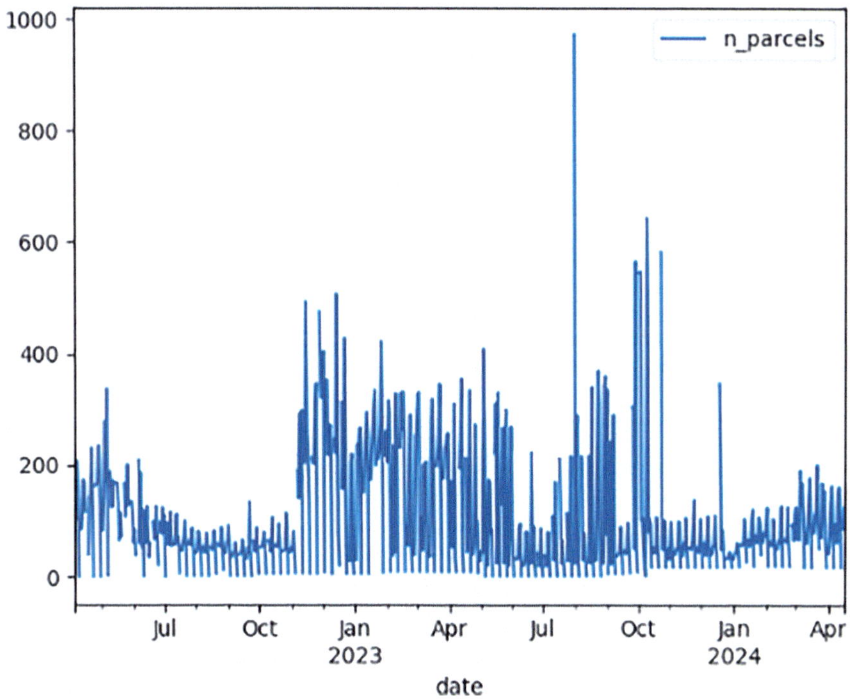

Fig. 2 Parcel delivery count April 2022 to April 2024

data can be represented as tensors in three dimensions, where the first dimension represents the number of samples, the second dimension represents time steps, and the third dimension represents the number of features. Then, sequence data is taken as input, and feature extraction is carried out through one or more convolution layers.

The convolution layer can extract the spatial features of the input data. And the pooling layer is added after the convolution layer to reduce the dimension of the feature map and retain important feature information. The pooling layer can be downsampled in spatial dimensions, such as maximum pooling or average pooling, and the pooled feature map is converted into a one-dimensional vector and input into the layer. In this way, it can use the spatial features extracted by the convolution layer and combine the timing information of the sequence data for modelling. In the layer, multiple units can be set up to model timing information, capture long-term dependencies in sequence data and learn timing patterns in the data.

After that, one or more fully connected layers can be added to output the final predicted results. The appropriate activation function and loss function can be selected according to the specific task requirements. It uses labelled data to train the model and perform hyper-parameter tuning, such as learning rate, batch size, and network structure. The model can be optimized using conventional optimization algorithms. The trained model is evaluated using validation sets or test sets, and various evaluation metrics can be used to evaluate the performance of the model.

5 Analysis

The statistical analysis of the number of parcels on different dates in the dataset is performed: a histogram of the number of parcels by dates is plotted in Figs. 3, 4, 5 and 6, and a geospatial data example of Post and Pedal deliveries is visualized in Fig. 7.

The first experiment carried out to explore daily patterns in the dataset was to cluster days in 10 patterns using clustering. Figure 5 shows the results of the experiment per day, and Fig. 4 shows the results per weekday.

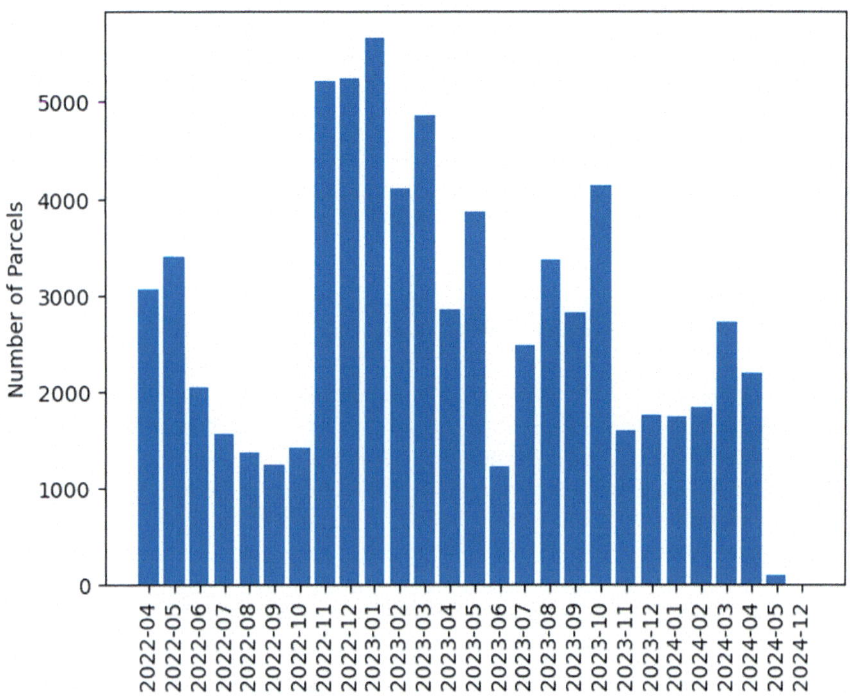

Fig. 3 Number of parcels 2022–2024

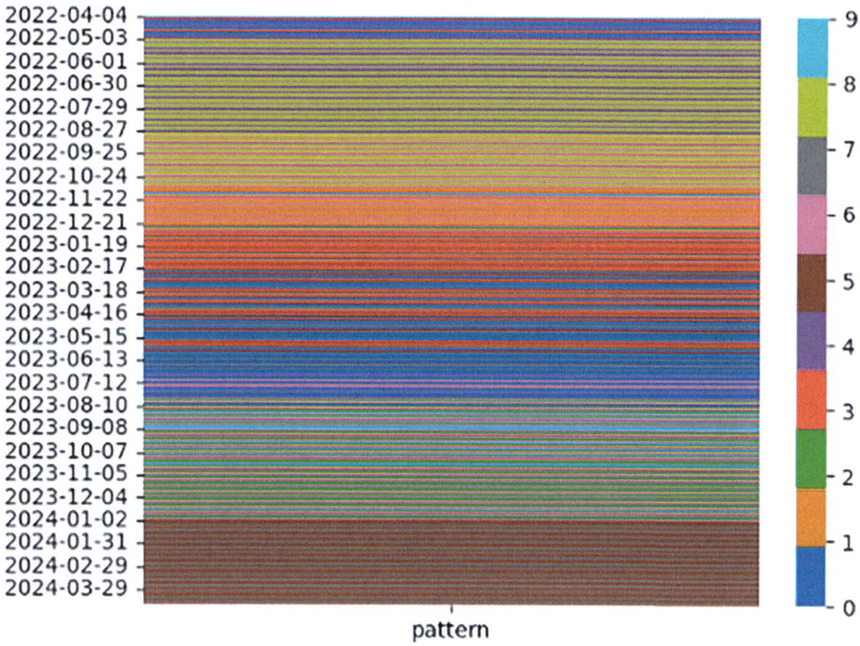

Fig. 4 Daily patterns using clustering

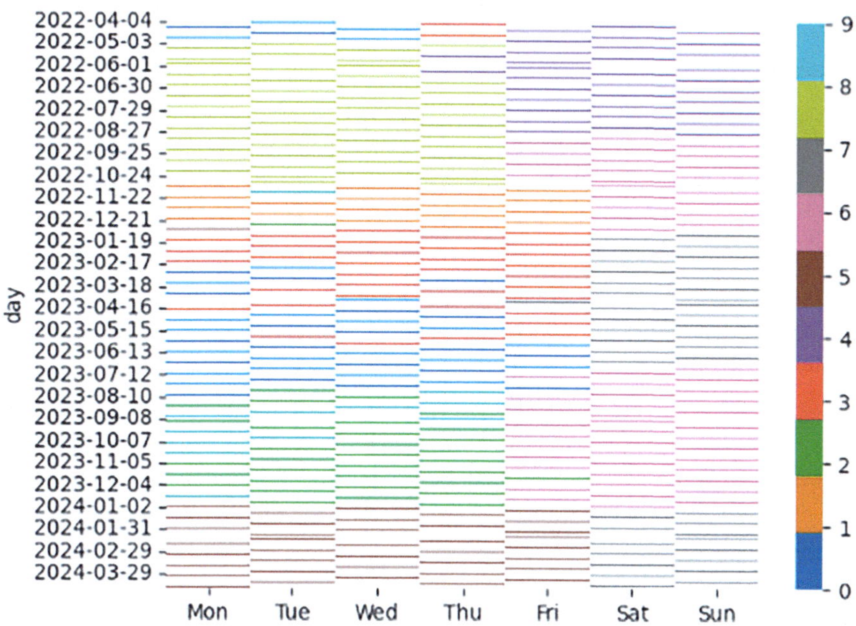

Fig. 5 Weekly patterns using clustering

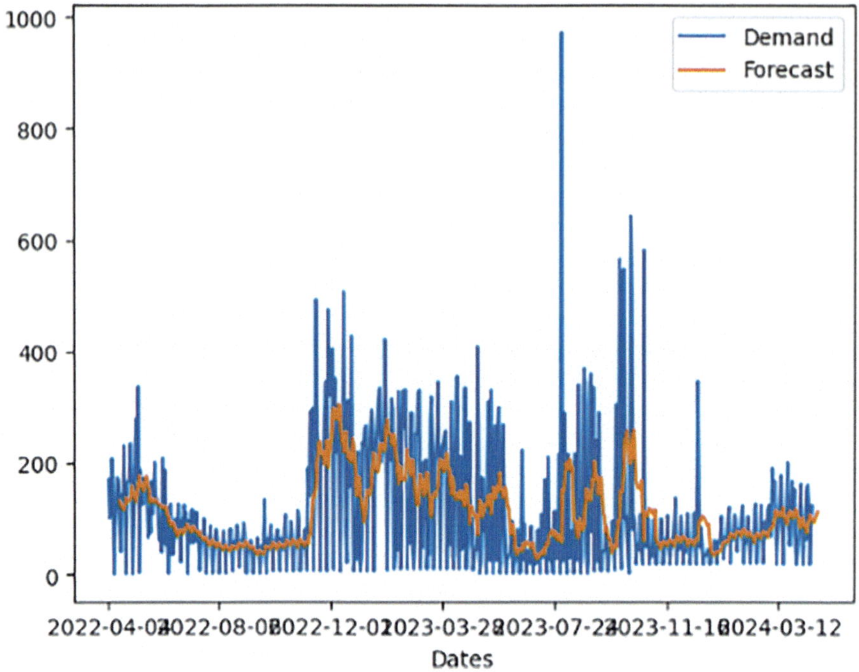

Fig. 6 Prediction results with a simple moving average

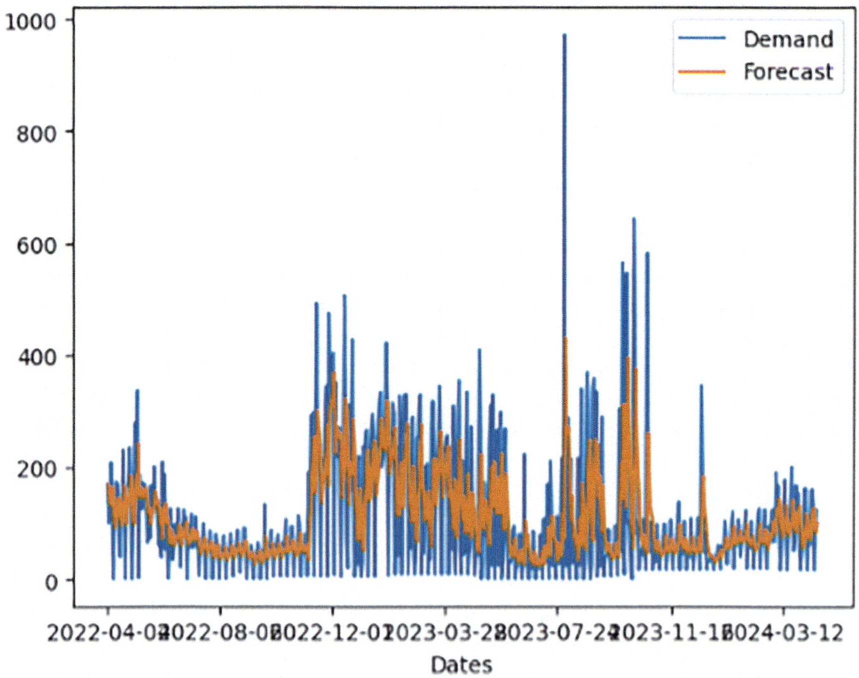

Fig. 7 Prediction results with exponential smoothing

Similar patterns on Fridays and weekends are observed versus Monday to Thursday, especially in 2022, and from July to December 2023.

Next, moving averages with and without exponential smoothing were explored. Moving averages smooth time series data by removing random variations. As equal weights are assigned to all historical data points, this technique is usually preferred when trends and seasonality are absent, which seemed to be the case in P&P data. As this technique neglects actual ups and downs in data, results need to be carefully interpreted and customized to the actual reality of the problem.

Figure 5 shows the output of a simple moving average using the previous 10 days to each forecasting horizon of 5 days. The plot compares real data in blue to the forecast in orange, which is conservative as expected and does not fit well to days with very high or low demand.

Figure 6 shows the output of exponential smoothing using an alpha of 0.4, which seems to be a promising method due to the high variability of the data. It should be noted though that high peaks are not anticipated using this method.

Two machine learning techniques have been tested as the most popular regression methods for time-series forecasting: random forest and XGBoost regressors. The dataset has been split into train and test using 01-03-2024 as a turning point. Both algorithms use the past 10 days as features for the forecast and predict 5-time horizons. Figure 8 shows the compared output of these two methods to the truth data for 1–5 horizons for each timestamp.

Finally, we have tested a deep learning (DL) framework even though the amount of data is quite limited for DL methodologies. We have used a long short-term memory (LSTM) architecture with 36 units and Tahn activation. Figure 9 shows the output of the forecast.

As DL architecture are data-hungry algorithms, we decided not to split the dataset. To have a more accurate interpretation of these results, and as the project lasts long enough to be able to collect more data, we expect that in the future once more data are available they can be used to analyze the accuracy of all these methods.

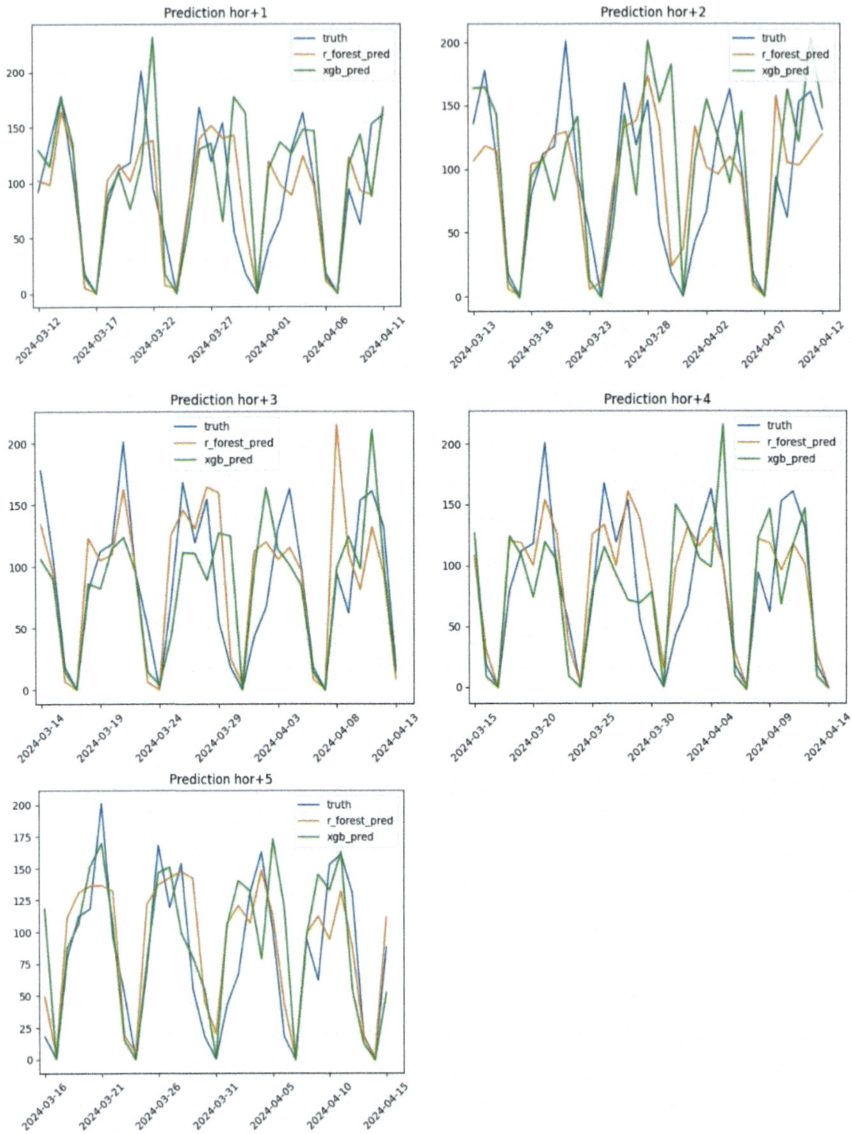

Fig. 8 Comparison of results using random forest and XGboost regressors at 5 time horizons

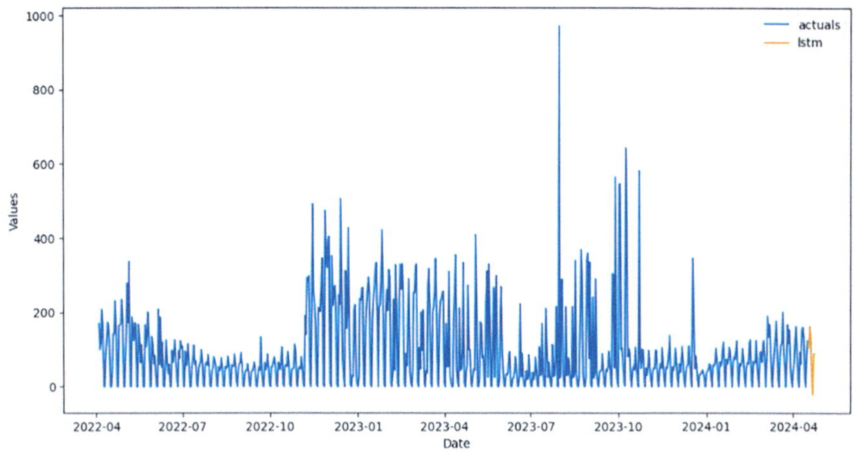

Fig. 9 LSTM prediction results

6 Conclusions and Future Work

In our work, we have designed and studied some logistics demand datasets with real-life characteristics, and then tested different efficient demand models and analyzed their performance on our designed datasets, based on our best understanding of our project. We also had a detailed summary of our main inquiries about the current and desired states of demand prediction within the project, and then conducted an in-depth discussion with our logistics partner Pedal & Post, our project team, and other partners.

A seven-day forecast is generated from historical delivery data. A single day's forecast volume is shown in Fig. 10. Figure 11 shows a small section of Fig. 10, where the postcode boundaries and labels can also be seen. There is much further work to be demonstrated on the detailed representation of delivery volume predictions in geographical information systems. We have been in discussion with stakeholders in the GreenLog project about the forms that this GIS data is presented. These results will be part of the final living laboratory operational reports for the project, available after Q4 2026.

As with all time-series data, there is a combination of low and high frequency variance. For a prediction duration of one day, with a learning window for all data of several months or years, the high-frequency variance data has essentially appeared in the prediction output as an error. As was discussed, simple variance smoothing with a moving average or exponential fitting does remove this high-frequency variation but does not improve prediction error. Even with the machine learning algorithms discussed, it is apparent that longer duration, low-frequency variances are more accurately predicted. These long-duration fluctuation—trends—are useful for seasonal prediction over a period of weeks or months.

The more complex algorithms applied show an improvement in error reduction at the daily prediction period. It is shown that to attempt prediction of parcel delivery volumes for the following day, it is necessary to use these machine learning algorithms to reduce error to a usable level. Prediction error of a few percent is

Fig. 10 Predicted delivery volume overlaid on Oxford postcode boundary polygons

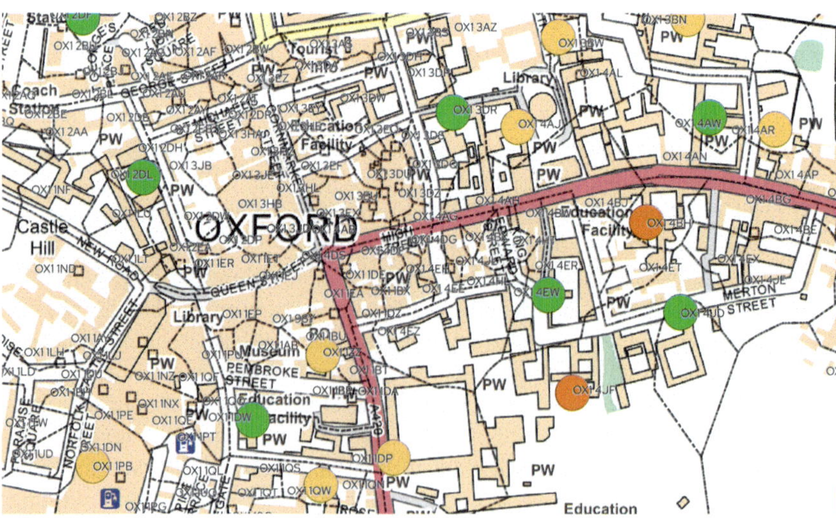

Fig. 11 Detail of central Oxford showing delivery volume predictions and postcode boundary polygons

demonstrated and is agreed with the LSPs to be both a useful predictor, and an improvement on the human judgement of delivery couriers and managers on site.

The LSP partner, Pedal & Post, has used the prediction data to plan staffing and vehicle usage. Seasonal trends, which are anecdotal knowledge within the courier team, have been confirmed by data analysis. These trends are now part of the business knowledge, which increases competitiveness within the sector and internal efficiency.

Oxford City Council has a remit to decrease traffic volumes and therefore pollution in the city centre and increase the safety and convenience of non-vehicular journeys for the population. These KPIs are being achieved by decreasing delivery traffic volumes and increasing the efficiency and availability of last-mile delivery companies using electric vehicles such as cargo bikes. The ability of these cargo bike couriers to efficiently cover large areas of Oxford is enhanced by siting multiple micro consolidation centres within, or just outside, the couriers' routes. The couriers can be supplied from multiple sites, which receive just one large delivery per day from national logistics companies. The location of these sites can now be accurately selected by using the delivery volume prediction data. As well as fixed micro consolidation centres, when an area's delivery volume is greater than the capacity of a single cargo bike courier, the data provides the option of locating temporary, mobile or consolidation sites. Where a courier may require more capacity for a delivery route than is available in one cargo bike, a second bike can meet them en route carrying more parcels. In this way, one courier can provide a greater capacity when an area has a temporary increase in volume. The volume prediction data allows short-term decisions on the location and capacity of these mobile delivery centres.

In future trials, the mobile delivery centres will be fully autonomous to further increase the efficiency of the last-mile delivery solution. Although currently autonomous vehicle cannot be used on public highways in the UK, solutions are being sought which will provide this whether through legislation or using remotely supervised semi-autonomous vehicles.

In the next step, based on our discussion with our project partners, we will further improve our logistics demand models and generated demand datasets in order to make it further closer to the real and complex environment of our project, and also carry on more investigations in order to gain better understanding about the operational details, demand patterns, forecasting needs, data limitations, and metrics, and further refine our demand prediction model effectively, and continue to maintain close communication with our project partners and obtain a further insight of both the current operational state and the aspirations post-project completion, facilitating a more targeted and efficient approach to demand prediction modelling in our logistics demand project, and we further apply our built model on real-life big dataset and test our prediction model on the real-life P&P dataset.

Moreover, we will do a detailed analysis of the P&P dataset and target the most impactful information from inside the large-scale dataset, and adjust our prediction model to further improve its performance on the real-life large-scale demand dataset.

We have run several preliminary experiments on the available data from P&P to get insights into the problem that needs to be tackled and start having an overview of different ways to provide insights, which may be useful for delivery services. First, we need to confirm that all data gathering from P&P is automated so that the implementation of a potential analytics service can be done. Then, once data providers allow communication of data feeds, a protocol needs to be established which complies with GDPR standards. Once data verification is successful, experiments need to be re-run to have a complete view and analysis of their outcome.

In this preliminary report, we have observed data inconsistencies which can be interpreted as data gaps from manual interference or merging of data sources or actual shifts in demand that could be recurrent. In case the second hypothesis applies, collecting more data from historical periods could prove this trend or reject it, and hence a decision on whether peaks should be removed needs to be taken.

With respect to statistical and machine learning methodologies, preliminary results show that traditional regression techniques such as random forest or XGBoost are adequate to tackle aggregating demand forecasting for 5-day time horizons. Moving averaged and deep learning are discarded due to their limited capabilities or demanding requirements.

References

Abbasimehr H, Shabani M, Yousefi M (2020) An optimized model using LSTM network for demand forecasting. Comput Ind Eng 143:106435

Abraham AV, Sasidharan P, Tejas SJS, Manohara M, Muthu R, Naidu RC (2022) Predicting energy consumption using LSTM and CNN deep learning algorithm. In: 2022 7th international conference on environment friendly energies and applications (EFEA), Bagatelle Moka MU, Mauritius, pp 1–6

Adamowski J, Chan HF (2011) A wavelet neural network conjunction model for groundwater level forecasting. J Hydrol 407:28–40

Berbiu J, Ocvirk E, Careviu D, Loncar G (2017) Application of neural networks and support vector machine for significant wave height prediction. Oceanologia 59:331–349

Chen J, Li Y, Zhang S (2023) Fast prediction of urban flooding water depth based on CNN–LSTM. Water 15:1397

Devapogu K, Venkatesh M, Vullam N, Babu VD, Naidu CR, Lakshmanarao A (2025) A novel approach to farm weather prediction with hybrid CNN, LSTM, and attention mechanisms. In: 2025 IEEE International Conference on Interdisciplinary Approaches in Technology and Management for Social Innovation (IATMSI), pp 1–7

Dou Z, Sun Y, Zhang Y, Wang T, Wu C, Fan S (2021) Regional manufacturing industry demand forecasting: a deep learning approach. Appl Sci 11(13):6199

Etemad-Shahidi A, Mahjoobi J (2009) Comparison between M5 model tree and neural networks for prediction of significant wave height in lake superior. Ocean Eng 36:1175–1181

Liu S, Hu Y (2025) Air quality prediction based on factor analysis combined with transformer and CNN-BILSTM-ATTENTION models. Sci Rep 15:20014

Malekmohamadi I, Bazargan-Lari M, Kerachian R, Nikoo M, Fallahnia M (2011) Evaluating the efficacy of SVMs, BNs, ANNs, and ANFIS in wave height prediction. Ocean Eng 38:487–497

Na N (2024) Research on supply chain demand prediction model based on LSTM. Proc Comput Sci 243:313–322. ISSN 1877-0509

Toncharoen R, Piantanakulchai M (2018) Traffic state prediction using convolutional neural network (2018). In: 15th international joint conference on computer science and software engineering (JCSSE), Nakhonpathom, Thailand, pp 1–6

Wang J, Liu G, Liu L (2019) A selection of advanced technologies for demand forecasting in the retail industry. In: 2019 IEEE 4th International Conference on Big Data Analytics (ICBDA), pp 317–320

Zhao X (2024) Exploring the performance of the CNN-LSTM model in stock prediction. In: Highlights in business, economics and management, vol 45. Darcy & Roy Press, pp 903–907

The Future of Urban Deliveries? Perceptions and Intentions About Delivery Robots and Drones in Europe

Shulei Zhang, Paulo Anciaes, and Emmanouil Chaniotakis

1 Introduction

The demand for home deliveries has been growing rapidly in most countries, especially since the COVID-19 pandemic. This has increased pressure on delivery networks, not only due to the rise in the number of deliveries but also because of changing expectations from customers regarding delivery times and flexibility. In many countries, delivery companies have also struggled to recruit enough drivers. While new solutions have been found for deliveries in dense areas, such as deliveries by cargo cycle, these also bring new challenges related to long working hours, poor safety conditions, and time pressures imposed by companies on delivery workers.

The growth in home deliveries has also had negative effects on urban environments. In particular, delivery van traffic has been a major contributor to increased road congestion in many cities. The circulation of these vehicles is especially problematic in city centres, due to limited parking availability and growing demands for other uses of kerbside space, such as bicycle and e-scooter parking, ridesharing vehicle stops, and outdoor cafés—uses that have also grown during and after the COVID-19 pandemic. These problems are exacerbated in older areas with narrow roads and streets.

S. Zhang · E. Chaniotakis
MUSA Lab, Bartlett School of Environment, Energy, and Resources, University College London, London, UK

P. Anciaes (✉)
Centre for Transport Studies, University College London, London, UK
e-mail: p.anciaes@ucl.ac.uk

A. Melkonyan-Gottschalk, Y. Li (eds.), *Urban Logistics Transformation*, Greening of Industry Networks Studies 15,
https://doi.org/10.1007/978-3-032-08387-6_5

Addressing these problems requires increasing the efficiency of urban deliveries, i.e. reducing the time required to complete deliveries, the amount of road space used, and the emissions generated. Small autonomous vehicles are a possible solution, as they do not require drivers and use less road space (or do not use road space at all). These vehicles can potentially enable faster and cheaper deliveries, compared with conventional vehicles. Being small and electric vehicles, they could also reduce the negative impacts of deliveries on congestion and emissions.

An example is delivery robots. These are small autonomous vehicles that can cover short distances using road footways, using sensors to safely interact with pedestrians and cross the road. They have been deployed mainly in enclosed spaces such as business or industrial parks, hospitals, or university campuses, with a growing number of services using public roads. The main use of these vehicles has been to make food or grocery deliveries. While they are a potentially efficient method, questions have been raised about how safe they are for pedestrians, as well as the possibility of theft of packages and vandalism of the vehicles. There is also a substantial gap in regulations governing the operation of these vehicles on public roads.

Another example is delivery drones. These are small aerial devices that can carry light parcels. They have been tested in several locations around the world and can be particularly useful for emergency deliveries of medical supplies or food, or to make deliveries in remote locations with difficult road access. Delivery drones could also reduce road congestion (although they may shift congestion to airspace). They also allow for faster deliveries as they can travel faster and in straight paths. Despite these potential benefits, the use of drones for deliveries in cities remains limited due to the many uncertainties the public has about them, such as the risk of crashing, falling, or dropping parcels, as well as noise, privacy issues, and the possibility of hacking. Regulations are even more of a problem than in the case of delivery robots.

While the technology has been proven to be feasible, and gains in efficiency and environmental quality are likely to happen, there is a risk that customers will not adopt autonomous delivery vehicles. This would affect demand for using them and thus make them commercially unviable. For example, customers may be concerned about the cost of deliveries using these vehicles. There are also uncertainties regarding business models. For example, individuals could own their own delivery robots or simply receive deliveries by robots owned by delivery companies or by the supplier of the products ordered. Customers may also not be receptive to using these vehicles due to concerns about the reliability of the delivery, privacy, and theft of the package. The concerns people might have about these vehicles as a citizen, such as those listed in the previous paragraphs, can also impact their likelihood of using them.

It is also uncertain whether the possibility of using delivery robots and drones could lead to a reduction in the number of trips people make (e.g. for shopping), which would have benefits in terms of reduced road congestion and emissions. However, individuals could respond to the possibility of using delivery robots and drones by making more deliveries (which would lead to more delivery trips, even if made by small autonomous vehicles) or more trips for purposes other than shopping (e.g. leisure).

This chapter explores these questions by analysing perceptions about delivery robots and drones, intentions to make online orders delivered by these vehicles, and potential impacts on delivery and travel behaviour. We use results from a European project to examine people's acceptance, as potential customers, of delivery robots and drones, and their main concerns about using these vehicles. We also compare results across eight countries and gender and age groups. This is important because the socio-economic and cultural context, and the current dominant travel and delivery patterns in each country, may shape people's perceptions. Likewise, age and gender may be linked to different experiences and beliefs about travelling or about technology in general.

In the next section, we review existing literature. We then use a multi-method approach to analyse data from three activities: an 8000-participant survey across the eight countries about intentions and perceived impact of the two types of vehicles (Sect. 3), a smaller survey in the United Kingdom focusing on concerns about using these vehicles (Sect. 4), and a demonstration of a delivery robot in the Netherlands, to gauge how people respond to seeing this vehicle in operation and whether that changes their intentions to use one (Sect. 5). In Sect. 6, we synthesise the results, compare them with previous literature, and discuss their implications for the delivery industry, broader society, and policy-makers.

2 Previous Knowledge

The development and real-world deployment of delivery robots and drones has been accompanied by a rapid growth in relevant academic literature, with studies conducted in a variety of countries both in the Global North and Global South. The demand side of the problem has been approached mainly in terms of people's acceptance of robots or drones as potential customers, i.e. their intentions to use the vehicles to deliver packages ordered online. In most studies, intentions were modelled based on the perceptions of survey respondents as potential customers and as citizens interacting with robots in public places or concerned with drones. Most studies have used structural equation modelling using latent variables (constructs) from established models of technology acceptance, such as the Technology Acceptance Model and the Unified Theory of Acceptance and Use of Technology.

Table 1 shows the constructs found to have either a direct effect on intentions to use delivery robots and drones or an indirect effect via attitudes towards these vehicles.

As shown in the table, customer acceptance of delivery robots and drones has been linked to the two key constructs of the Technology Acceptance Model: perceived usefulness and ease of use. It has also been linked to several constructs from the Unified Theory of Acceptance and Use of the Technology, such as performance expectancy, social influence (specified in some studies as "subjective norms"), hedonic motivation (i.e. pleasure derived from the experience of using the technology), effort expectancy, and price value (modified in some studies as price

Table 1 Evidence on the factors associated with customer acceptance of delivery robots and drones

Factor	Delivery robots	Delivery drones
From the technology acceptance model		
Perceived usefulness	Esmaili et al. 2025, Yuen et al. 2022	Leong and Koay 2023, Koh et al. 2023, Leon et al. 2023, Badu-Baiden et al. 2025
Ease of use	Montero-Vega et al. 2025, Esmaili et al. 2025, Yuen et al. 2022	Leong and Koay 2023, Koh et al. 2023, Badu-Baiden et al. 2025
From the unified theory of acceptance and use of the technology		
Performance expectancy	Kapser and Abdelrahman 2020, Oulmakki et al. 2024, Zibarzani et al. 2024, AlKheder et al. 2023	Osakwe et al. 2022, Chi and Hanh 2023, Kellermann et al. 2023
Social influence (or subjective norms)	Kapser and Abdelrahman 2020, Oulmakki et al. 2024, Montero-Vega et al. 2025, Zibarzani et al. 2024, AlKheder et al. 2023, Ju et al. 2024, Esmaili et al. 2025, Yuen et al. 2022	Schmidt and Saraceni 2024, Mathew et al. 2023, Nunkoo et al. 2024, Badu-Baiden et al. 2025
Hedonic motivation	Kapser and Abdelrahman 2020, Esmaili et al. 2025, AlKheder et al. 2023	Liébana-Cabanillas et al. 2025, Schmidt and Saraceni 2024, Mathew et al. 2023
Effort expectancy		Mathew et al. 2023
Price value (or price sensitivity)	Kapser and Abdelrahman 2020, Montero-Vega et al. 2025, Zibarzani et al. 2024, AlKheder et al. 2023	
Other factors		
Trust in technology	Oulmakki et al. 2024, AlKheder et al. 2023, Ju et al. 2024	Leon et al. 2023, Abbasi et al. 2024
Personal innovativeness	Oulmakki et al. 2024, AlKheder et al. 2023, Ju et al. 2024	Chi and Hanh 2023
[−] Perceived risk	Montero-Vega et al. 2025, Kapser and Abdelrahman 2020, AlKheder et al. 2023, Ju et al. 2024, Koh et al. 2024	Koh et al. 2023, Badu-Baiden et al. 2025, Nunkoo et al. 2024, Chi and Hanh 2023, Kellermann et al. 2023
Personal norms	Ganjipour and Edrisi 2023	
[−] Environmental concerns	Esmaili et al. 2025, Ganjipour and Edrisi 2023	Nunkoo et al. 2024
[−] Security concerns	Zibarzani et al. 2024, Esmaili et al. 2025	
[−] Privacy concerns	Zibarzani et al. 2024, Koh et al. 2024	Koh et al. 2023, Leon et al. 2023
Relative advantage	Ju et al. 2024, Ganjipour and Edrisi 2023	Schmidt and Saraceni 2024, Nunkoo et al. 2024
Perceived value	Koh et al. 2024	
Compatibility with lifestyle	Koh et al. 2024, Ganjipour and Edrisi 2023	Osakwe et al. 2022

Table 1 (continued)

Factor	Delivery robots	Delivery drones
Perceived behavioural control (or self-efficacy)	Yuen et al. 2022	Badu-Baiden et al. 2025, Osakwe et al. 2022
Observability	Ganjipour and Edrisi 2023	
[−] Complexity	Ganjipour and Edrisi 2023	

Note: [−] denotes factors with a negative association with intention to use

sensitivity). There is little evidence on the role of facilitating conditions and habit, the other two constructs of the model. In the latter case, this is because only a few customers have had experience of receiving deliveries using these vehicles.

Other studies have used constructs from extended versions of either the two above-mentioned models or from other models. Examples of significant constructs associated with intention to use delivery robots or drones include trust in technology; personal innovativeness; perceived risk (of anything going wrong with using the technology); personal norms; environmental, security, or privacy concerns; relative advantage of the technology compared with alternatives; perceived value; compatibility with lifestyle; perceived behavioural control (also specified as "self-efficacy", i.e. confidence in the ability to use the technology); observability (i.e. the extent to which processes and outcomes of the technology are observable); and complexity (of using the technology).

The effects of these factors are moderated by socio-economic and other personal characteristics. For example, most of the studies listed in Table 1 found that the willingness to adopt robots or drones increases with education and decreases with age, with more mixed results regarding gender. Intention to use is also positively associated with the level of knowledge and familiarity people have with delivery robots (Pani et al. 2020, Montero-Vega et al. 2025) and drones (Liébana-Cabanillas et al. 2025). The study of Pani et al. (2020) also found that the intention to use robots decreases with satisfaction with existing delivery methods (Pani et al. 2020).

Factors also vary from country to country. As an example, Montero-Vega et al. (2025) showed that in Asia intentions to use delivery robots were more strongly associated with social influences than with cost and perceived risks, compared to Europe and the United States. Within each country, the factors influencing intentions also differ between urban and rural areas (Liébana-Cabanillas et al. 2025).

While there is extensive research on factors explaining acceptance, there are fewer studies quantifying the potential size of the market for deliveries with autonomous vehicles. This depends on the number of customers willing to use delivery robots and drones, on the proportion of deliveries shifted from conventional delivery methods, and on the number of new deliveries, i.e. deliveries that were not previously made by any type of vehicle. The limited evidence there is suggests a modest market potential for delivery robots and drones. For example, in a small-sample study in France, Oulmakki et al. (2024) found that participants were, on average, neutral towards delivery robots. In the study by Pani et al. (2020) in the US, 61% were willing to pay extra for having their packages delivered by a delivery robot.

However, this study did not control for the characteristics of deliveries (e.g. cost, time) made by delivery robots and alternative vehicles. Another study that controlled for those characteristics found that participants were willing to pay US$6.70 for same-day human delivery but only US$1.90 for autonomous delivery (Forsythe et al. 2024).

In the case of drones, a study in a Chinese city, Zhao et al. (2024) estimated that even considering the best possible conditions, drone delivery could reach only 21% of the urban delivery market. A study in Portugal showed that only 25% of participants had an intention to use delivery drones, while 36% did not intend to use them (some of them due to concerns about the security of the package and legal aspects) and 39% had not formed an opinion about the matter (Bafouni-Kotta et al. 2023). In Germany, Merkert et al. (2022) found that, on average, people preferred conventional delivery, only choosing drones if they could deliver their packages faster and cheaper. Even in the most promising market (express deliveries), drone deliveries with the same price as conventional deliveries would only capture 30% of the market.

Another gap in the literature is the impact of delivery robots and drones on travel behaviour. While studies have analysed possible travel time gains due to increased efficiency in freight distribution, they have not quantified the impacts of these time gains (and of the fact that delivery orders could substitute trips) on the number of trips people make for shopping or other purposes.

3 Eight-Country Survey on Intentions and Effects of Delivery Robots and Drones

3.1 Methods

We ran an online survey in eight European countries (Cyprus, France, Germany, Greece, the Netherlands, Poland, Spain, and the United Kingdom), involving 7941 participants. The survey aimed to assess participants' intentions to use autonomous vehicles and the resulting effect on delivery and travel behaviour, comparing findings across countries, genders, and age groups.

The questionnaire covered both passenger and delivery vehicles. This chapter focuses only on the latter. Each participant was randomly presented with information about either delivery robots or drones. They were presented as "use cases", defined in terms of ownership of the vehicle, delivery procedure, and possible time savings for travel and parking. The delivery robot was specified as a private vehicle: the individual buys and maintains the robot and can send it to pick up or deliver packages (products or food). The drone was specified as a service provided by a company which owns the vehicle and delivers packages. In both cases, participants were reminded that they could save time currently used for travelling and looking for parking.

Participants were then asked about: (1) their likelihood of using delivery robots and drones (on a 5-point scale from "highly unlikely" to "highly likely"), (2) the expected change in the monthly number of delivery orders and weekly number of trips if they could use those vehicles (expressed by moving sliders on a scale between −20 and +20), and (3) the expected change on delivery costs (on the same 5-point scale). Information was also collected about demographic characteristics and current travel and delivery order behaviour.

Participants were recruited through market research companies. In each country, quotas were imposed on sex, age groups (18–34, 35–64, 65+) and regions (using the European Union's Classification of Territorial Units for Statistics classification, Level 1). Data were collected during January–May 2024. 51% of the overall sample was male and 48% was female. The sample mean age was 46.5 years, with 28% of participants in the 18–34 age group, 52% in the 35–64 group, and 20% in the 65+ group. The Cyprus sample was imbalanced, having only 1/3 male participants and 2% of people aged 65+. 55% of the overall sample had a university degree. 21% were not aware of autonomous vehicles, 50% had only heard about these vehicles, 23% were aware, and 6% were well aware. 6% never make delivery orders, 31% make them a few times per year, 48% a few times per month, 14% a few times per week, and 2% almost every day.

3.2 *Likelihood of Using Delivery Robots and Drones*

Figures 1 and 2 show the participants' stated intentions to use delivery robots and drones. Among the whole sample, more participants said they were (somewhat or highly) unlikely to use these vehicles (robot: 50%, drone: 43%) than those who said

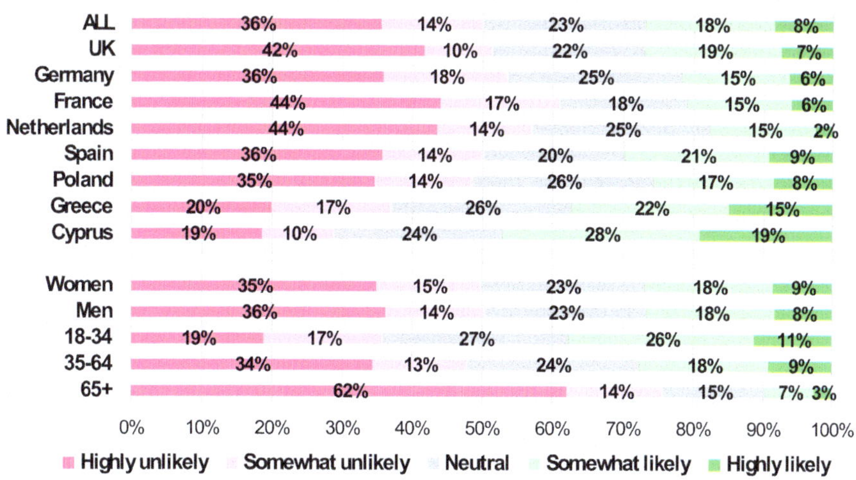

Fig. 1 Likelihood of using delivery robots, by country, gender, and age

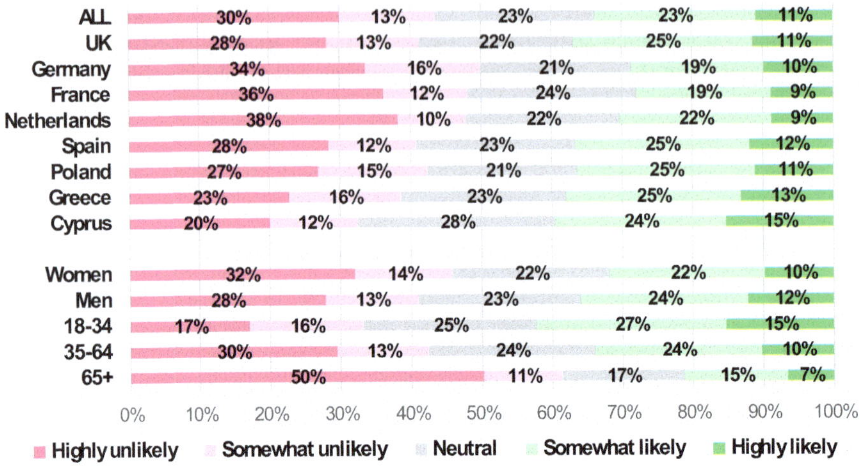

Fig. 2 Likelihood of using delivery drones, by country, gender, and age

they were (somewhat or highly) likely (robot: 26%, drone: 34%), with 23% being neutral (in both cases). Intentions were slightly stronger towards using delivery drones: 11% of participants stated they were highly likely to use them, and another 23% were somewhat likely. The respective numbers for the delivery robot were 8% and 18%.

As shown in Fig. 1, the intention to use delivery robots was markedly higher in Greece and Cyprus than in the other six countries. Intentions were also above average in Spain and Poland. Women and men had similar intentions. The stated likelihood of using a delivery robot decreased sharply with age. Among the 65+ age group, 62% said they were highly unlikely to use a delivery robot, and another 14% said they were somewhat unlikely, that is, a total of 76% expressing negative intentions. Only 10% had positive intentions among the 65+ group. Intentions were more positive among the 36–64 group and even more positive among the 18–34 group. However, even among the 18–34 group, 36% expressed negative intentions.

There were less pronounced cross-country differences in intentions regarding the delivery drone (Fig. 2). However, participants in Germany, France, and the Netherlands had more negative intentions than those in the other five countries. Again, women and men had similar intentions. The stated likelihood of using a delivery drone decreased with age. Among the 65+ age group, 50% said they were highly unlikely to use a drone, and another 11% said they were somewhat unlikely, that is, a total of 61% expressing negative intentions. This number is lower than in the case of the delivery robot. 22% had positive intentions among the 65+ age group. Intentions were more positive among the 36–64 group and even more positive among the 18–34 group. However, even among the 18–34 group, 33% expressed negative intentions.

3.3 *Impact on the Number of Delivery Orders*

Table 2 shows the average change in the number of monthly delivery orders people would make if they could use delivery robots and drones. The results are disaggregated by country, gender, and age. On average, robots and drones would increase delivery orders by 0.2 and 0.4, respectively. This is a positive but small impact. Although data were not collected for the current number of monthly delivery orders people make, the 0.2 and 0.4 increases can be compared with the frequency of orders: 66% of the sample make deliveries at least "a few times per month" (with 16% making deliveries at least a few times per week). The 0.2 and 0.4 increases due to the availability of delivery robots and drones are therefore small, in the context of the current delivery order behaviour.

The overall positive change in the number of delivery orders is driven mainly by the results in Greece and Cyprus, where orders would increase by 2.3–2.7 per month. The number is also positive in Spain and (in the case of drones) in Poland. In the other four countries (the Netherlands, France, Germany, and the United Kingdom), the availability of delivery robots or drones would lead to a reduction in the number of deliveries. On average, women reported more positive changes in orders delivered by robots, but men reported more changes in orders delivered by drones. Age was inversely related to the change in the number of delivery orders by both vehicles. While the effect on the number of orders among the 35–64 group was neutral (in the case of robots) or positive but small (in the case of drones), the effect among the 18–34 group was positive and the effect among the 65+ group was negative (and almost symmetrical to the effect among the 18–34 group).

Table 2 Average expected change in the number of monthly delivery orders

	Delivery robot	Delivery drone
All	0.2	0.4
United Kingdom	−0.7	−0.6
Germany	−0.1	−0.1
France	−0.9	−0.4
Netherlands	−0.9	−1.0
Spain	0.6	0.4
Poland	0.0	0.9
Greece	2.3	2.7
Cyprus	2.7	2.3
Women	0.3	0.2
Men	0.0	0.6
18–34	1.9	1.9
35–64	0.0	0.2
65+	−1.9	−1.6

3.4 *Impact on Delivery Costs*

Opinions about the impacts of the two vehicles on delivery costs were expressed by participants on a 5-point scale from "reduce significantly (50% reduction or more)" to "increase significantly (50% increase or more)". The results (Figs. 3 and 4) are almost identical for the two vehicles. 62% of the sample thought that there would be no change. Almost equal proportions thought that costs would increase or increase significantly (18–19%) and decrease or decrease significantly (19–20%).

The proportion of the sample who thought that costs would decrease was markedly higher in Greece and Cyprus than in other countries. In Germany, there was a larger proportion of participants who thought there would be no change, compared with other countries. The opinions of men and women were similar. The propensity to think that costs would remain unchanged increases with age: 48–50% of participants in the 18–34 age group thought costs would remain unchanged, compared with 63–65% among the 35–64 group and 75–76% among the 65+ group. Age was related both to a lower propensity to think costs would increase and a lower propensity to think they would decrease.

3.5 *Effect on the Number of Trips*

Table 3 shows the average change in the number of trips people would make (for all purposes) if they could use robots and drones to order deliveries. On average, the availability of a delivery robot would have no impact on the number of trips. The

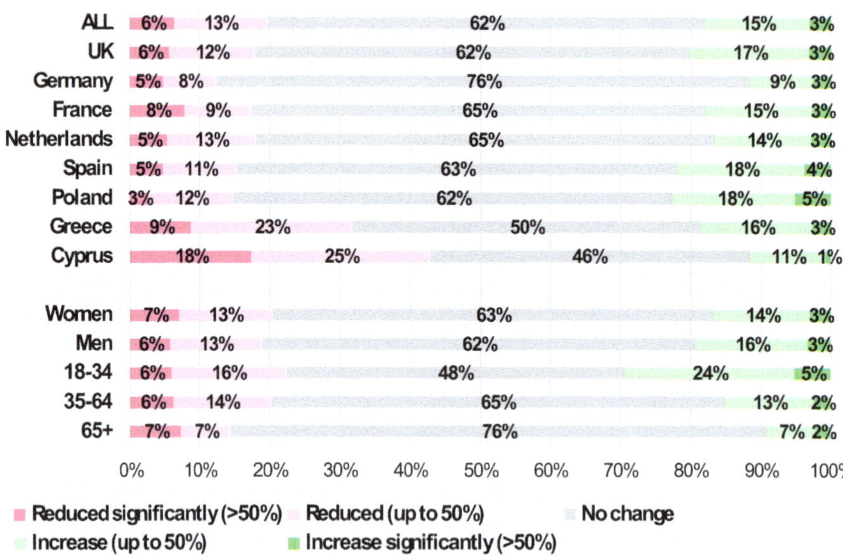

Fig. 3 Impact of delivery robots on delivery costs

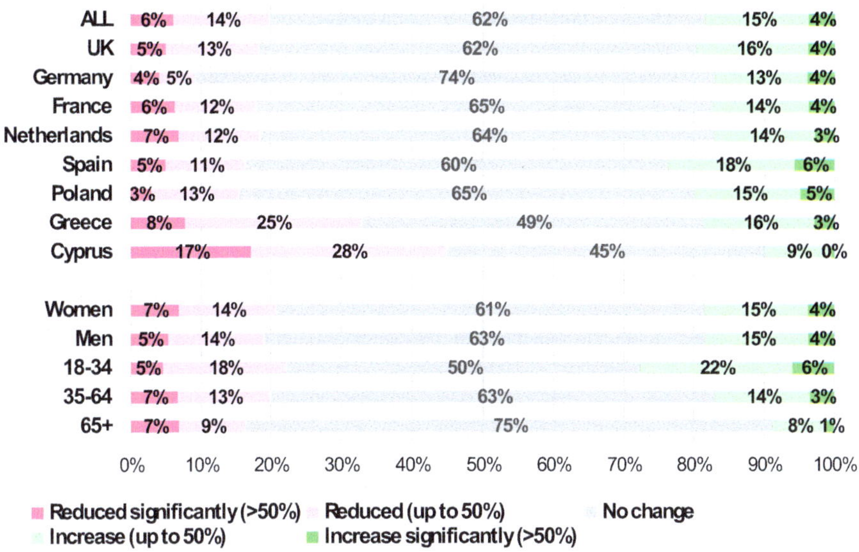

Fig. 4 Impact of delivery drones on delivery costs

Table 3 Average impact of delivery robots and drones on the number of passenger trips per week

	Delivery robot	Delivery drone
All	0.0	0.2
United Kingdom	−1.0	−1.4
Germany	−0.6	−0.3
France	−0.8	−0.2
Netherlands	−0.9	−1.2
Spain	0.5	0.7
Poland	0.7	1.3
Greece	1.5	1.8
Cyprus	1.7	2.2
Women	0.1	0.2
Men	−0.1	0.3
18–34	1.4	1.9
35–64	0.0	0.0
65+	−1.7	−1.4

availability of a drone would have a small impact of an extra 0.2 trips per week. The availability of drones would increase the number of trips slightly more, or decrease less, than delivery robots do in almost all countries, genders, and age groups.

The availability of robots or drones would increase the number of trips in Cyprus (1.7–2.2) additional trips per week in the robot and drone cases, respectively), followed by Greece (1.5–1.8), Poland (0.7–1.3), and Spain (0.5–0.7). It would decrease the number of trips in the other countries. The effect on trips made by men and women is similar. The availability of delivery robots or drones would have no effect on the number of trips made by the 35–64 group but would have a positive impact on trips made by the 18–34 group (1.4–1.9) and a negative impact on trips made by the 65+ group (−1.7 and −1.4).

4 UK-Only Survey on Attitudes and Intentions About Delivery Robots and Drones

We ran an online survey in the United Kingdom involving 700 participants. The survey complemented the eight-country survey by exploring not only customer intentions but also their attitudes regarding autonomous delivery vehicles, including delivery robots and drones.

Participants were first introduced to the two vehicles and then asked, for each one, if they agreed or disagreed with a series of statements, on a 5-point scale. The statements included: (1) opinion about the chance of deliveries having problems (being stolen, delayed, damaged by someone, damaged by the vehicle, injuring someone, delivering to the wrong address, or failing to deliver in bad weather), (2) opinion about the convenience, speed, and punctuality of the delivery, compared with a delivery by human-driven vehicles, and (3) intention to use the vehicle to order and return goods. Participants were then asked if they agreed or disagreed with a series of statements about autonomous delivery vehicles in general, covering privacy, hacking, and technology failure.

As in the eight-country survey, participants were recruited through a market research company and quotas were imposed on sex, age groups and regions. Data were collected during June–July 2024. 48% of the sample were male and 51% female. 29% were in the 18–34 group, 49% in the 35–64 group, and 22% in the 65+ group. 29% had a university degree. 7% were not aware of autonomous vehicles, 44% had only heard about these vehicles, 29% were aware, and 20% were well aware. Participants made an average of 6.1 deliveries per month for household and personal items, 4.4 for clothes, 4.1 for supermarket orders, and 3.3 for other items. Their main factor affecting delivery options were delivery cost (48%) and time (41%).

Figure 5 shows the results of the questions about attitudes towards delivery robots. The proportions agreeing or strongly agreeing that the robot would cause problems varied between 40% and 61%. The main concerns were failing in bad weather and the robot or the contents being damaged by someone (61%), followed by the contents being stolen (60%), delivery to a different address (55%), the robot injuring someone (49%), the delivery being less on-time than delivery with a driver (47%), the robot damaging the package (44%), and the delivery taking too long (40%).

Smaller proportions agreed or agreed strongly with the positive statements, i.e. using the robot delivery would be more convenient (39%) and faster (35%). 31%

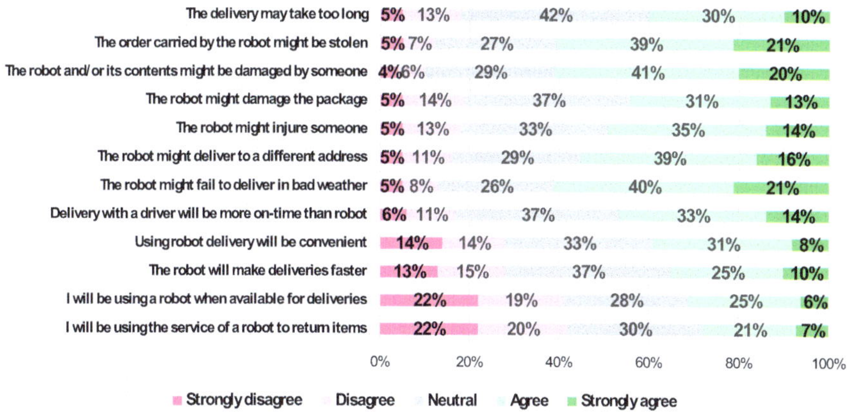

Fig. 5 Customer attitudes towards delivery robots

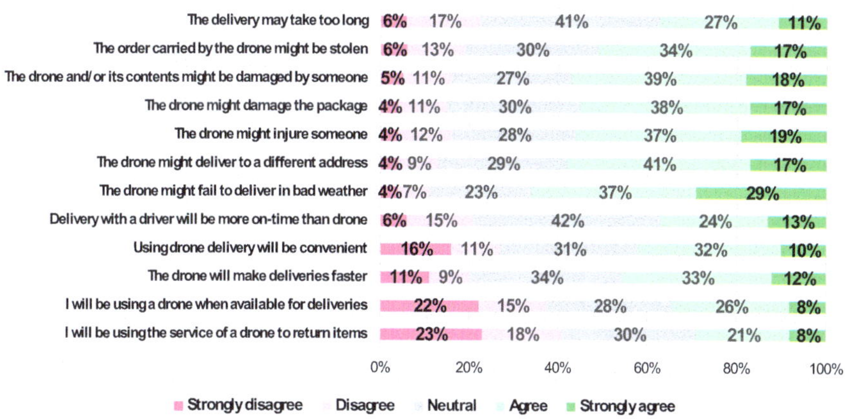

Fig. 6 Customer attitudes towards delivery drones

and 28% agreed/strongly agreed that they would use a delivery robot to order and return goods, respectively. These proportions were lower than the proportions of participants stating they disagreed or disagreed strongly (41% and 42% for ordering and returning, respectively).

Figure 6 shows the results of the questions about attitudes towards delivery drones. The proportions agreeing or strongly agreeing that the drone would cause problems varied between 38% and 66%. The main concern was failing in bad weather (66%), followed by delivering to a different address (58%), the robot or the contents being damaged by someone (57%), the robot injuring someone (56%), someone damaging the package (55%), the contents being stolen (51%), and the delivery taking too long (38%), and taking more time than a delivery with a driver (37%).

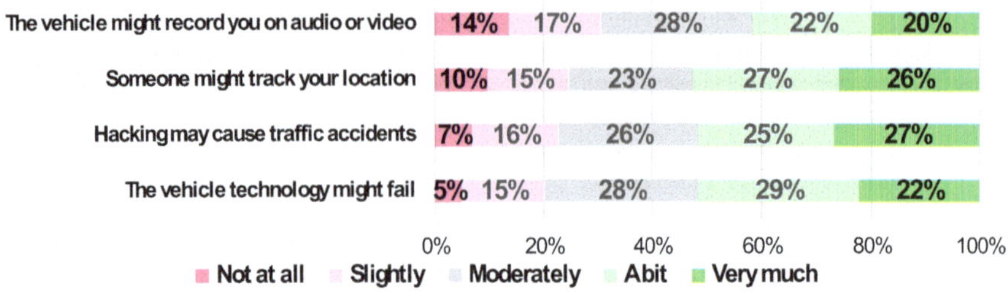

Fig. 7 Customer concerns about autonomous delivery vehicles

Smaller proportions agreed or agreed strongly with the positive statements, i.e. using the drone delivery would be more convenient (42%) and faster (45%). 34% and 29% agreed/strongly agreed that they would use a delivery drone to order and return goods, respectively. These proportions were lower than the proportions of participants stating they disagreed or disagreed strongly (37% and 41% for ordering and returning, respectively).

Figure 7 shows the participants' general concerns about autonomous delivery vehicles. 42% were concerned "a bit" or "very much" that the vehicle would record them on audio and video, 53% were concerned that someone would track their location, 52% that hacking might cause accidents, and 51% that the vehicle technology might fail. The proportions who were only slightly or not at all concerned are much smaller.

5 Demonstration of a Delivery Robot

We invited 35 individuals to a demonstration of autonomous vehicles in Helmond, a city in the South of the Netherlands with a population of 95,940. The objectives of the demonstration were to assess participants' attitudes and intentions about autonomous vehicles after using and observing them. The demonstration included three autonomous vehicles, one of them being a delivery robot. Participants could observe how the robot moved in a real-world environment and reflect on the advantages and disadvantages of this type of vehicle to make deliveries. A demonstration is a useful approach to gathering data on attitudes and intentions about autonomous vehicles, as most people have not yet experienced using these vehicles. The inclusion of a delivery vehicle also brings value added, as few studies to date have reported how people perceive these vehicles after experiencing them.

The demonstration was held on 20 January 2024 in the Helmond Automotive Campus. Over an 8-hour period, eight groups of four participants observed the delivery robot (Fig. 8). The vehicle was a Macrostep Autonoom Delivery Robot, an electric vehicle with a container, with dimensions 2 x 1.1 x 1.7 m. Participants gathered in tents, where they were briefed on what was going to happen. They then

Fig. 8 Aspect of the demonstration of a delivery robot in the Netherlands

observed the delivery robot moving for 3 minutes, turning several times. At the end, participants had an opportunity to ask questions to the event organisers.

Participants were recruited through a market research company. Quotas were imposed on sex and age groups. 59% of the sample were male and 41% female. 18% of participants were in the 18–34 group, 56% in the 35–64 group, and 26% in the 65+ group. These proportions are not balanced but are reasonably aligned with those in the Helmond adult population. 62% of participants had a university degree. Only one of the 35 participants had previously seen an autonomous delivery vehicle.

Participants answered a questionnaire regarding their experience observing the delivery robot and using the other vehicles (not reported in this chapter). The questions included what participants liked and disliked about the vehicle (open-ended question), safety for pedestrians and cyclists, how deliveries made by this type of vehicles would compare with deliveries made by vehicles driven by humans (e.g. vans), three main concerns about ordering goods delivered by these vehicles (open-ended question), and intention to order goods delivered by this type of vehicles in the future.

We coded the open-ended questions about the aspects participants liked and disliked. The main aspects they liked about the delivery robot were its perceived safety (26% of all answers), the design (23%), and its multi-functionality (as a modular vehicle) (20%). They also liked that the vehicle was practical (17%), quiet (17%), compact (14%), environment-friendly (11%), funny (in its design and/or movement) (9%), and did not need a driver (9%). The main aspects that participants disliked were that the vehicle could be vandalised (14% of all answers) and it being slow (11%) and small (11%).

20% and 40% of participants thought that it would be "very safe" and "safe", respectively, for pedestrians to walk on roads used by delivery robots. 37% thought it would not be safe not unsafe. 23% and 29% thought that it would be "very safe" and "safe" for cyclists, respectively. 43% thought they would be not safe nor unsafe. None of the 35 participants thought it would be unsafe for pedestrians, and only one (3%) thought it would be unsafe for cyclists.

The large majority (71%) thought that the delivery robot would be cheaper than human-driven vehicles, while only 3% thought they would be more expensive. The others thought that the two types of vehicles would be similar or did not know. 37% thought delivery robots would be safer (from accidents), compared with 6% who thought they would be more dangerous. On the negative side, only 9% thought they would be more secure from crime, compared with 51% who thought they would be less secure. In addition, only 21% thought delivery robots would be faster, compared with 35% who thought they would be slower.

We coded the open-ended questions about the main concerns participants had about delivery robots. The main concern was related to crime: the fact that goods can be stolen from the vehicle (mentioned in 46% of the answers). The other concerns were accessibility to the front door (23%), delivery time (17%), vandalism (14%), and delivery failures in general (9%).

In the intention question, 68% of participants said they would use a delivery robot and 32% said "maybe" they would use it. No participant said they would not use the delivery robot.

6 Synthesis and Discussion

This chapter analysed perceptions and intentions of individuals in eight European countries about the use of delivery robots and drones. It also provided additional detail about the perceptions of residents in the United Kingdom and residents in Helmond, a city in the Netherlands, where participants observed a delivery robot.

The various data collection activities revealed only a modest interest in the use of delivery robots and drones. This was evident, for example, in the eight-country survey, as more people stated they were unlikely to use those vehicles to order deliveries than those stating they were likely to use them. We found that only 26% of participants were likely to use robots and 34% were likely to use drones (Figs. 1 and 2). These figures are comparable with those of previous studies, conducted in a variety of countries, which showed that robots and drones could reach only 20–30% of the market (Zhao et al. 2024, Bafouni-Kotta et al. 2023, Merkert et al. 2022).

The study also added to the previous literature, providing evidence on the impacts of delivery robots and drones on delivery orders and travel behaviour. We found that the expected effect of the availability of the two vehicles on the number of delivery orders people make was small, and the effect on the number of trips was even smaller. Opinions were split regarding delivery costs: almost the same proportion of participants thought that delivery costs would increase and decrease.

However, there were considerable differences in results across countries. For example, participants in Cyprus and Greece showed more interest in using the two vehicles and in using them to make more delivery orders than they currently do. They also had a higher propensity to think delivery costs would decrease and expected to increase the number of trips they make, for all purposes. The intentions and attitudes in the four countries in the sample that have higher income per capita (Germany, France, the Netherlands, and the United Kingdom) were less optimistic, with a lower stated likelihood of using the vehicles, weaker belief that delivery costs would decrease, and a perceived negative impact on delivery orders and the number of trips. The existence of cross-country differences is consistent with the previous literature (e.g. Montero-Vega et al. 2025), which also found differences (in other sets of countries). However, previous studies have not found patterns related to income per capita.

There were only a few differences between the intentions and attitudes of men and women. However, age was highly relevant. The older the age group, the lower the stated likelihood of using the vehicles and of using them to increase the number of delivery orders; the greater the tendency to think delivery costs will remain unchanged, and the lower the expected effect on the number of trips made.

It is not possible to know the motivations behind the lack of strong interest in the two vehicles in the large eight-country survey. For example, the small effect on delivery orders could be due to the low likelihood of using the vehicles, but it could also be explained by the belief that these vehicles are more efficient, which would allow customers to combine orders into a single delivery. There was no evidence that the lack of strong interest was due to an expectation that delivery costs would increase, as the proportion of participants saying they were not likely to use delivery robots and drones was much higher than the proportion who thought that delivery costs would increase if they could use those vehicles.

The UK-only survey confirmed the limited interest in the two vehicles, again showing a higher proportion of participants saying that they were unlikely to use the vehicles, compared with those saying they were likely to use them. The survey also provided some clues about the factors that might contribute to the lack of strong interest in the two vehicles. For example, the majority of the sample agreed that the two vehicles could fail and that the goods could be stolen or damaged. There were also substantial concerns about hacking and data privacy. These results are consistent with several previous studies, which, as shown in Table 1, found that customer acceptance of delivery robots or drones was negatively related to factors such as performance expectancy, perceived risk, and concerns about security and privacy.

We found that some of these concerns are partly mitigated when individuals have the experience of observing the vehicles. In the demonstration of a delivery robot in the Netherlands, participants mentioned several concerns, especially about security, but they could also identify several advantages of using these vehicles. They also believed that deliveries would be cheaper and safer.

Overall, our analyses showed some scepticism among individuals in Europe about the implementation of delivery robots and drones. This has several implications for the delivery industry. People's concerns about autonomous vehicles affect their likelihood of using them, reducing the potential market and the financial viability of a wide deployment of the vehicles. This means that companies may find it difficult to achieve the necessary economies of scale to justify the large investments to buy and operate the vehicles. There could also be resistance from residents, who may protest against robots using footways or drones using airspace in their area. This may lead policy-makers to impose tight regulations on the use of the vehicles, which would increase costs or even prevent their use in densely populated areas, which are precisely the areas where automated vehicles could bring more efficiency gains and serve a larger market.

To mitigate these market and societal risks, companies could adopt hybrid solutions, giving customers the option to choose their preferred type of vehicle, or deploying autonomous vehicles only in areas where acceptance is higher. They could also implement marketing strategies to inform the public about how the vehicles work and to reassure them about their safety and security. This could be achieved, for example, through public trials and demonstrations, such as the one conducted in our study, where members of the public could observe the vehicles and ask questions about how they would operate if they were deployed in their area.

The findings of this chapter are also useful for the development of the vehicles (e.g. their design and operation). Efforts must be focused on designing vehicles that can deal with unexpected situations, such as extreme weather events, that can cause their failure. The design of vehicles and associated systems also needs to ensure that packages are delivered in good condition and are not stolen along the way. Safeguards are also needed to prevent the system from being hacked.

The lack of strong intentions to use autonomous vehicles for deliveries could also indicate broader societal resistance to the widespread application of automated processes to aspects of daily life. This resistance could be linked to concerns about the impacts of automation on social interaction, jobs, urban land use, and human dependence on technology. If similar patterns are found regarding the adoption of automation in other sectors, this could signal that the transition to an economy relying on automation is happening too fast.

Finally, the findings are also useful for the application of public policies to address the potential impacts of the vehicles. Acceptance of delivery robots and drones will depend on the regulations applied to address concerns such as the use of road and air space and the collection and use of customer data. Governments would also need to impose strict safety and security standards, as well as implement measures to ensure that robots are safe to be in environments shared with pedestrians, and that drones are safe to be used in densely populated areas.

Acknowledgements This work is an output of the Move2CCAM project, which was funded by the European Union. Views and opinions expressed are, however, those of the authors only and do not necessarily reflect those of the European Union or the European Climate, Infrastructure and Environment Executive Agency. Neither the European Union nor the granting authority can be held responsible for them. Participation of partners in the United Kingdom was funded by UK Research and Innovation under the United Kingdom government's Horizon Europe funding guarantee grant numbers 10044500 (University College London).

References

Abbasi GA, Rodriguez-López ME, Higueras-Castillo E, Liébana-Cabanillas F (2024) Drones in food delivery: an analysis of consumer values and perspectives. Int J Log Res Appl. https://doi.org/10.1080/13675567.2024.2407496

AlKheder S, Bash A, Al Baghli Z, Al Hubaini R, Al Kader A (2023) Customer perception and acceptance of autonomous delivery vehicles in the State of Kuwait during COVID-19, Technological Forecasting and Social Change 191:122485

Badu-Baiden F, Chiu W, Hagan E, Hodibert VA (2025) Unmanned deliveries: a complexity theory approach to understand consumer acceptance and intention in drone food delivery in Africa. Int J Hosp Manag 130:104234

Bafouni-Kotta A, Villanueva J, Kalakou S, Dias JG (2023) Upcoming drone delivery services: A perceptions-based clustering of citizens in Lisbon. Technol Soc 73:102250

Chi NTK, Hanh NT (2023) The drone delivery services: an innovative application in an emerging economy. Asian J Ship Logist 39:39–45

Esmaili A, Rejali S, Aghabayk K, Mohammadi A, De Gruyter C (2025) Autonomous delivery vehicle acceptance: the moderating role of perceived risk of theft. Transp Policy 162:406–423

Forsythe CR, Harper CD, Michalek JJ (2024) Bringing home the bacon: Estimating willingness to pay for autonomous grocery delivery across US households. Transp Res Interdiscip Perspect 26:101118

Ganjipour H, Edrisi A (2023) Consumers' intention to use delivery robots in Iran: an integration of NAM, DOI, and TAM. Case Stud Trans Policy 13:101024

Ju C, Wang S, Hu Z (2024) The impact of individual innovativeness and social influence on consumer intention to adopt autonomous delivery vehicles. J Knowl Econ. https://doi.org/10.1007/s13132-024-02407-9

Kapser S, Abdelrahman M (2020) Acceptance of autonomous delivery vehicles for last-mile delivery in Germany–Extending UTAUT2 with risk perceptions. Trans Res C Emerg Technol 111:210–225

Kellermann R, Biehle T, Mostofi H (2023) Modelling public attitude towards drone delivery in Germany. Eur Transp Res Rev 15:38

Koh LY, Lee JY, Wang X, Yuen KF (2023) Urban drone adoption: Addressing technological, privacy and task–technology fit concerns. Technol Soc 72:102203

Koh LY, Xia Z, Yuen KF (2024) Consumer acceptance of the autonomous robot in last-mile delivery: a combined perspective of resource-matching, perceived risk and value theories. Transp Res A Policy Pract 182:104008

Leon S, Chen C, Ratcliffe A (2023) Consumers' perceptions of last mile drone delivery. Int J Log Res Appl 26:345–364

Leong MK, Koay KY (2023) Towards a unified model of consumers' intentions to use drone food delivery services. Int J Hosp Manag 113:103539

Liébana-Cabanillas F, Rodríguez-López ME, Ghazanfar AA, Higueras-Castillo E (2025) A behavioral study of food delivery service by drones: insights from urban and rural consumers. Int J Hosp Manag 127:104098

Mathew AO, Chowdhury S, Devpura S, Lingappa AK (2023) Factors influencing technology acceptance of drones for last-mile food deliveries: an adaptation of the UTAUT2 Model. Human Behav Emerg Technol. https://doi.org/10.1155/2023/7399080

Merkert R, Bliemer MC, Fayyaz M (2022) Consumer preferences for innovative and traditional last-mile parcel delivery. Int J Phys Distrib Logist Manag 52:261–284

Montero-Vega M, Estrada M, Prouvier M, Siebeneich A (2025) Autonomous delivery robots: differences in consumer's acceptance in the USA, Europe, and Asia. J Urban Mobil 7:100110

Nunkoo R, Pillai R, Sivathanu B, Rana NP (2024) Drivers and inhibitors of consumers' adoption of AI-driven drone food delivery services. Int J Hosp Manag 123:03913

Osakwe CN, Hudik M, Říha D, Stros M, Ramayah T (2022) Critical factors characterizing consumers' intentions to use drones for last-mile delivery: does delivery risk matter? J Retail Consum Serv 65:102865
Oulmakki O, Verny J, Janjevic M, Khalfalli M (2024) Empirical study on consumer's acceptance of delivery robots in France. Int J Log Res Appl 27:2049–2079
Pani A, Mishra S, Golias M, Figliozzi M (2020) Evaluating public acceptance of autonomous delivery robots during COVID-19 pandemic. Transp Res Part D: Transp Environ 89:102600
Schmidt S, Saraceni A (2024) Consumer acceptance of drone-based technology for last mile delivery. Res Transp Econ 103:101404
Yuen KF, Koh LY, Anwar MHDB, Wang X (2022) Acceptance of autonomous delivery robots in urban cities. Cities 131:104056
Zhao B, Suo Y, Tang L, Li C, Fu M, Huang L (2024) Urban air mobility for time-sensitive goods with explicit customer preferences: A case study on Chengdu. J Air Transp Manag 118:102613
Zibarzani M, Abumalloh RA, Nilashi M (2024) Adoption behavioural intention of robots in last mile food delivery: the importance of environmental friendliness and moderating impacts of privacy and security concerns. Res Transp Bus Manag 55:101146

Digital Twins for City Logistics: Towards a Roadmap

Lóri Tavasszy, Rodrigo Tapia, Michiel de Bok, Merve Cebeci, Finn Niklas Winkelmann, and Önder Gürcan

1 Introduction

Digital Twins are one of the key new applications of computing technology in urban management and one of the key ingredients of the smart and sustainable city paradigm. Smarter cities can significantly contribute to sustainable freight transport by curbing traffic pollution, CO_2 emissions, noise annoyance, social inequity and other externalities (Tavasszy et al. 2025). The main function of a digital twin is to support

L. Tavasszy (✉)
Delft University of Technology, Delft, The Netherlands
e-mail: L.A.Tavasszy@tudelft.nl

R. Tapia
Roma Tre University, Rome, Italy

Significance BV, The Hague, The Netherlands
e-mail: rodrigojavier.tapia@uniroma3.it

M. de Bok
Delft University of Technology, Delft, The Netherlands

Significance BV, The Hague, The Netherlands
e-mail: M.A.deBok@tudelft.nl

M. Cebeci
Panteia BV, Zoetermeer, The Netherlands
e-mail: M.S.Cebeci@tudelft.nl

F. N. Winkelmann
TNO, The Hague, The Netherlands
e-mail: finn_niklas.winkelmann@tno.nl

Ö. Gürcan
NORCE AS, Bergen, Norway
e-mail: ongu@norceresearch.no

A. Melkonyan-Gottschalk, Y. Li (eds.), *Urban Logistics Transformation*, Greening of Industry Networks Studies 15,
https://doi.org/10.1007/978-3-032-08387-6_6

public and private stakeholders with the management of logistics processes inside the city, including the planning and execution of these processes, their facilitation with infrastructure and their regulation (see also Table 1). Digital twins aim to make cities more responsive, flexible and inclusive in identifying pressing problems and solutions (Tapia and Tavasszy 2024). This chapter explores the current and future approaches towards the design of digital twins for sustainable urban logistics systems.

In its simplest form, the digital twin is not more than a shadow of the system, reactive in its behaviour to data produced by the real system. In its more advanced form, it contributes to the control of system behaviour by actuating actions, responding to analysis of states, predictions of future states and prescriptive analysis to determine optimal control signals. "Smartness" then resides in the sensors that monitor the system (possibly interconnected as IoT), the digital representation of subsystems (simulation), the chained algorithms for problem interpretation, forecasting and system optimisation (opportunities for AI) and the visualisation of the system for stakeholders (VR/AR). Clearly, much technical advancement is possible to help digital twins develop from the simple implementations of today as simulation models, to advanced controllers of the urban freight transport system. In addition, urban logistics systems are complex, with many stakeholders, technologies, and processes fraught with uncertainty and volatility, and urgent problems with management of efficiency and external impacts.

Logistics management decisions can relate to many aspects (Riopel and Langevin 2005 count 48 fields of decision making) and many different stakeholders (public and private) are relevant. DTs act as decision support tools in decision cycles, which can be conveniently pictured as control loops or—in management language—Deming cycles (PDCA). We take this framing as a starting point and posit that all

Table 1 DT functions and positioning in decision types

	Decision	Measurement	Evaluation	Decision generation	Actuation
Operational	Dynamic access, traffic management	Cameras, loop detectors, GPS, trucks	Delivery satisfaction, costs, emission levels	Scenario-based algorithms in traffic management centres	Speed management, access restrictions, enforcement
Tactical	Time windows, vehicle restrictions; planning of curb use; reservations	Planning systems, capacity utilisation	(infrastructural, environmental) capacity utilisation patterns	Local policy design, implementation and enforcement teams	Local traffic and urban design plans
Strategic	Infrastructure, hub development	Land use, transport, environmental statistics	Long-term cost-effectiveness, sustainability	Political, policy departments, development actors	Land-use plans, construction and maintenance

logistics decisions, with different degrees of formality, follow such cycles. Applications of DTs for city logistics can be envisaged in strategic, tactical and operational decision cycles. Strategic decisions concern investments and heavy governance arrangements related to institutions and their governance (e.g. law). These cycles are slow and typically revolve over years. Tactical decisions are related to the reservation of asset capacity for operations, and its codification in fleet management plans, service schedules and sourcing and delivery contracts. These decisions will typically be updated on a yearly basis. Operations concern the execution of logistics services, with high-frequency decisions, measured in days or even seconds. DTs in the public policy spheres will aim to support or regulate processes at these levels (Table 1).

Clearly, the nature of and requirements for DTs will differ between these contexts. Especially in the strategic context, sophisticated simulation and optimisation models have existed since the second half of the previous century. So far, they have developed mostly as digital shadows of systems, however. Actuation happens through years-long processes of policy preparation, political decision-making and policy implementation. The different models used to support these management processes emulate and predict the behaviour of the system, using observations, which, again, move at a slow pace, typically built on public statistics. The framing of these models as digital twins (more accurately now: digital shadows) has provided a new way of looking at these models. It has opened up an innovation perspective in two new directions. Firstly, the closing of control loops (or decision-making cycles) by making the sensing and actuation activities consistent with each other. This begs the question whether the correct data are acquired to produce decisions and (ex-post) evaluate them. Secondly, it connects long and short-term cycles within the same paradigm. This opens up possibilities to use strategic data and models for tactical and operational decisions and vice versa.

This chapter reports on recent advances in digital twin development for urban logistics systems. Several recent EU projects have brought new insights about the positioning of DTs in city logistics management, the alternative trajectories that DT technology can take in this context and the possible impacts of applying DTs. While many studies and experiments have propelled our technological DT capabilities further, there has been less attention to alternative experiences with this technology in city logistics. This chapter aims to contribute by framing the DT functions in the urban logistics context, developing an innovation pathway for DTs based on this framing and positioning recent DT use cases.

Within the scope of the complete and multi-dimensional framing of DTs, this chapter takes a narrow approach in two respects. Firstly, our focus lies on the representation capabilities of DTs in the context of city logistics, in contrast to the sensing and control capabilities. Such a DT is often called a digital shadow of the physical system. The main difference with a fully bi-directional DT is that the sensing and actuation of the system are not fully automated. Secondly, we limit ourselves to strategic decisions related to policies of public institutions, specifically supporting infrastructure and the permanent regulation of its use. Broader discussions of DTs for short-cyclic urban management, the bi-directional, control-focused

DTs and their respective positioning along the lines of Table 1 can be found elsewhere in the literature (Nag et al. 2025; Le et al. 2024; Tavasszy and Quak 2023; Batty 2018).

It is worthwhile to note that the scientific literature on digital twins for city logistics is still nascent. Recent work which addresses the use of DTs for urban freight policy purposes can be found in (Tavasszy et al. 2025; Belfadel et al. 2023; Marcucci et al. 2020; Abouelrous et al. 2023; Tapia and Tavasszy 2024; Jeong et al. 2022; Nag et al. 2025; Grotto et al. 2024; Taniguchi et al. 2024; de Bok et al. 2024). In this chapter, we contribute by presenting results from European projects that have focused on two very different architectures: a centralised, single-form DT which is used in similar way for different cities and cases (de Bok et al. 2024; Gürcan et al. 2025) and a federated approach with a library of DTs, where model components are chosen to be applied depending on the policy question (Belfadel et al. 2023). The second, even though experimental, is interesting as it shows the direction for the evolution of DTs according to a generic maturity model (Tapia and Tavasszy 2024), from sectoral to multi-sectoral DTs. This model shows that sectoral city DTs can be seen as systems-of-systems.

The next section provides an overview of these approaches. Section 2 provides examples of applications. Section 3 reiterates main learnings from the projects and concludes the chapter.

2 Digital Twins for City Logistics: Three Cases

2.1 Alternative Architectures

Here, we introduce the approaches for urban DTs as implemented in recent EU projects. A general architecture for the DT/PT duplet is shown here, based on (Belfadel et al. 2023). The physical world communicates with the DT by feeding it with relevant data obtained from measurements and sensors (lower arrow) and recording recommendations for policy and decision making (upper arrow). Note that the storage system includes the model library, which can also be reduced to an extreme case of a single model. The existence of such a library with the resulting degrees of freedom marks the difference between a centralised and a distributed architecture. In the centralised case, one model will be prescribed for all cases as the simulation engine for the DT. In the distributed case, a whole library of models can be used, and the model management system has the additional task of configuring the specific model system required to answer a question. In other words, parts of the system thought to be less relevant will be switched off in the DT. This approach has the advantage of efficiency and flexibility, but may also reduce the transparency of the applications and their validity, if pragmatic choices are made in the selection of models (Fig. 1).

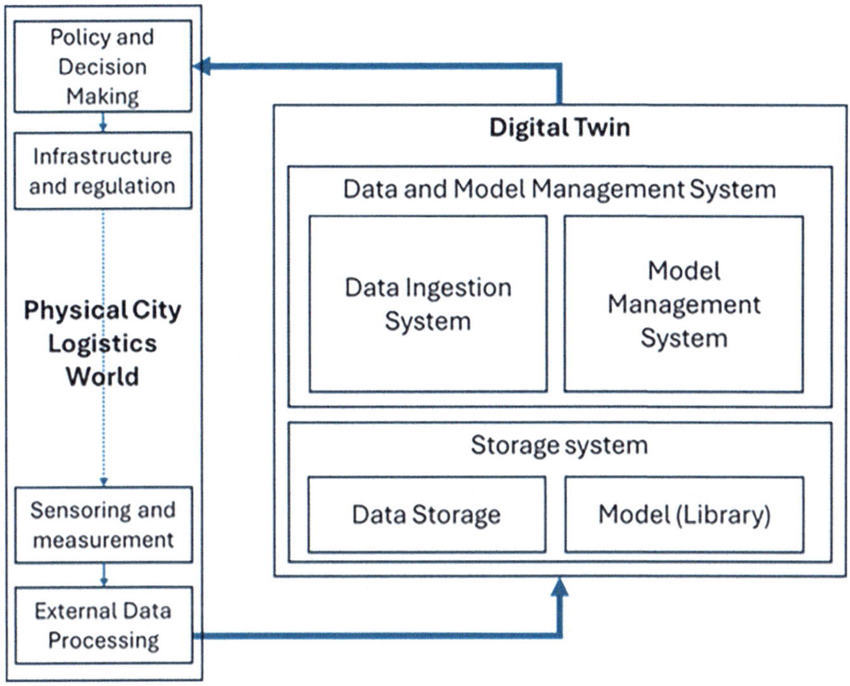

Fig. 1 Basic architecture of DT in relation to the physical world (Adapted from (Belfadel et al. 2023))

Recent EU projects that have contributed to DTs for city logistics are HARMONY, LEAD and URBANE. One of the objectives of these projects was to create models for DTs. In LEAD, the emphasis was on a library which could allow many existing models to be used, which led to the distributed approach. In HARMONY and URBANE, a centralised approach was chosen due to the innovative nature of the models, which were developed within the project. HARMONY created the Tactical Freight Simulator, an agent-based simulation system for urban logistics (de Bok et al. 2021a). The model development for URBANE focused on the joint application of ABMs for social-network-driven learning behaviour and urban freight simulation, respectively HUMAT (Jager et al. 2025) and MASS-GT (de Bok et al. 2025a). Our focus in this chapter is on the URBANE model. In addition to these EU project results, we provide a third example of a unique DT development in The Netherlands, where a freight simulation system was integrated into a full-fledged urban digital twin toolbox, to strengthen its capabilities for transport analyses.

Here, we describe in more detail the following three DTs: the URBANE development of coupled models together forming a DT to be applied for various cities, the LEAD project that created a DT with a whole library of models, and the broader city DT toolbox which included MASS-GT.

2.2 *URBANE: A Marriage Between Agent-Based Models*

2.2.1 Method

In URBANE (Franklin et al. 2025), the MASS-GT model has been further developed by integrating components of consumer decision-making behaviour using the model HUMAT, which simulates decisions of individuals who are part of a social network and allows the learning process from decisions to propagate through the social network (Tapia et al. 2023). One of the living labs of the project was situated in the city of Helsinki, Finland. The overall ambition of this Living Lab is to achieve carbon neutrality by implementing the Carbon-Neutral 2035 Action Plan. The Helsinki LL proposed testing the concept of micro hubs in the city, specifically focusing on innovative LMD options such as robot deliveries using autonomous delivery vehicles (ADVs), cargo bikes and teleoperation. The LL included multiple piloting locations. To explore the acceptability of ADVs, the integrated modelling framework given here is calibrated and used for such a use case (Gürcan et al. 2024) (Fig. 2).

In this use case, while the HUMAT modules simulate consumer choices between the ADV service and home delivery options over time, MASS-GT is used to generate household demand and a vehicle routing model (VRP) in the calibrated modelling framework. This collaboration between the two models can be extended to encompass the full range of capabilities of MASS-GT and include simulation of logistics decisions. Conceptually, this architecture allows a simulation of social and logistics networks to inform decision makers. As agent-based simulations can be relatively easily connected to practice, this approach allows us to connect to data

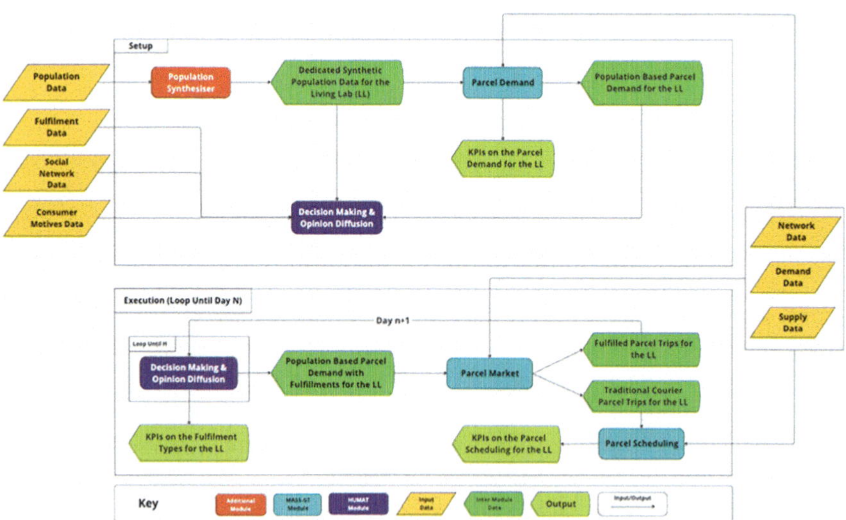

Fig. 2 Integrated modelling framework in URBANE

sources which monitor daily delivery operations, as well as to advise or steer management decisions related to daily operations. In URBANE, these applications and extensions have been out of scope.

2.2.2 Application

Simulations with the URBANE ABM were carried out by implementing what-if scenarios in a digital twin of Helsinki. The Parcel Demand module of MASS-GT estimated the demand based on various socio-demographic characteristics and the household structure of the population. Demand data provided by an LSP company were used to populate the system. The Parcel Demand module generated 220 parcels for the study area. Next, detailed simulations with HUMAT were conducted to show the service and queuing times under different parameters (Fig. 3).

Figure 4 shows the performance of an ADV over 31 days by comparing the number of parcels to deliver and parcels delivered under various capacity and customer arrival rates.

The arrival rates of customers show that higher ADV capacities (30 and 300 parcels) do not significantly improve the delivery rate. The main reason for this outcome is the time window constraint (i.e. the period within which the ADV must visit a location, e.g. between 11:30 and 12:45) and the capacity constraint of the ADV. The analysis shows that increasing ADV capacity alone does not ensure higher delivery efficiency, especially under higher demand scenarios. Increasing the width of the time windows as well as the ADV's waiting time at each delivery

Fig. 3 Demand setting in the Helsinki LL

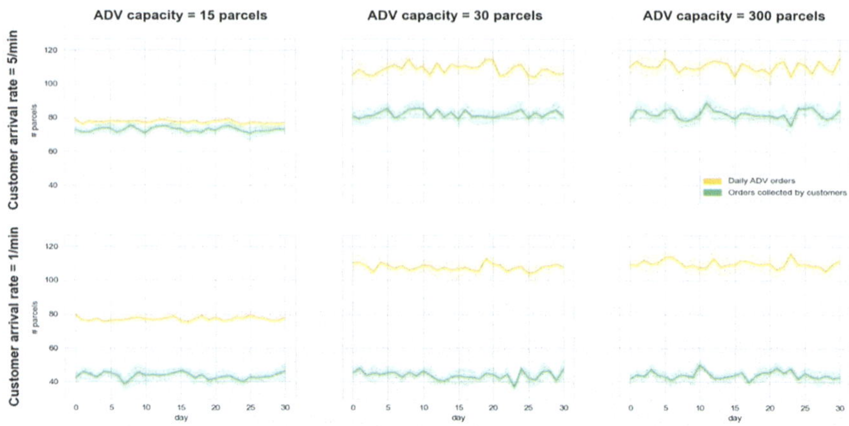

Fig. 4 Average parcels delivered by ADV

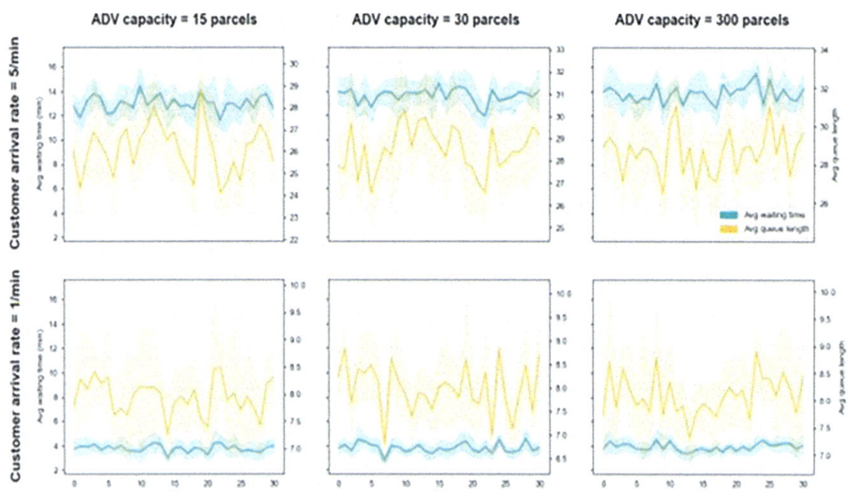

Fig. 5 Average waiting time and queue length with ADV delivery

location will improve the delivery rate. This, however, will result in an increase in the ADV's total operating time. For a given delivery scenario, the locations' time windows and waiting times should be optimised such that a high delivery rate is achieved while ensuring that the ADV is capable of operating for the required length of time. Figure 5 shows the average waiting time and average queue length over a 31-day period of ADV delivery.

As expected, when the customer arrivals are higher (5 customers per minute), the average waiting time and average queue length become higher since there are more customers picking up their parcels from the ADV, creating higher waiting times and a longer queue.

2.2.3 Discussion

Together, the DT-like implementation of the two models analysed the deployment of autonomous vehicles for direct parcel deliveries to consumers. The simulation suggests that while increasing the capacity of ADVs can reduce the number of trips, this benefit is most effective when coupled with optimised delivery schedules. The environmental impact of ADVs is also significant, with the potential for substantial reductions in CO_2 emissions in regions where renewable energy sources are prevalent. However, in areas dependent on fossil fuels, the environmental benefits are less pronounced. By integrating social networks into the simulation of delivery services, the model captures how consumer preferences can be influenced by peer influence.

2.3 LEAD Project: A Model Library-Driven DT

2.3.1 Method

The approach of the LEAD project to develop DTs was to establish a set of Living Labs with the objective of obtaining different use cases in detail that could be combined in a single platform. This way, the DT could develop a deep understanding of the public authorities' and private sector requirements. The multiplicity of use cases and models needed to approach them requires a flexible modelling framework, besides having the flexibility to adopt further use cases. This was done by developing a model library (ML) that is model and ontology agnostic. This means that the ML can include different model types, with different assumptions, timescales, coding languages and frameworks in a flexible and modular way. The communication between models is done by a set of connectors (similar to APIs) that make the relevant data transformations to make the outputs of a model compatible with another one. By being ontology agnostic, the ML allows the easy inclusion of new models that may use different terminologies or conceptual frameworks, provided they generate the relevant connections with existing models.

The main disadvantage of this decentralised approach compared with a centralised approach is that there is not an internal consistency guarantee, relying on the users to be sure that they deploy the set of models that do not have conflicting assumptions. This makes the approach more reliant on either already developed use cases, or an expert user that is able to understand the strengths and limitations of each model from the ML. The Meta Model (M2) of the ML helps users select models and assists developers in creating them. The M2 organises and categorises the models according to their aggregation and temporal scope, and by their type of model. In the M2, the connectors are mapped by the relationships between different models.

The temporal scales used in the M2 are strategic, where long-term models are included, such as population synthesis and echelon models; tactical, where medium-term (such as a week or a full day) models are included, such as tour formation; and operational, where short-term models such as routing are included. The level of aggregation dimension can be either zonal aggregation or a lower-level disaggregated parcel and shipment level. These two dimensions are related to the types of logistical decisions as described in Table 1. The M2 also distinguishes different types of models, such as optimisation models, demand models, ABMs, network models and impact assessment models. More details on the ML structure and the LEAD DT can be found in (Belfadel et al. 2023). Figure 6 shows the M2 of the LEAD library.

2.3.2 Applications

A large number of use cases were applied in the LEAD project, all supported by the LEAD DT.

- Madrid: Study the impact of implementing an Urban Consolidation Centre in a parking lot at the entrance of the limited traffic zone. The LL consisted of the city transport authority, a logistic service provider, a shipper, a vehicle routing service provider and academic partners.
- The Hague: Study the impact of connecting different logistical networks and enhancing collaboration across logistical service providers. The LL consisted of

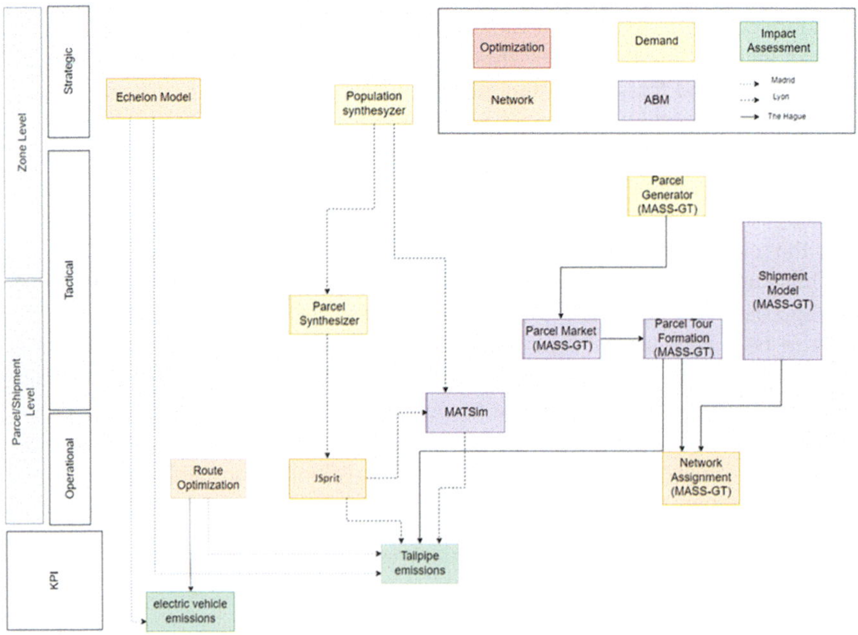

Fig. 6 Meta-model of the DT model library in the LEAD project (Abouelrous et al. 2023)

the city authorities, a digital infrastructure company, parcel locker companies, a crowdshipping company and academic partners.

- Lyon: Study the impact of cargo bikes and autonomous last-mile distribution vehicles in the city centre. The LL consisted of city authorities, a technology provider, a carrier and academic partners.
- Oslo: Study the impact of implementing crowdshipping for last-mile delivery. The LL consisted of a crowdshipping company, public authorities and academic partners.
- Budapest: Study the impact of the location of Urban Consolidation Centres outside the city centre. The LL consisted of the city authorities, a shipper and academic partners.
- Porto: Optimise delivery routes for electric delivery vans, considering grid restrictions and characteristics. The LL consisted of a technology provider, a shipper, a carrier and academic partners.

Here, we provide some illustrative outcomes of the living lab in The Hague, The Netherlands. Here, the LEAD DT has been used to test the effect of crowdshipping in the urban environment. The main question asked in this simulation is the effect of pricing on the total emissions and congestion. The KPIs obtained are GHG emissions by car and vans, total distance travelled by cars and vans and the expected revenue by the crowdshipping company. The first set of simulations changed the remuneration given to crowdshipping carriers, assuming a flat rate. From all the simulations we registered the crowdshipping platform remuneration and selected the one with the maximum (7 euros) for a more in-depth analysis. In this case, it is assumed that the platform would set the prices that maximise its own revenue. Next, flows were simulated for all the parcels in The Hague. It can be seen in Fig. 7 how these parcels are fulfilled across the multiple carriers and, in the case of crowdshipping, which mode they take. Since crowdshipping is only available for Local-to-Local flows (i.e. flows that start and end in the area), they are restricted to these flows that represent around 4% of the total demand. From these, it can be seen that circa two-thirds of the crowdshipping is done by car.

2.3.3 Discussion

The DT approach has also been proven as an interesting element throughout the LL activities. Some of the KPIs were not possible to measure in the physical implementation, since the pilots were done in an open system (i.e. not in a controlled environment). By using data collected from the pilots and the DT approaches, the DT was able to simulate large-scale implementations. Through the DT, the possible effects of crowdshipping were measured and signalled a warning regarding the potential detrimental effects it can have if it attracts car travellers (Cebeci et al. 2024; Tapia et al. 2023). This procedure can become a key in the future, when public authorities are able to simulate the effects of a new logistic innovation and be aware of the

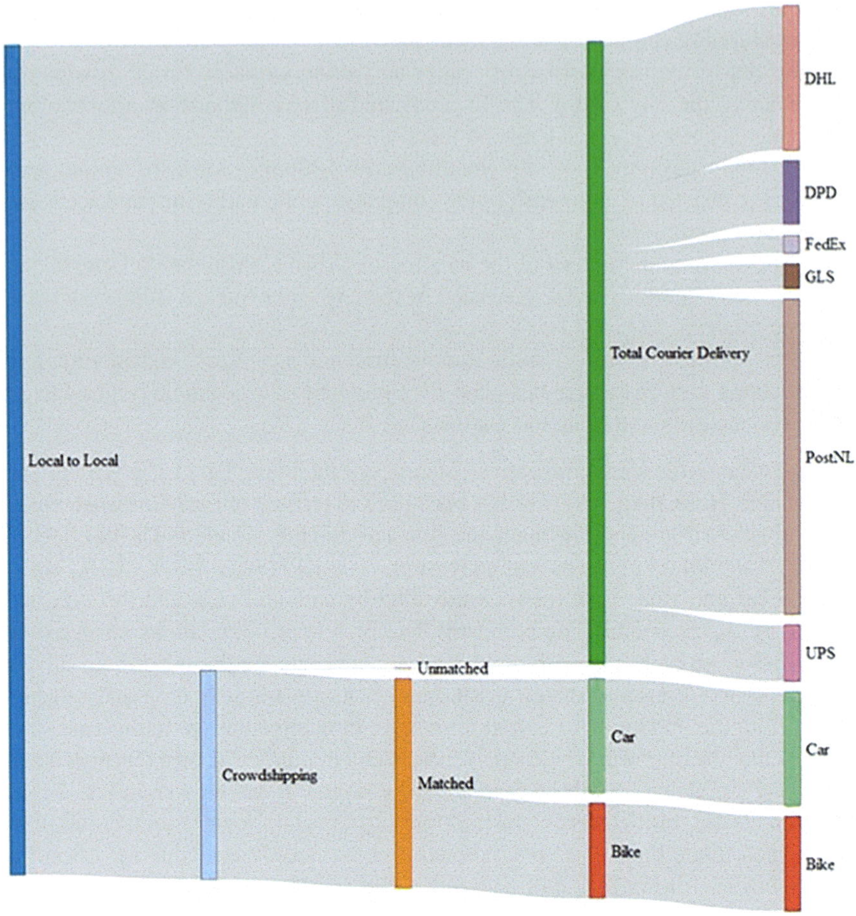

Fig. 7 Parcel fulfilment in LEAD LL

potential impacts before they happen and be able to avoid possible negative externalities.

2.4 MASS-GT Integration into DT Toolbox

Finally, we present an example of how the MASS-GT simulation model was integrated into an existing digital twinning framework for Amsterdam (Winkelmann et al. 2025). It shows how an agent-based simulator can be added as a module to describe the city's logistic activities. In a test application, it is shown how the simulator capabilities of MASS-GT are used for strategic policy application, in this case, the introduction of a ZE zone.

2.4.1 Method

The method integrates the MASS-GT city logistic simulator into an existing DT toolbox. The starting point is the TNO Digital Twin, an extensive digital twinning framework that focuses on scenario-based analysis in the mobility domain (Lohman et al. 2023). The tool consists of several models, which simulate a variety of travel modes including cars, freight, bicycles and public transportation. These travel modes are assigned to a transport network using an assignment model. For the Amsterdam case study, the Verkeersmodel Amsterdam traffic model, VMA 4.5, is integrated into the DT to predict the mobility demand (Gemeente Amsterdam 2024). Motorised traffic is assigned using a (multi-user class) volume averaging (car) or an all-or-nothing (freight) assignment algorithm. The Digital Twin can be used in an interactive fashion to adjust urban characteristics, to allow users (e.g. city authorities or municipalities) to actively simulate a variety of policy scenarios. Example cases include the impact of road closures (city-wide), speed reduction, population growth or increased parking prices. All simulated scenarios can be evaluated using a variety of indicators. Moreover, a number of models exist to compute air quality, emissions and noise pollution, but also further indicators related to mobility (e.g. vehicle kilometres, modal split), health or accessibility.

MASS-GT was integrated into the toolbox as an additional module to evaluate the city's logistic developments. MASS-GT is an agent-based model for urban freight transportation (de Bok et al. 2025b). The model utilises a commodity-based approach to distinguish between the number of agents in the freight transportation system and models their behaviour according to their respective sector. This agent-based approach is integrated with commodity data and synthesised firm locations to simulate representative freight transportation patterns through logistics choice models. The demand is simulated at the shipment level to accurately reflect the decision-making process within eight logistics segments: food (general cargo), miscellaneous (general cargo), temperature controlled, parcels and express, facility logistics, waste logistics, construction logistics and dangerous goods. Furthermore, shipments originate from supply nodes and are transported to consumers, either directly or via one or more logistics nodes, such as distribution centres or trans-shipment terminals. MASS-GT makes various decisions regarding vehicle type, distribution channel and shipment size. Once all shipments are generated, they are organised into tours and assigned to specific times of day. Additionally, a parcel module operates in parallel, generating similar choices for parcel-specific traffic while considering unique specifications of parcel logistics. Similarly, an additional service demand module generates the tours for construction and service vans.

2.4.2 Application

The integrated DT toolbox is applied in a case study to analyse the impacts of a planned zero-emission zone (ZEZ) for city logistics traffic in Amsterdam, the Netherlands. The case study analyses scenarios with two different potential sizes for

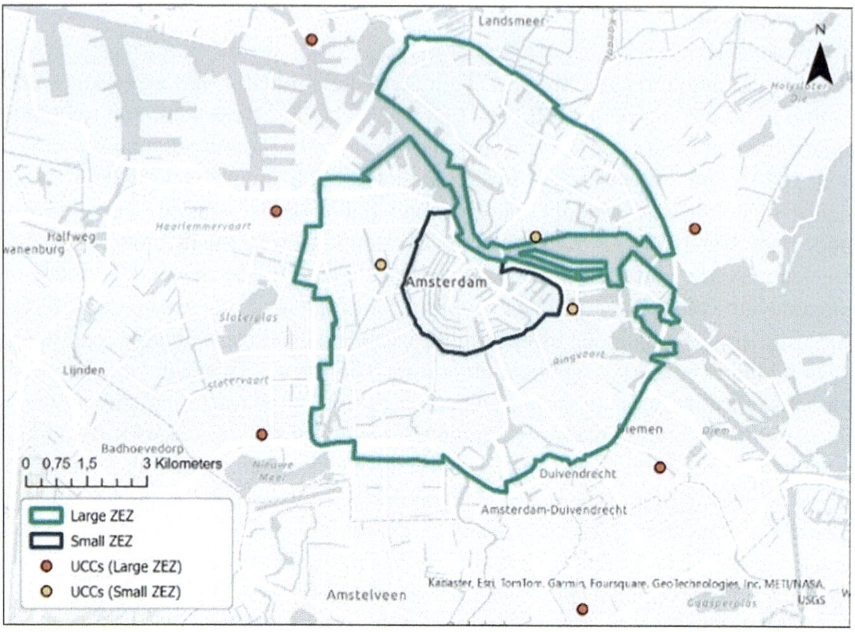

Fig. 8 Small and large zero-emission zone (ZEZ) scenarios and their respective UCCs

the ZEZ (see Fig. 8). The smaller example follows the S100 orbital road, and the larger example follows the A10 highway. In both examples, the (ring) road itself (i.e. S100 and A10) is not considered to be part of the ZEZ. The MASS-GT module simulates the behavioural impacts: existing transports with conventional vehicles are either replaced by zero-emission vehicles (electric trucks or vans) or they are redirected to urban consolidation centres (UCC) on the edge of the zero-emission zone, and the last-mile is organised from these UCCs using zero-emission vehicles. Due to the scope of the research, it is assumed that all last-mile transportation from the UCC is done by electric light commercial vehicles (ELCV). The behavioural parameters are based on previous empirical studies (City of Rotterdam 2019; de Bok et al. 2021b).

The objective of the case study is to analyse the spatial distribution of impacts. Outputs confirmed that in the scenarios analysed, the neighbourhoods of lower socio-economic status bear the largest part of negative effects, while higher income neighbourhoods, often located closer to city centres and therefore ZEZs, experience an increase in air quality (see Fig. 9). One example of this is the neighbourhood of Slotermeer, located in the west of Amsterdam, close to the middle UCC in the large ZEZ scenario. The average income in Slotermeer is €26.350 (CBS 2023) and from the relative NO_2 chart in Fig. 6, a large increase in emissions can be seen throughout the neighbourhood. In contrast, the neighbourhood of Jordan, located on the western side of the city centre of Amsterdam, has a population with an average income of €45.500 (Antosz et al. 2019) and is shown to experience a large decrease in

Fig. 9 Difference in total traffic intensity for the large ZEZ scenario (left; green = reduction, red = increase) and difference in NO_2 emissions (right; green = reduction, red = increase)

overall NO_2 emissions. This warrants attention, since the placement of a UCC will determine the location of these negative externalities, potentially causing emissions to be shifted to lower socio-economic neighbourhoods.

2.4.3 Discussion

This study exemplifies how simulation models are developing into dynamic toolboxes, evolving incrementally from digital shadow to a digital twin. This study shows how an agent-based simulator for urban freight traffic (MASS-GT) is integrated into a digital twin framework (TNO Digital Twin), which brings together existing policy support tools for traffic analysis, emission modelling and urban environmental impact assessment. Such a framework allows the assessment of impacts across different aspects with high spatial granularity. The use case shows how diverse the spatial distribution of these impacts may be.

Another aspect of the modular structure of a DT toolbox is that the data formats and interfaces between modules are automated, which allows more efficient updates with new observed data compared to conventional (freight) transport models, e.g. once every 4 years. We consider the presented Digital Twinning toolbox as a logical step from a digital shadow towards a digital twin.

3 Concluding Remarks

Together, the cases from the different EU and non-EU projects suggest a clear evolutionary line for DTs for urban freight, which has materialised so far. Individual models that provide answers to policy questions evolve further into coupled models, which fulfil complementary roles and together are larger than their sum. Finally, libraries of models can create flexible digital twins, which boost the applicability of the DT for a multitude of cases. DTs that already contain such a library can continue to absorb models to elevate their capabilities. This is only the start of a larger change, however. DTs will evolve further into the future, stepping beyond their current function of digital shadows of systems and becoming more instrumental for urban freight decision-making and management. As next steps, DTs for city logistics will focus on the following:

- Explicit connections to city sensors and databases—this can build on digitalised connected assets in IoT-like structures, and on aggregation processes for data that help to monitor longer-term, stable developments.
- Explicit connections to actuators in the system, be it policy processes, investment planning processes or operational management systems. Today, the impact of DTs is still indirect, largely without automation or pre-programmed connections to decisions.
- Once these connections are in place, DTs will allow faster and automated decision-making for long-cycled and short-cycled decisions by closing the control loop and installing (partial or full) autonomous decision-making and management.

The DTs presented here are a prerequisite for the above-mentioned development. It seems likely that the above-mentioned evolution of closing the control loop will occur domain by domain, or sector by sector, as also in the physical world control across domains and sectors is still an important challenge. Therefore, we position these projects as cornerstones for a roadmap which contains different lines of development that, when converged, will eventually lead to cross-domain, city-level DTs with full control capabilities (Fig. 10).

We argue that such an evolution is necessary for targeted progress, as it sets a clear agenda for R&D and valorisation of technology. This should help to drive the next steps forward in applied computing technology and decision-making processes, ultimately allowing convergence towards fully functional digital twins for city logistics.

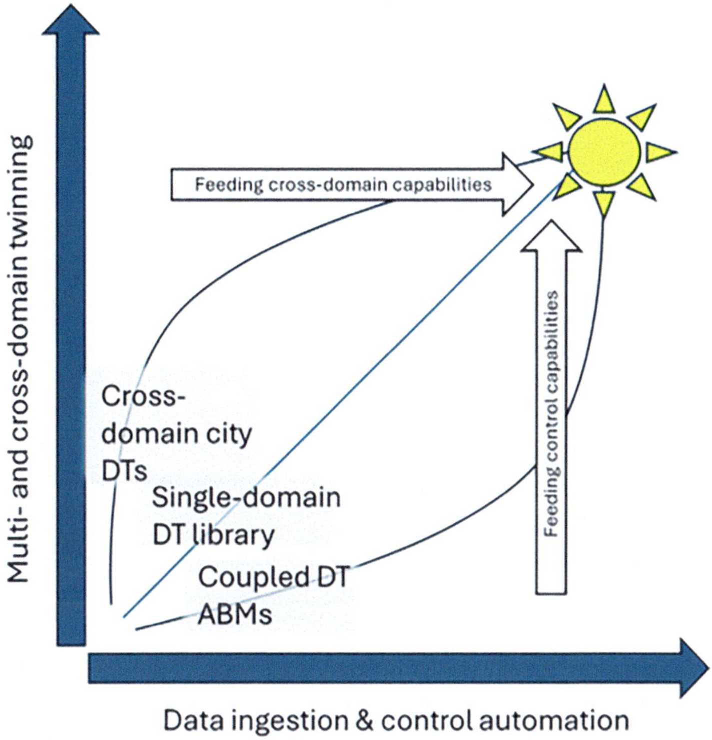

Fig. 10 DT roadmap with cases as cornerstones

References

Tavasszy LA, Browne M, Piecyk M (eds) (2025) Freight transport planning. Elsevier, Amsterdam

Belfadel A, Hörl S, Tapia RJ, Politaki D, Kureshi I, Tavasszy L, Puchinger J (2023) A conceptual digital twin framework for city logistics. Comput Environ Urban Syst 103:101989

Marcucci E, Gatta V, Le Pira M, Hansson L, Bråthen S (2020) Digital twins: a critical discussion on their potential for supporting policymaking and planning in urban logistics. Sustainability 12(24):10623

Abouelrous A, Bliek L, Zhang Y (2023) Digital twin applications in urban logistics: an overview. Urban Plan Transp Res 11(1):2216768

Tapia RJ, Tavasszy L (2024) Digital twins for freight planning. In: Tavasszy LA, Browne M, Piecyk M (eds) Freight transport planning. Elsevier, Amsterdam

Jeong DY, Baek MS, Lim TB, Kim YW, Kim SH, Lee YT et al (2022) Digital twin: technology evolution stages and implementation layers with technology elements. IEEE Access 10:52609–52620

Nag D, Brandel-Tanis F, Pramestri ZA, Pitera K, Frøyen YK (2025) Exploring digital twins for transport planning: a review. Eur Transp Res Rev 17(1):15

Grotto A, Casas PFI, Zubaryeva A, Sparber W (2024) Formalizing sustainable urban mobility management: an innovative approach with digital twin and integrated modeling. Logistics 8(4):117

Taniguchi E, Thompson RG, Qureshi AG (2024) Recent developments in urban freight analytics for collaborative city logistics. Transp Res Proc 79:3–12 de Bok M, Giasoumi S, Tavasszy L, Thoen S, Nadi A, Streng J (2024) A simulation study of the impacts of micro-hub scenarios for city logistics in Rotterdam. Res Transp Bus Manag 56:101186

Gürcan Ö, Antosz P, Szczepanska T, Falck V, Cebeci M, de Bok M, Tavasszy LA, Tapia R (2025) Urbane city logistics modelling framework and agent-based models, D3.2, urbane consortium

Le TV, Li X, Le TA (2024) National Digital Twins for logistics and supply chain systems: outlooks, components, enablers, and architecture

Tavasszy L, Quak H (2023) Computing technology and its applications in urban logistics. In: Monios J, Budd L, Ison S (eds) The Routledge handbook of urban logistics. Routledge

Batty M (2018) Digital twins. Environ Plan B Urban Analy City Sci 45(5):817–820

de Bok M, Tavasszy L, Kourounioti I, Thoen S, Eggers L, Nielsen VM, Streng J (2021a) Application of the HARMONY tactical freight simulator to a case study for zero emission zones in Rotterdam. Transport Res Rec

Jager W, Antosz P, Li T, Polhill G, Szczepanska T, Wang S (2025) HUMAT: an integrated framework for modelling individual motivations, social exchange and network dynamics. J Artif Soc Soc Simul 28(1):4

de Bok M, Tavasszy L, Thoen S, Eggers L, Kourounioti I (2025a) MASS-GT: an empirical model for the simulation of freight policies. Simul Model Pract Theory 103140

Cebeci MS, Tapia RJ, Nadi A, Bok MD, Tavasszy L (2024) Does crowdshipping of parcels generate new passenger trips? evidence from The Netherlands. Transp Res Rec 2678(6):360–375

Tapia RJ, Kourounioti I, Thoen S, de Bok M, Tavasszy L (2023) A disaggregate model of passenger-freight matching in crowdshipping services. Transp Res A Policy Pract 169:Article 103587. https://doi.org/10.1016/j.tra.2023.103587

Winkelmann F, Nandakumar A, Spruijtenburg D, de Bok M (2025) Exploring local impacts of urban freight policies with a digital twin for Amsterdam. In: Proceedings of the 13th International Conference on City Logistics

Antosz P, Jager W, Polhill G, Salt D, Alonso-Betanzos A, Sánchez-Maroño N et al (2019) Simulation model implementing different relevant layers of social innovation, human choice behaviour and habitual structures. SMARTEES Deliverable

CBS (Centraal Bureau voor de Statistiek) (2023) Kerncijfers per postcode. https://www.cbs.nl/nl-nl/dossier/nederland-regionaal/geografische-data/gegevens-per-postcode Accessed on 29 Jan 2025

City of Rotterdam (2019) Roadmap ZECL: moving towards Zero Emission City logistics (ZECL) in Rotterdam in 2025. City of Rotterdam, Rotterdam

de Bok M, Tavasszy L, Kourounioti I, Thoen S, Eggers L, Mayland Nielsen V, Streng J (2021b) Application of the HARMONY tactical freight simulator to a case study for zero emission zones in Rotterdam. Transp Res Rec 2675(10):776–785. https://doi.org/10.1177/03611981211012694

de Bok M, Tavasszy L, Thoen S, Eggers L, Kourounioti I (2025b) MASS-GT: an empirical model for the simulation of freight policies. Simul Model Pract Theory 142. https://doi.org/10.1016/j.simpat.2025.103140

Gemeente Amsterdam (2024) Uitgangspunten Verkeersmodel Amsterdam 5.0 Microsoft Word - UitgangspuntenVMA5.0_1_0.docx

Lohman W, Cornelissen H, Borst J, Klerkx R, Araghi Y, Walraven E (2023) Building digital twins of cities using the inter model broker framework. Futur Gener Comput Syst 148:501–513. https://doi.org/10.1016/j.future.2023.06.024

Franklin R, Fergadiotou I, Kampa M, Antosz P, Gürcan Ö et al (2025) The URBANE innovation transferability platform: learnings for decarbonizing last-mile delivery networks. In: Nathanail EG, Gavanas N, Adamos E (eds) Climate crisis and resilient transportation systems. The 7th Conference on Sustainable Mobility (CSuM 2024), Lecture notes in intelligent transportation and infrastructure. Springer, Cham. https://doi.org/10.1007/978-3-031-82818-8_29

Gürcan O, Szczepanska T, Falck V et al (2024) A generic modelling framework for last-mile delivery systems. In: Proceedings of the 19th Social Simulation Conference (SSC'2024), Cracow Poland, 16–20 September 2024

Sustainable E-commerce

Sustainability Score of Last-Mile Delivery Strategies: Conceptual Approach and Application in European Cities

José Manuel Vassallo, Natalia Sobrino, and Laura Garrido

1 Introduction and Background

Last-mile delivery (LMD), i.e. the last stage of the goods supply chain, represents one of the major logistics challenges in urban environments. LMD involves many actors and stakeholders with diverging goals and perspectives (shippers, carriers, logistics operators, governments, regulators, customers, etc.), and is characterised by high fragmentation, use of smaller vehicles (usually light-duty ones), low load factors and high degree of empty running, short routes, low speed driving, short effective driving time and long vehicle idle times (Bosona 2020). It is also subject to increasingly demanding customer service levels, dynamic demand, a high level of widespread destinations, infrastructure limitations (dedicated parking and loading space in cities) and regulatory issues (vehicle sizing, access and timing restrictions, congestion pricing) (Corti et al. 2024; Mangiaracina et al. 2019; Olsson et al. 2019). All the above reduce the efficiency of LMD while increasing its complexity and cost and making it a significant source of road congestion, pollutant emissions and inefficient use of urban space (Allen et al. 2017). In a context of increasing urbanisation and population growth, where e-commerce continues to expand, these challenges intensify, putting additional pressure on urban infrastructures and generating negative impacts on the quality of life of cities.

To address this issue, various LMD strategies have been implemented or proposed. Several authors have attempted to classify measures aimed at reducing the negative impacts of LMD according to different criteria. For instance, Russo and Comi (2010) differentiated between measures related to (i) material infrastructure, (ii) immaterial infrastructure, (iii) equipment and (iv) governance; Ranieri et al.

J. M. Vassallo (✉) · N. Sobrino · L. Garrido
Transport Research Centre (TRANSyT), Universidad Politécnica de Madrid, Madrid, Spain
e-mail: josemanuel.vassallo@upm.es

A. Melkonyan-Gottschalk, Y. Li (eds.), *Urban Logistics Transformation*, Greening of Industry Networks Studies 15,
https://doi.org/10.1007/978-3-032-08387-6_7

(2018) classified them into (i) innovative vehicles, (ii) proximity stations or points, (iii) collaborative and cooperative urban logistics, (iv) optimisation of transport management and routing and (v) innovations in public policies and infrastructures; and Boggio-Marzet (2022) proposed four categories depending on whether the measures were related to operation, fleet, governance or infrastructure. The following are some of the measures included in the aforementioned classifications, grouped by their functional focus. Those related to vehicle fleet-related strategies include the adoption of low-carbon delivery fleet configurations such as electric vans, cargo bikes or tricycles, autonomous vehicles and other non-motorised modes, as well as the development of recharging infrastructure. Access management measures encompass temporary access restrictions, environmental regulations or access limitations based on vehicle size or load. Urban space and infrastructure-based approaches involve dedicated parking, freight curbside management and the implementation of consolidation facilities such as micro-hubs, mobile depots or Urban Consolidation Centers (UCCs). Service innovation includes crowdshipping, collaborative logistics models or the integration of parcel deliveries into public transport systems. Finally, customer-focused solutions cover self-service parcel collection methods such as in-store pick-up, proximity stations or parcel lockers.

While these solutions can contribute to reducing the environmental footprint of freight urban transport, they also have drawbacks or limitations. For example, compared to internal combustion engine vehicles, electric vehicles have a higher purchase price and a lower technological and operational performance (require charging infrastructure, have a lower range), which discourages their adoption by logistics operators (Quak et al. 2016). In turn, key challenges in establishing UCCs include their high cost, the complexity of coordinating multiple transport operators to use a shared facility, governance-related difficulties and the need to identify and categorise goods from various companies, among other issues (Gonzalez et al. 2023a, b). This diversity of options and contexts underscores the need for robust analytical frameworks capable of assessing their impacts holistically through the lens of sustainability, ensuring that environmental, social and economic dimensions are coherently addressed.

The concept of sustainability has gained momentum since the term "sustainable development" was defined by the Brundtland Commission in 1987 as "meeting the needs of the present without compromising the ability of future generations to meet their own needs" (Brundtland 1987). Sustainability is usually assessed following the triple bottom line (TBL) approach (Elkington 1998), which establishes three interdependent dimensions: economic, social and environmental (Alhaddi 2015). Applied to urban logistics, a sustainable solution must fulfil the present needs of customers and businesses while safeguarding the environment and ensuring long-term social and economic welfare (Gudmundsson et al. 2016). That includes reducing the environmental footprint of transportation and delivery, supporting economic development and ensuring access to goods and services for all members of society now and in the future (Lyons and McDonald 2022).

The United Nations (UN) 2030 Agenda and the Sustainable Development Goals (SDGs) reinforce the need to promote more inclusive, safe, resilient and sustainable cities (SDG 11), as well as to take urgent action on climate change (SDG 13). In this context, urban freight distribution —particularly in the last mile— emerges as a key component in achieving these goals. However, despite its relevance, tools to assess the sustainability of urban logistics strategies remain limited and, in many cases, inadequate to reflect the specificities of different urban areas.

In this context, there is a need for a comprehensive sustainability assessment of LMD strategies that considers not only logistics efficiency but also their environmental and social impacts. However, as it will be shown in the next section, there is a lack of methodologies to compare solutions in different scenarios and cities that take into account the city context of the area where the solutions are to be implemented. This chapter presents a sustainability assessment method that comes up with sustainability ratings of LMD strategies, incorporating expert stakeholders' opinions and considering the specific geographical and social context of the urban area under study.

2 Overview of Sustainability Assessment Methods and Existing Challenges

2.1 Overview of Sustainability Assessment Methods

Sustainability assessment has evolved considerably in recent decades, from approaches focused exclusively on environmental impact to integrative frameworks that simultaneously consider economic, social and environmental dimensions. This evolution responds to the acknowledgement of contemporary problems —such as climate change, inequality or biodiversity loss— that cannot be addressed from a single perspective. Assessing the sustainability of policies, technologies and practical solutions has become a fundamental tool for decision-making in various sectors, including urban planning, mobility and logistics.

There are numerous methods and techniques that can be used for assessing the sustainability of urban logistics solutions. These can be broadly classified into economic and multi-criteria techniques (Holguín-Veras et al. 2015).

Economic analyses are often employed in policy and planning to assess both the cost-effectiveness of a given option and to identify which alternative yields the highest overall benefits (Browne and Ryan 2011). These techniques quantify the impacts, either costs or benefits, of a solution in economic units, using economic valuation techniques to monetise non-market costs and benefits. Discount factors are often applied to the benefits and costs incurred over a time horizon to produce economic performance indicators such as net present value, internal rate of return or benefit-cost ratio of the solution. Different economic approaches, such as the cost-benefit analysis (CBA) or the social cost-benefit analysis (SCBA), have been used to evaluate urban logistics solutions (i.e. Van Duin et al. 2008; Kin and Macharis 2015; Gonzalez-Feliu 2016). While the classic CBA only considers the monetary costs (investment and operational) and financial benefits of a solution, its extended version, the SCBA, also takes into account the social and environmental costs and benefits to the extent that they can be quantified.

Although economic analysis is commonly applied to other transportation policy initiatives, its application in the context of urban logistics remains relatively limited (Gonzalez-Feliu 2017). The first study that applied CBA to urban logistics was the one by Van Duin et al. (2008), where a CBA was conducted for assessing the mid-term feasibility of various scenarios of UCC development. A few more authors have discussed the potential of this method to evaluate city logistics solutions but incorporating, to a lesser or greater extent, societal/environmental impacts (Gonzalez-Feliu et al. 2013; Holmgren 2018). For instance, de Langhe (2014) proposed the inclusion of social costs/benefits, such as emissions and other externalities, besides private costs/benefits when assessing the feasibility of rail for urban freight distribution. Gonzalez-Feliu (2016) also proposed and applied an SCBA-based analysis framework for assessing the potential deployment of a freight tramway. The framework included social and environmental non-monetary costs (pollution and noise of railway works, potential decrease of customers for retailing and service activities around works) and benefits (GHG emissions, pollution and noise savings, congestion decrease), although only GHG and pollution savings were considered in the final assessment.

Kin and Macharis (2015) also applied an SCBA to evaluate a private UCC already in operation and compare it with the previous situation, where transport operators delivered the goods themselves throughout the designated urban area. The analysis considered direct effects (capex, opex and revenues) and external effects (air pollution, climate change, noise, accidents, congestion and marginal infrastructure costs), which were monetised using willingness to pay valuation. However, indirect effects (e.g. service level, employee satisfaction, quality of life) could not be included given that their monetisation was not possible.

Multi-criteria decision-making (MCDM) methods are tools designed to help decision-makers rank or select the best solution among a set of alternatives, which have different performances with respect to several key indicators that can be both quantitative and qualitative (Mardani et al. 2015). These methods allow options to be compared and assessed against multiple, often heterogeneous and even conflicting criteria, such as economic, social or environmental (Watróbski et al. 2017). A key component of these methods is the incorporation of the preferences of the different actors involved in the process, which is achieved by assigning scores and weights. Multi-criteria weights represent the relative degree of importance given to each criterion within the decision process. By quantifying the importance, the weights allow the different alternatives to be compared in a structured way, giving greater importance to the factors that are considered most relevant to the objectives of the analysis.

In contrast to CBA approaches, MCDM methods have been largely used to assess the sustainability of city logistics solutions. Researchers such as Jamshidi et al. (2019), Alvarez Gallo and Maheut (2022, 2023) and Corti et al. (2024) have carried out comprehensive literature reviews on the use of MCDM approaches to assess the sustainability of last-mile solutions. These studies identify MCDM methods as key tools for evaluating the sustainability of urban freight logistics strategies (i.e. Deveci et al. 2022; Tadić et al. 2018), the best location for a city logistics project or service (i.e. Wang et al. 2023; Rao et al. 2015) or the suitability of low-carbon delivery fleets (i.e. Saha et al. 2023; Watróbski et al. 2017). In their literature reviews, the authors also examined the MCDM techniques applied to sustainable logistics initiatives in urban environments, and the relevant criteria and sub-criteria mostly used for decision-making in this field. For instance, Alvarez Gallo and Maheut (2023) found that the Analytical Hierarchical Process (AHP) and its fuzzy variant (F-AHP) have been the preferred methods for evaluating sustainable urban logistics alternatives (Silva et al. 2023; Katsela and Pålsson 2019; Janjevic et al. 2019; Awasthi and Chauhan 2012). However, they could not establish any direct link between the decision problem being addressed and the specific MCDM methods applied.

While some studies apply a single MCDM method to make an assessment, many academic contributions propose hybrid models that combine MCDM methods to leverage their complementary strengths. For instance, in the context of urban logistics strategy decision-making, Tadić et al. (2014) proposed a hybrid model integrating fuzzy Decision Making Trial and Evaluation Laboratory (DEMATEL), fuzzy analytic network process (FANP) and fuzzy VIKOR models, while Krstić et al. (2021) proposed a hybrid MCDM model combining fuzzy Delphi, fuzzy FARE and fuzzy VIKOR methods. On their part, Watróbski et al. (2017) proposed a hybrid approach that combines PROMETHEE II and fuzzy TOPSIS to support the selection of electric freight vehicles in urban logistics contexts.

Another commonly used method to assess urban logistics solutions is Multi-Actor Multi-Criteria Analysis (MAMCA), which emphasises the importance of involving different stakeholders in the decision-making process at a very early stage (Macharis et al. 2010). For instance, Aljohani and Thompson (2020) used MAMCA to evaluate and rank different configurations of green delivery fleets for consolidated deliveries. A more recent example is provided by the study of Jardas et al. (2024), who used MAMCA to assess different potential last-mile delivery strategies involving UCCs and environmentally friendly vehicles.

Some academic works have made a hybrid use of economic and MCDM methods to improve decision-making for sustainable city logistics solutions. For instance, Balm et al. (2014) developed an evaluation framework comprising three methods —SCBA, business model analysis and MAMCA— to assess innovative urban freight transport solutions in the context of the STRAIGHTSOL project. On their part, Velickovic et al. (2025) applied a combination of SCBA and AHP methods for assessing the potential of different strategies involving a UCC in mitigating urban freight transport negative externalities, taking into account their financial viability.

Finally, there are different global sustainability indices and sustainability reporting frameworks in the field of logistics and supply chain that might be useful for practitioners to incorporate sustainability principles into the assessment of urban logistics strategies. Some examples reviewed by Zhang (2024) include the UN

SDGs framework for sustainable development, the Global Reporting Initiative (GRI) Standards framework for sustainable reporting, the SmartWay programme, the Global Logistics Emissions Council (GLEC) framework and the European Sustainability Reporting Standards (ESRS) for Road Transport Sector (currently under development). These frameworks address different perspectives of sustainability depending on their purpose. While the UN SDGs and GRI Standards adopt an index and indicator-based approach that encompasses economic, environmental and social dimensions, the ESRS for road freight transport sets its sustainability reporting using ESG (Environmental, Social, Governance) indicators, and SmartWay and the GLEC framework concentrate exclusively on the environmental dimension of sustainability (Zhang 2024).

Some of the above-mentioned frameworks have been used for assessing the sustainability performance of city logistics. For instance, Andruetto et al. (2022) proposed a sustainability performance assessment framework of Key Performance Indicators (KPIs) that combined SDGs with existing indicators in the city logistics literature. Then, Andruetto (2022) applied the aforementioned framework to evaluate qualitatively the impacts of four urban logistics solutions —consolidation, electrification, cargo bikes and automation— on sustainability, based on results from the literature. Another framework developed for measuring the sustainability performance of city logistics is that of Zhang (2024), who complemented the GRI indicator framework with literature and industry-based indicators. The author then assessed the sustainability performance of road transport in city logistics in different European capitals using secondary data at the city level.

2.2 Existing Challenges

Although economic and multi-criteria evaluation methods have been widely used to assess policies and intervention strategies in urban contexts, their application to last-mile distribution still poses some methodological and practical challenges. Both methods have the potential to provide valuable information for decision making, but have significant limitations when applied to complex, dynamic and multistakeholder urban logistics systems.

The CBA approach has important methodological limitations when confronted with highly complex urban contexts. First, the quantification of non-market impacts objectively and comparably is complicated. Many key impacts associated with urban logistics —such as noise, use of public space, health effects or social aspects— are difficult to monetise, which may lead to a systematic underestimation of social and environmental externalities (Browne and Ryan 2011). Second, CBA does not consider who gains and who loses from an intervention. Distributional analysis, central to social sustainability, is often absent or underdeveloped in this type of assessment (Holmgren 2018). Finally, the establishment of the time horizon and, especially, the discount rate has been a controversial issue, since it can have a large impact on the results (Mouter 2018; Almansa Sáez and Calatrava Requena 2007). CBA tends to apply discount rates that reduce the weight of future impacts, which can be problematic in decisions related to climate change or long-term urban transformation. Ultimately, while economic methods provide valuable support in decision-making processes, their ability to capture all dimensions of sustainability is limited (Bueno et al. 2015).

Multi-criteria methods have been promoted as a more flexible and qualitatively rich alternative to address complex decisions with multiple dimensions. Unlike CBA, MCDM allows the simultaneous incorporation of environmental, social, economic, technical, operational, political or legal criteria (Alvarez Gallo and Maheut 2023), without the need to quantify them in monetary units. This approach is particularly suitable for the evaluation of LMD solutions, where multiple stakeholders with divergent interests are involved (Jamshidi et al. 2019) and aspects that are difficult to compare with each other are assessed (Browne and Ryan 2011).

However, MCDM methods also face significant challenges in their practical application. First, the definition of relevant criteria and sub-criteria and their relative weighting is often subjective and context-dependent. While this allows for adaptability, it also introduces risks of bias and high dependence on expert judgment or the participatory process used. Regarding the latter, the quality and legitimacy of an MCDM largely depend on a robust participatory process, where all stakeholders are sufficiently represented and are allowed to express their priorities. In practice, many studies fail to sufficiently involve different stakeholders (Alvarez Gallo and Maheut 2023), which can compromise the validity and usefulness of the results.

Also, although MCDM allows specific cases to be assessed in detail, its comparative applicability between different urban contexts remains limited. Differences in criteria structure, sustainability issues, local values or data availability make it difficult to create homogeneous frameworks that allow comparability between cities and strategies. Finally, given the qualitative nature of many social and environmental criteria, some MCDM frameworks largely rely on qualitative indicators (Aljohani and Thompson 2020), which are often difficult to measure in an objective way.

The sustainability assessment method proposed in this chapter offers a practical approach to bridging the methodological divide between traditional economic analysis and stakeholder-centred approaches such as MAMCA. While CBA adopts a strictly economic perspective and often disregards non-magnetisable impacts —particularly key social dimensions critical to sustainability— MAMCA approaches that rely exclusively on stakeholder opinions may fail to account for the contextual variability of sustainability challenges across different urban settings. The proposed framework addresses these limitations by integrating both monetary and non-monetary impacts of LMD solutions, incorporating stakeholder input and aligning with the specific characteristics of the urban environment.

3 Method for Setting the Sustainability Scores of Last-Mile Delivery Strategies

This section describes a new method for assessing the sustainability of last-mile delivery (LMD) strategies. The approach, inspired by the Sustainability Tool for the Appraisal of Road Projects (STAR) Method developed by Bueno Cadena and Vassallo Magro (2015), is designed to support decision-makers and logistics operators in identifying the most sustainable LMD solution. The method follows a multi-criteria framework and comprises three steps, as illustrated in Fig. 1. In this context, the most sensitive criteria within each urban area —those with the highest level of importance— should be assigned greater weights, as they are interpreted as constraints that provide valuable support for the sustainability of the proposed strategy.

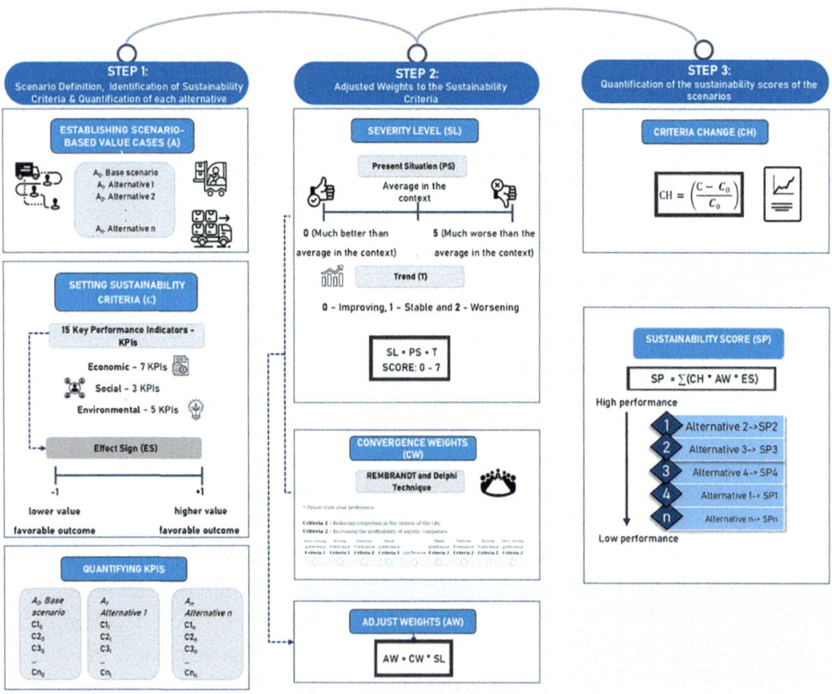

Fig. 1 Methodological framework for setting the sustainability score of last-mile delivery strategies

3.1 Step 1: Scenario Definition, Identification of Sustainability Criteria and Quantification of each Alternative

This step consists of defining and characterising the scenarios of the value-case solutions to be assessed compared to the baseline (do-nothing) scenario. In addition, a set of sustainability criteria measured through KPIs are defined. These KPIs will be quantified for all the LMD solution scenarios, including the baseline. Here, a more detailed explanation of this step is provided.

Scenario-based value case definition. A set of scenario-based value cases, each representing a different LMD solution, will be defined to compare with the baseline scenario (the business-as-usual, BAU). As previously mentioned, these solutions may involve different low-carbon vehicle fleet configurations, parcel lockers, UCCs or crowdshipping, among others. Each alternative influences sustainability criteria (reducing air pollution, increasing operational efficiency or improving the quality of life in the city) in a different way. Through the development of these scenario-based cases, it is possible to evaluate the variation of all the KPIs of each solution compared to the BAU scenario.

Identification and quantification of sustainability criteria. For each specific policy or case study, a range of sustainability criteria (in the form of KPIs) are set to characterise the well-known sustainability triple bottom line dimensions: economic efficiency, environmental preservation and social considerations (Alhaddi 2015). The way of measuring each KPI within the assessment framework should be clearly defined to provide a comprehensive understanding of its meaning. An Effect Sign (ES) is also applied to each KPI based on its impact (positive or negative) on sustainability. Specifically, $ES = +1$ is assigned when a higher value represents a more favourable outcome (i.e. "more is better"), and $ES = -1$ is assigned when a lower value indicates the appositive (i.e. "more is worse"). For instance, regarding energy consumption or delivery costs, "more is worse", while, for job creation, "more is better". The signs ensure that all KPIs align consistently with the sustainability objectives and allow for accurate comparison and assessment of different scenarios. Many criteria can be considered to evaluate LMD sustainability, with numerous academic studies exploring the selection of these KPIs (i.e. Giret et al. 2018). However, the method can be easily adjusted to the KPIs that can be effectively measured, given data availability constraints.

For the application that is provided in the Chapter of this book, the selection made by Gonzalez et al. (2023a) using a Consequence Analysis Framework (Licata 1978) and grounded in consensus among expert stakeholders within the LEAD project (EC 2023) is adopted. As a result, a total of 15 criteria were selected, aligning with the triple bottom line of sustainability:

- Social criteria*:* job creation, quality of the jobs and neighbourhood quality of life.
- Environmental criteria: energy consumption, GHG emissions, NOx emissions, PM emissions and noise pollution.

- Economic criteria: average delivery cost, congestion, urban storage and parking space, financial internal rate of return (FIRR), shop retail benefits, delivery time and delivery reliability within the time windows.

After the identification and definition of criteria, the next step is to quantify the KPIs for each scenario. Depending on their nature, the criteria can be measured on either quantitative or qualitative scales (Bueno Cadena and Vassallo Magro 2015; Vassallo and Bueno 2021). Data sources could be obtained from different channels, such as direct data from logistics operators, simulation-based or survey-based.

3.2 *Step 2: Assignment of Adjusted Weights to the Sustainability Criteria*

As part of the multi-criteria framework, this step involves obtaining Adjusted Weights (*AW*) to determine the relative impact of each criterion (i) on sustainability in each specific context. The weighting coefficients ($AW_{i,k}$) assigned to the sustainability criteria in each city or urban context (k) are based on the combination of expert stakeholders' preferences (Convergent Weights —CW_i) and the sensitivity of each sustainability criteria in the urban area in which the LMD strategies are implemented (Severity Level —$SL_{i,k}$) (see Eq. 1). The procedure combines the REMBRANDT system (Olson et al. 1995), the Delphi method for the *Convergent Weights* and the *Severity Level* to characterise the context (Vassallo and Bueno 2021). The following is a summary of the main steps for obtaining CW_i and $SL_{i,k}$. The work by Gonzalez et al. (2023b) provides a detailed description of the weighting framework.

$$AW_{ik} = CW_i \cdot SL_{ik} \forall i,k \in C \tag{1}$$

Convergent Weights (CW_i) is the weighting coefficient corresponding to criteria i from the general opinions of expert stakeholders without knowing where the solution is implemented (with both theoretical and practical backgrounds). Bueno Cadena and Vassallo Magro (2015) proposed a structured four-step process to obtain the CW_i based on the REMBRANDT technique and Delphi method as follows:

- Design a questionnaire that includes pairwise comparisons among the previously identified sustainability criteria list. Experts should be asked to compare the importance of different sustainability criteria based on a −8 to +8 scale known as the REMRANDT scale.
- Select experts in sustainability to complete the survey. Application of the questionnaire. Experts are selected to collect all the views of the urban logistics landscape. A minimum of 30 responders is required to obtain trustworthy results.
- Survey each expert through a matrix of preferences using the REMBRAND technique described by Olson et al. (1995).

- Conduct a statistical test to evaluate the experts' consensus level. The statistical test consists of a simulation based on a cross-validation technique.
- Obtain Convergent Weights. Firstly, the creation of a matrix of criteria comparisons, where δ represents the median importance of criterion i compared to criterion j, after getting the consensus. Secondly, the creation of a matrix using the geometric scale values calculated by raising e. The preference ratio (γ) used in this process was ln, as recommended by Olson et al. (1995). Next, the estimation of the geometric mean for each criterion i is conducted. Finally, a normalisation of each criterion is made by dividing its geometric mean by the sum of all geometric means. This approach allows criteria weights to be derived based on expert judgments and ensures a comprehensive evaluation of the criteria.

Severity Level ($SL_{i,k}$) is the weighting coefficient that considers the sensitivity of the sustainability criteria i to the urban context of the city k where the LMD strategies will be implemented. The $SL_{i,k}$ is a composite indicator that comprises both the *Present Situation* and *Trend* for the selected sustainability criteria in each urban context (Eq. 2).

$$SL_{ik} = PS_{ik} + T_{ik} \forall i,k \in C \tag{2}$$

- The *Present Situation* ($PS_{i,k}$) for each criterion i in city context k is estimated on a score ranging from 0 (the attribute is much better or moderately better than the average in context) to 5 (the attribute is much worse than the average in context). This value compares the current situation of an attribute for a specific urban context j with the average current situation of that attribute for the average value for other similar contexts. For instance, considering the KPI unemployment rate, the value of $PS_{i,k}$ will be much higher in a city with a high unemployment rate compared to a city with a low unemployment rate.
- The *Trend* ($T_{i,k}$) measures whether a certain sustainability criterion i is expected to improve or worsen in the future in a specific urban context k. The T can be categorised as "improving", "stable" or "worsening", and allocates 0 points, 1 point and 2 points, respectively.

Finally, the *Severity Level Index* varies in a range from 0 to 7. As the *SL* increases, the criterion becomes more context sensitive.

3.3 Step 3: Quantification of Sustainability Scores of the Scenarios

The third step quantifies the Sustainability Performance (SP_{ijk}) of each scenario A_j. This will help select the option with the best sustainability score. The process starts by calculating the change of KPI_i (CH_i) relative to the business-as-usual scenario (A_0), according to Eq. 3. The use of a relative calculation framework allows for the normalisation of performance across criteria with differing units and magnitudes, ensuring methodological consistency by focusing solely on deviations from the BAU scenario.

$$CH_{ij} = \frac{\left(C_{ij} - C_{i0}\right)}{C_{i0}} \forall i \in C, \forall j \in A \tag{3}$$

Finally, for a given city k the Sustainability Performance SP_{jk} of each scenario j is calculated by multiplying the change rates (CH_{ij}) times the corresponding adjusted weights (AW_{ik}) with its effect sign (ES_i) of each criterion, following Eq. 4.

$$SP_{jk} = \sum_{i} CH_{ij} AW_{ik}\ ES_i \forall i, k \in C, \forall j \in A \tag{4}$$

The best alternative is the one with the highest score. However, the analyst should also provide a descriptive comparison of each criterion across alternatives. After breaking down sustainability performance, these results should be presented alongside the overall evaluation.

4 Application to Case Studies in European Cities

To illustrate the applicability of the sustainability assessment method described in Sect. 3, this study examines its application to three European cities—Madrid, The Hague and Porto—which participated in the European LEAD research project (EC 2023) developing different pilots. In fact, European cities are increasingly recognising the need to adopt sustainable LMD solutions in response to growing environmental, operational and urban planning challenges. For example, Madrid, as Spain's capital and major logistics hub, has advanced policies like the "Madrid 360 Strategy" (Ayuntamiento de Madrid 2023a) and Low Emission Zone (LEZ) initiative (Ayuntamiento de Madrid 2023b) to reduce vehicle emissions, promote cleaner freight distribution and support innovations such as micro hubs for e-cargo bikes. Similarly, The Hague is tackling high car dependency and freight intensity through LEZs (The Hague 2022), urban renewal plans and emerging delivery models like crowdshipping (Balm 2022), aiming to cut down emissions and reducing congestion while fostering social inclusion. Facing rising e-commerce demand and limited logistics space, Porto enforces access regulation (Porto 2023) and prioritises strategic site allocation to enhance air quality and efficiency. These cities illustrate the urgent and context-specific need for sustainable LMD strategies that align with broader urban sustainability goals across diverse metropolitan areas.

4.1 Description of LMD Strategies to Be Evaluated in Different Cities

In Madrid, the sustainability solution analysed is the implementation of a UCC in an underground car parking at the city centre, enabling cross-docking from vans to low-carbon vehicles for B2C deliveries in the central district. The strategy involves a freight operator company and the public entity in charge of managing public infrastructure, such as the underground parking. In the BAU scenario, the logistics operator makes direct delivery (without cross-docking) using diesel internal combustion engine cargo vans from its logistics warehouse located 25 km away from the city centre. In the UCC scenario during the early hours, cargo vans travel from the warehouse to the UCC located in the underground parking, where parcels are cross-docked to smaller electric cargo vehicles for last-mile delivery within the central district in Madrid's LEZ. Table 1 includes the five scenarios studied for the case of Madrid. The main objective is to assess and validate which scenario, including UCC and different configurations of low-carbon vehicle fleets for B2C deliveries, presents the most sustainable solutions for the city centre.

Table 1 Cities' scenarios definition

City	Scenarios	Description	Delivery type
Madrid	S_Mad_0	**BAU—Diesel Van:** Direct delivery from a platform 25 km away using diesel cargo vans	B2C
	S_Mad_1	**BAU—Electric Van**: Direct delivery from a platform 25 km away using 100% electric small cargo vans	
	S_Mad_2	**UCC—Hybrid van + e-trikes:** Platform to city-centre UCC (underground parking) with hybrid vans, then last-mile by electric three-wheeled trikes	
	S_Mad_3	**UCC—Small e-Van + e-trikes:** Platform to UCC with small electric vans, then last-mile by electric three-wheeled trikes	
	S_Mad_4	**UCC—Large e-Van + e-trikes:** Platform to UCC with large electric vans, then last-mile by electric three-wheeled trikes	
The Hague	S_Hag_0	**BAU**—No services: No crowdshipping or parcel lockers available	B2C & C2C
	S_Hag_1	**Crowdshipping app**: Widely used crowdshipping app in the Hague	
	S_Hag_2	**Parcel lockers + bikes:** Public parcel lockers served by bikes in a neighbourhood	
	S_Hag_3	**Hag_1 + Hag_2 combined:** Both crowdshipping and parcel lockers in use	
Porto	S_Por_0	**BAU**—Business as usual scenario	B2B
	S_Por_1	**75% EV Fleet:** Deliver with 75% of e-vehicles fleet	
	S_Por_2	**50% EV Fleet:** Deliver with 50% of e-vehicles fleet	
	S_Por_3	**25% EV Fleet**: Deliver with 25% of e-vehicles fleet	

In The Hague, the sustainability solution analysed is implementing crowdshipping in the city and parcel lockers in a neighbourhood or the combination of both. The crowdshipping service is assumed to be available for B2C and C2C parcel deliveries by a local pool of bringers using an app service. The parcel lockers are implemented in public spaces in a neighbourhood for B2C and are fulfilled by bikes. Table 1 includes the four scenarios created for the case of The Hague.

In Porto, the sustainability solution analysed is the implementation of an electric vehicle fleet for B2B deliveries for an important logistics operator. Table 1 includes the three scenarios created for the case of Porto.

4.2 Selection and Measurement of Sustainability Criteria

As the first step in the process and following the definition of the scenarios to be evaluated, sustainability criteria for each scenario were selected. As outlined in Sect. 3.1, a total of 15 criteria—aligned with the triple bottom line of sustainability—were ultimately chosen within the framework of the LEAD project, under which each city developed and assessed its respective scenarios. Table 2 shows the measurement of the sustainability criteria for each scenario in the three cities, specifying the associated data sources used to quantify them. The criteria were gathered from three different sources: (i) Logistics Service Providers: logistics operators, logistics services apps, etc. involved in the LEAD project pilots; (ii) Digital Twin Simulations: criteria were also estimated using the Digital Twins developed to simulate the previous LMD scenarios and evaluate the operational impact of various interventions as part of the LEAD project; (iii) Survey-based: additional criteria-related data were obtained through participant responses in several LEAD project workshops. In those sessions, stakeholders completed surveys after reviewing the proposed scenarios, providing qualitative insights and feedback. Detailed scenario description and the calculation of the sustainability criteria are available in the project deliverables (see results in EC 2023).

Table 2 compares different LMD scenarios in Madrid, The Hague and Porto, focusing on their social, environmental and economic impacts. In **Madrid**, all the proposed scenarios lead to significant environmental improvements, notably in emissions and energy use. These gains also enhance neighbourhood quality of life, although they come with a slight decrease in economic profitability. **The Hague** stands out for the significant improvement in delivery times and strong environmental performance. However, these advantages are counterbalanced by higher operational costs. Job creation can be substantial in some scenarios. It is worth noting that several KPIs in The Hague, such as job quality and delivery reliability, remain constant across scenarios because they could not be measured or differentiated (see figures in grey). **Porto**, on the other hand, offers a well-rounded balance. It achieves high levels of job creation and quality, along with strong perceived benefits for local shops and very reliable delivery performance. While its environmental outcomes are not as strong as in the other cities, Porto compensates with low delivery costs and solid social benefits. Like The Hague, some indicators in Porto do not vary across scenarios due to limitations in data availability. Additionally, it is worth

Table 2 Measurement of sustainability criteria across last-mile delivery scenarios and cities

City	KPI	*Social.* Job creation (full-time employees) [a]	*Social.* Quality of the jobs [1–5] [c]	*Social.* Neighbourhood quality of life [1–5] [c]	*Environmental.* Energy consumption (MJ/delivery) [b]	*Environmental.* GHG emissions (gCO_2/delivery) [b]	*Environmental.* NOx emissions (gNO_2/delivery)[b]	*Environmental.* PM emissions (gPM/delivery) [b]	*Environmental.* Noise pollution (% population affected) [b]	*Economic.* Average delivery cost of the business model (€/delivery) [a]	*Economic.* Congestion (m^2*h/day) [a]	*Economic.* Urban storage & parking space (m^2) [a]	*Economic.* Financial Internal Rate of Return (%) [a]	*Economic.* Shop retail benefits [1–5] [c]	*Economic.* Delivery time (min/delivery) [a]	*Economic.* Delivery reliability within the time windows (%) [a]
Madrid	S_Mad_0	5	3.0	3.0	1.33	356.960	0.440	0.030	59.73	2.3	300.65	45.07	4.0	3.0	5.0	0.8
	S_Mad_1	5	3.0	4.4	0.07	0.030	0.000	0.000	49.13	2.4	300.65	45.07	3.0	3.0	5.0	0.8
	S_Mad_2	10	3.6	4.4	0.28	63.610	0.000	0.010	34.96	2.5	180.56	245.07	2.0	3.8	4.0	0.9
	S_Mad_3	10	3.6	4.4	0.07	0.020	0.000	0.000	34.96	2.6	180.56	245.07	1.0	3.8	4.0	0.9
	S_Mad_4	10	3.6	4.4	0.07	0.010	0.000	0.000	34.96	2.4	180.56	245.07	2.0	3.8	4.0	0.9
The Hague	S_Hag_0	10	3	3	0.089	663,308	0,00244	0,00109	100	9.2	343.5	1	1	3	1	1
	S_Hag_1	48	3	4.1	0.08165	608,540	0,00224	0,00100	100	5.95	310.5	1	1	3.8	1	1
	S_Hag_2	10	3	4.1	0.08848	659,477	0,00243	0,00108	100	9.2	308.3	1	6	3.8	1	1
	S_Hag_3	48	3	4.1	0.08165	608,540	0,00224	0,00100	100	5.95	308.3	2	6	2.7	1	1
Porto	S_Por_0	31	3.5	4.7	32,438	150,44	0,0027	0,0166	70*	2.05	610.03	1	775000**	3.5	6,34	94
	S_Por_1	32	4.1	4.9	1892	93,39	0,0001	0,0056	66.24*	1.25	189.70	1	390577**	3.5	6,52	94
	S_Por_2	33	4.1	4.8	24,515	116,34	0,0016	0,0083	67.49*	1.62	333.11	1	547383**	3.5	6,61	94
	S_Por_3	32	4.1	4.7	27,507	135,93	0,0021	0,0138	68.75*	1.84	471.01	1	639791**	3.5	6,52	94

Data sources: [a]logistic operator, [b]digital twin tool and [c]survey-based
Magnitudes: *dB; **investment €

noting that for Porto, the noise pollution criterion and the FIRR are expressed using different magnitudes compared to those used in the other two cities. This does not pose a problem, as the third step of the methodology includes a weighted change relative to scenario 0, ensuring consistency in comparisons. Overall, each city emphasises different priorities—Madrid leans towards environmental gains, The Hague towards delivery efficiency and Porto towards a more balanced integration of social and economic outcomes.

4.3 *Estimation of Weights*

The procedure for calculating the adjusted weights for each of the three cities—Madrid, The Hague and Porto—is summarised in the next section, according to the methodological framework described in Sect. 3.2. The adjusted weights reflect both the intrinsic relevance of each KPI for experts' stakeholders and the contextual needs of each city. A more comprehensive explanation of the weighting framework, including data sources, normalisation techniques and weighting algorithms, can be found in Gonzalez et al. (2023b).

4.3.1 Stakeholders' Opinions: Convergent Weights

To assess the relative importance of sustainability criteria in last-mile delivery, a structured expert survey was conducted using a pairwise comparison methodology. Experts were asked to evaluate trade-offs between different sustainability indicators without reference to any specific urban context. The questionnaire employed the REMBRANDT scale, which ranges from −8 (indicating a strong preference for one criterion) to +8 (for the opposite), allowing for graded and meaningful judgments (Karbassi Yazdi and Farshid 2017; Olson et al. 1995). To support clarity in responses, the survey included practical examples—for instance, demonstrating how prioritising job creation over job quality reflects a societal preference for employment volume rather than work conditions. The survey was conducted online from May to June 2022 and was disseminated to experts through the LEAD network of partners. A collaborative data collection approach was adopted, whereby each project partner shared the survey with their respective stakeholders, including public authorities, logistics providers, academics and consultants. A total of 54 valid responses were collected, predominantly from European experts (Vassallo et al. 2023). The respondents represented a balanced mix of public, academic and industry sectors, with over 60% having more than 5 years of experience in last-mile logistics. This broad engagement ensured a robust and multidimensional foundation for weighting sustainability criteria.

Based on the methodology previously detailed in Sect. 3.2, the weights for each sustainability criterion were calculated and normalised to a standardised scale from 0 to 100. The results, illustrated in Fig. 2, reveal that stakeholders assign the highest

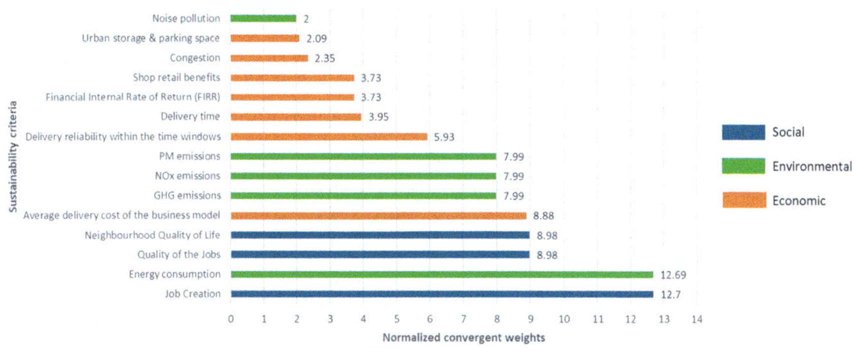

Fig. 2 Experts' normalised convergent weights for each criterion

importance to job creation in the social dimension, energy reduction in the environmental dimension and lower delivery costs in the economic dimension. Notably, the analysis highlights the strong emphasis experts place on the social aspects of last-mile delivery, indicating their central role in sustainability evaluations.

4.3.2 City Context: Severity Level

The Severity Level Index, detailed in Sect. 3.2, is calculated by selecting and comparing specific attributes for each sustainability criterion, either quantitatively or qualitatively. For quantitative attributes, data were gathered from official reports or databases at the city or country level. Qualitative criteria, in their turn, were assessed by project representatives of each city. These representatives, consisting of public authorities and logistics companies, evaluate the significance of various sustainability factors in their respective cities. For example, when evaluating urban and parking space, representatives are asked to rate the importance of these issues in their city compared to cities with similar characteristics on a scale from 1 to 5 (ranging from "Not at all important" to "Extremely important"). Similarly, for criteria such as average delivery costs, representatives compare the delivery costs in their city to those in other cities with similar features.

The Severity Level (*SL*) for each city is determined by adding the Present Situation (*PS*) value and the Trend (*T*) value. If the attribute is numerical, the *PS* is adjusted on a 0–5 scale, comparing the city's value to the average in the context of all the cities. For qualitative criteria, the *PS* value is directly provided by the city representatives.

Table 3 presents the SL_{ik} for each criterion i and city k. For detailed information on how to reach the SL, see Gonzalez et al. (2023b). This approach allows for a nuanced understanding of how each city is performing on each criterion in relation to others, considering both the present situation and future trends. The results reveal that each city faces distinct challenges, emphasising the importance of considering local contexts in the analysis. For instance, Madrid stands out due to its social

Table 3 Severity level index (Source: Gonzalez et al. 2023b)

		Madrid			The Hague			Porto		
Component	Criteria	PS	T	SL	PS	T	SL	PS	T	SL
Social	Job creation	5	0	***5***	1	0	***1***	2	1	***3***
	Quality of the jobs	5	1	***6***	1	1	***2***	3	1	***4***
	Neighbours' quality of life	3	2	***5***	1	0	***1***	1	1	***2***
Environmental	Energy consumption	5	0	***5***	3	0	***3***	5	0	***5***
	GHG emissions[1]	3	0	***3***	3	0	***3***	3	0	***3***
	NOx emissions	2	0	***2***	5	0	***5***	3	1	***4***
	PM emissions	1	1	***2***	1	0	***1***	1	1	***2***
	Noise pollution	1	0	***1***	4	2	***6***	1	2	***3***
Economic	Average delivery cost of the business model	3	1	***4***	3	0	***3***	3	2	***5***
	Congestion	1	1	***2***	4	1	***5***	2	0	***2***
	Urban storage & parking space	4	1	***5***	4	0	***4***	4	2	***6***
	Financial internal rate of return (FIRR)	2	2	***4***	2	0	***2***	2	1	***3***
	Shop retail benefits	4	2	***6***	4	0	***4***	4	0	***4***
	Delivery time	3	0	***3***	3	0	***3***	3	1	***4***
	Delivery reliability within the time windows	3	1	***4***	3	2	***5***	3	0	***3***

PS: Present situation, T: Trend and SL: severity level
[1] GHG emissions are a global criterion that is equally important for all cities.

sustainability issues. Compared to the other cities, Madrid has the highest unemployment rate and the poorest job quality, with trends for these factors expected to worsen. In contrast, other cities exhibit more stable social conditions, with less pronounced social sustainability concerns. When it comes to environmental sustainability, the results are mixed across cities. Both Madrid and Porto consider energy consumption as an important criterion for improvement relative to the average. Surprisingly, PM and NO_x emissions hold medium importance across most cities, although The Hague specifically is more sensitive to reducing nitrous oxide emissions. While CO_2 emissions are included as a criterion, they do not have a local impact, as climate change is a global issue, so no distinct weights are assigned to this criterion at the local level. Economic sustainability also varies among cities, with each city identifying different areas of concern. Madrid and Porto reported challenges with the lack of storage and urban parking, with no expected improvements in the following years. Additionally, Madrid highlights the impact of these issues on retail profits, making it an important economic criterion to consider in the analysis. This variation in concerns across cities underscores the need to tailor sustainability strategies to local realities and priorities.

4.3.3 Adjusted Weights

The analysis highlights how context-specific differences across cities influence the weighting of sustainability criteria. These differences across cities were used to adjust the experts' original weights, as explained in Sect. 3.2. Table 4 presents the final normalised adjusted weights per city and criterion, based on a methodology developed by Bueno Cadena and Vassallo Magro (2015). Each row represents a different *AW* per criteria, with values normalised to allow comparison across cities. *AW* Madrid values range roughly from 0.49 to 15.59. *AW* The Hague values vary more widely, from about 2.74 up to 14.68. *AW* Porto values fall between approximately 1.33 and 17.88. For example, the highest value in the table appears in Porto (17.88), while Madrid has some of the highest values close to 15.59. The Hague shows a few high values as well, reaching 14.68. This kind of normalised data allows comparison of criteria across the three cities on a similar scale, making it easier to identify which city has higher or lower priorities in specific cases.

Madrid shows a strong sensitivity to social criteria, particularly job creation. Energy consumption also receives significant weight, reflecting its environmental priorities. While experts had emphasised emissions such as GHGs, NO_x and PM, the local context of Madrid led to higher emphasis on economic criteria like average delivery cost and delivery reliability. Noise pollution, however, is less of a concern in Madrid. Overall, the city demonstrates a balanced profile for implementing LMD solutions that positively affect job creation, energy efficiency and delivery costs.

In contrast, The Hague places more emphasis on environmental sustainability. Its highest adjusted weights are for NOx emissions and energy consumption, followed by delivery reliability and average delivery cost. Unlike Madrid, noise pollution is a

Table 4 Adjusted weights (normalised) for each criterion and city—convergent weights and severity level

Criteria	Madrid	The Hague	Porto
Job creation	**15.59**	4.67	**10.74**
Quality of the jobs	**13.23**	6.60	10.13
Neighbourhood quality of life	11.03	3.30	5.06
Energy consumption	**15.58**	**13.98**	**17.88**
GHG emissions	5.89	8.81	6.76
NOx emissions	3.92	**14.68**	9.01
PM emissions	3.92	2.94	4.51
Noise pollution	0.49	4.40	1.69
Average delivery cost of the business model	8.72	9.78	**12.51**
Congestion	1.16	4.32	1.33
Urban storage & parking space	2.57	3.08	3.54
Financial internal rate of return (FIRR)	3.67	2.74	3.16
Shop retail benefits	5.50	5.48	4.21
Delivery time	2.91	4.36	4.46
Delivery reliability within the time windows	5.82	**10.88**	5.01

major concern in The Hague. On the other hand, financial indicators such as FIRR and PM emissions are less significant. Some social and economic factors—like job quality, retail benefits and urban infrastructure—receive similar medium-level weights. Therefore, sustainable LMD strategies in The Hague should focus on reducing NOx and energy use, improving reliability and lowering delivery costs.

Finally, in Porto, energy consumption receives the highest adjusted weight, followed by the average delivery cost of the business model. Social aspects such as job creation and job quality are also considered important, showing their role in achieving sustainability in last-mile delivery.

4.4 Sustainability Score

In the final stage of the methodology, the variation of each sustainability criterion from each scenario relative to the base case (Scenario 0) is calculated to allow for a consistent comparison across all criteria. The sustainability score of each scenario is then determined by multiplying each city's Adjusted Weights by the corresponding criterion variation rate. This calculation captures each scenario's contribution to sustainability, with the resulting values reflecting either a positive or negative impact, indicated by a sign change (+1 for positive, −1 for negative).

Table 5 provides a comparative overview of the **sustainability performance scores** for various **last-mile delivery (LMD) strategies** tested in **Madrid**, **The Hague** and **Porto**, based on a wide set of social, environmental and economic KPIs. Each strategy (Scenario 1 to 4 for Madrid, and 1 to 3 for The Hague and Porto) is evaluated relative to a base scenario (Scenario 0) using adjusted weights and variation rates. Positive scores indicate improved sustainability relative to the base case, while negative scores point to a reduction of sustainability compared to the baseline.

Madrid' solution leads significantly in sustainability across all its scenarios, with Scenario 4 (S_Mad_4: UCC large E-van + E-trikes) showing the highest score (41.8), indicating a strong performance in multiple criteria including energy consumption, social impact and delivery reliability and showing that fully electrified and consolidated urban logistics through **urban consolidation centres (UCCs)** and **electrified vehicle fleet** is the most sustainable approach for Madrid. On its part, The Hague shows moderate performance, with Scenario 3 (S_Hag_3: integrated crowdshipping + parcel lockers) being the most sustainable (39.7), reinforcing the value of **integrating multiple green delivery methods**. Finally, Porto's strategies focus on gradual electrification of the delivery fleet, but sustainability gains are relatively modest: Scenario 1 (S_Por_1: 75% e-vehicles) scores highest (25), showing that increased electric vehicle usage improves sustainability, particularly in emissions and energy use.

All in all, Madrid's UCC + E-Trike model (S_Mad_4) demonstrates the highest sustainability, thanks to a fully electric, consolidated system that addresses social, environmental and economic dimensions in a balanced way. The Hague benefits from combining crowdshipping and parcel lockers; however, its social and

Table 5 Sustainability scores for cities' scenarios

City	Madrid				The Hague			Porto		
Criteria	S_Mad_1	S_Mad_2	S_Mad_3	S_Mad_4	S_Hag_1	S_Hag_2	S_Hag_3	S_Por_1	S_Por_2	S_Por_3
Social. Job creation	0.0	15.6	15.6	15.6	17.7	0.0	17.7	0.0	1.1	0.0
Social. Quality of the jobs	0.0	2.6	2.6	2.6	0.0	0.0	0.0	2.0	2.0	2.0
Social. Neighbourhood quality of life	5.5	5.5	5.5	5.5	1.3	1.3	1.3	0.0	0.0	0.0
Environmental. Energy consumption	14.0	12.5	14.0	14.0	1.4	0.0	1.4	7.2	3.6	3.6
Environmental. GHG emissions	5.9	4.7	5.9	5.9	0.9	0.0	0.9	2.7	1.4	0.7
Environmental. NOx emissions	3.9	3.9	3.9	3.9	1.5	0.0	1.5	5.4	3.6	1.8
Environmental. PM emissions	3.9	2.7	3.9	3.9	0.3	0.0	0.3	3.2	2.3	0.9
Environmental. Noise pollution	0.1	0.2	0.2	0.2	0.0	0.0	0.0	0.2	0.0	0.0
Economic. Average delivery cost of the business model	0.0	−0.9	−0.9	0.0	3.9	0.0	3.9	5.0	2.5	1.3
Economic. Congestion	0.0	0.5	0.5	0.5	0.4	0.4	0.4	0.9	0.7	0.3
Economic. Urban storage & parking space	0.0	−11.3	−11.3	−11.3	0.0	0.0	−3.1	0.0	0.0	0.0
Economic. Financial internal rate of return	−0.7	−1.8	−2.9	−1.8	0.0	13.7	13.7	−1.6	−0.9	−0.6
Economic. Shop retail benefits	0.0	1.6	1.6	1.6	1.6	1.6	1.6	0.0	0.0	0.0
Economic. Delivery time	0.0	0.6	0.6	0.6	0.0	0.0	0.0	0.0	0.0	0.0
Economic. Delivery reliability within the time windows	0.0	0.6	0.6	0.6	0.0	0.0	0.0	0.0	0.0	0.0
Sustainability score	**32.6**	**37.1**	**39.9**	**41.8**	**29.1**	**17.1**	**39.7**	**25.0**	**16.1**	**9.9**

economic impacts may vary, as some criteria were not fully estimated, potentially affecting the overall evaluation. Similarly, Porto's approach—based on partial fleet electrification—shows incremental improvements, yet the absence of complete data for certain indicators suggests the need for more integrated and comprehensive solutions to improve its sustainability outcomes. These results illustrate how effective the different solutions are in each city for tackling their unique sustainability challenges.

It is important to note that the sustainability score reflects the relative performance of each scenario compared to the Business-As-Usual (BAU) baseline. Rather than being an absolute measure, it highlights how each intervention improves or worsens sustainability impacts in each city context. This comparative perspective allows decision-makers to identify the most promising strategies tailored to their specific urban challenges.

5 Applicability, Scalability and Future Developments

The high growth of urban freight delivery in most metropolitan areas in recent years, along with the need to make freight mobility more sustainable, has reinforced the need to have a method for measuring the sustainability of last-mile delivery strategies. However, this is not an easy task. Setting a quantitative rate to accurately measure sustainability faces crucial challenges. First, the concept of sustainability is very wide and touches upon different fields that until recently have followed separate methodological paths and used different metrics. Second, although combining different metrics and indicators is possible using multi-criteria methods, setting weights is still a big issue because sustainability is highly affected by the specific problems of urban and metropolitan areas where innovative solutions and strategies are implemented, and experts often have different opinions and views.

Although there have been some attempts to define sustainability ratings to assess the impact of LMD strategies, most of the methods defined are not scalable, as their implementation is limited to a single specific strategy and/or a predefined context where the solution is implemented. This chapter sets the basis for a sustainability assessment framework applicable to LMD strategies that can be easily adapted to the information available, the criteria selected to measure sustainability, and the city context where the LMD solution or strategy is implemented. One of the greatest advantages of the framework is its flexibility, which allows its replicability and scalability to be applied to any LMD strategy implemented anywhere. In addition, it facilitates the comparison of the effectiveness of different solutions in different places using the same sustainability criteria.

The framework can be implemented considering either a greater or shorter number of criteria, depending on the availability of data. It also allows for the combination of quantitative and qualitative criteria. Although the larger the number of criteria, the more precise the analysis will be, it is not always feasible to have quality data available for measuring all desirable KPIs. This method allows using either a larger or shorter number of criteria and even coming up with a result with missing values for a certain indicator.

One of the greatest differences of the framework compared to previous methods is stressing the impacts (both positive and negative) that are more critical in a certain urban context. For example, the creation/destruction of jobs will have a much greater sustainability impact (positive/negative) in a city with high unemployment rates than in a city with almost full employment. This adjustment allows the panel of experts to set the weights without the need to know the context, making the decision more objective and neutral.

The framework is useful for multiple end users (policymakers, planners, logistics operators, consultants or researchers) to prioritise urban logistics strategies based on sustainability, ensuring that key drivers are effectively included in the decision-making process according to the specific needs of the urban area where the solution is implemented.

Although the method has advantages, it also has limitations and constraints suitable for refinement and future research endeavours. A first limitation, which is clearly associated with the multi-criteria nature of the method, is subjectivity at the time of setting the weights. Although the Delphi and Rembrandt methods set a framework to ensure statistical confidence in the agreement of experts, setting the trade-offs among different sustainability criteria is still complicated, particularly if the units of different criteria are difficult to compare.

Another limitation refers to the fact that the method provides the sustainability variation of the LMD strategy analysed in relative terms by comparing with the baseline scenario, but it does not provide the gross variation. That means that, as it is currently outlined, the framework is not able to capture the size of the impact that the new solution or strategy is producing. Having a gross impact would be possible if, instead of measuring the variation of the performance indicator, the method separately measured both the solution that is evaluated and the baseline scenario, and the indicators are measured in the original units instead of being normalised. Future research work may focus on improving those aspects.

References

Alhaddi H (2015) Triple bottom line and sustainability: a literature review. Busi Manag Stud 1(2):6–10. https://doi.org/10.11114/bms.v1i2.752

Aljohani K, Thompson RG (2020) A multi-criteria spatial evaluation framework to optimise the siting of freight consolidation facilities in inner-city areas. Transp Res A Policy Pract 138:51–69. https://doi.org/10.1016/j.tra.2020.05.020

Allen J, Piecyk M, Piotrowska M, McLeod F, Cherrett T, Ghali K, Nguyen T, Bektas T, Bates O, Friday A, Wise S, Austwick M (2017) Understanding the impact of e-commerce on last-mile light goods vehicle activity in urban areas: the case of London. Transport Res D Trans Environ 61:325–338. https://doi.org/10.1016/j.trd.2017.07.020

Almansa Sáez C, Calatrava Requena J (2007) Reconciling sustainability and discounting in cost-benefit analysis: a methodological proposal. Ecol Econ 60(4):712–725. https://doi.org/10.1016/j.ecolecon.2006.05.002

Alvarez Gallo SM, Maheut J (2022) Protocol: systematic literature review of the application of the multicriteria decision analysis methodology in the evaluation of urban freight logistics initiatives. WPOM Work Pap Oper Manag 13:86–107. https://doi.org/10.4995/wpom.16780

Alvarez Gallo S, Maheut J (2023) Multi-criteria analysis for the evaluation of urban freight logistics solutions: a systematic literature review. Mathematics 11(19):4089. https://doi.org/10.3390/math11194089

Andruetto C (2022) Exploring electrification, consolidation, cargo bikes and automation using a sustainability performance assessment framework

Andruetto C, Martensson J, von Wieding S, Pernestal A (2022) Indicators for sustainability assessment in city logistics: perspectives of society and logistic service providers. In: 2023 Transportation Research Board annual meeting, 2022. https://doi.org/10.2139/ssrn.4034714

Awasthi A, Chauhan SS (2012) A hybrid approach integrating affinity diagram, AHP and fuzzy TOPSIS for sustainable city logistics planning. Appl Math Model 36:573–584. https://doi.org/10.1016/j.apm.2011.07.033

Ayuntamiento de Madrid (2023a) Estrategia de Distribución Urbana de Mercancías DUM. Madrid 360. https://www.madrid360.es/movilidad-sostenible/distribucion-urbana-de-mercancias/

Ayuntamiento de Madrid (2023b) Los Retos de La Movilidad Urbana. Or-denanza de Movilidad Sostenible. https://www.madrid360.es/movilidad-sostenible/ordenanza-de-movilidad-sostenible/

Balm S (2022) Using procurement power to accelerate Sustainable City logistics: lessons from change agents in The Netherlands. Sustainability 14(10):6225. https://doi.org/10.3390/su14106225

Balm S, Browne M, Leonardi J, Quak H (2014) Developing an evaluation framework for innovative urban and interurban freight transport solutions. Procedia Soc Behav Sci 125:386–397. https://doi.org/10.1016/j.sbspro.2014.01.1482

Boggio-Marzet A (2022) Strategies to improve efficiency and sustainability of last-mile delivery in urban settings: a holistic perspective. https://doi.org/10.20868/upm.thesis.72053

Bosona T (2020) Urban freight last mile logistics—challenges and opportunities to improve sustainability: a literature review. Sustainability 12(21):8769. https://doi.org/10.3390/su12218769

Browne D, Ryan L (2011) Comparative analysis of evaluation techniques for transport policies. Environ Impact Assess Rev 31(3):226–233. https://doi.org/10.1016/j.eiar.2010.11.001

Brundtland World Commission on Environment and Development (1987) Report of the world commission on environment and development. Retrieved from http://www.un.org/documents/ga/res/42/ares42-187.htm

Bueno Cadena PC, Vassallo Magro JM (2015) Setting the weights of sustainability criteria for the appraisal of transport projects. Transport 30(3):298–306. https://doi.org/10.3846/16484142.2015.1086890

Bueno P, Vassallo J, Cheung K (2015) Sustainability assessment of transport infrastructure projects: a review of existing tools and methods. Transp Rev 35(5):622–649. https://doi.org/10.1080/01441647.2015.1041435

Corti F, D'Alpaos C, Greco L (2024) Multicriteria evaluation of sustainability in last-mile logistics: a review. Valori e Valutazioni:125–147. https://doi.org/10.48264/vvsiev-20243608

De Langhe K (2014) Analysing the role of rail in urban freight distribution. In: Next generation supply chains: trends and opportunities, vol 18, p 223

Deveci M, Pamucar D, Gokasar I, Delen D, Wu Q, Simic V (2022) An analytics approach to decision alternative prioritization for zero-emission zone logistics. J Bus Res 146:554–570. https://doi.org/10.1016/j.jbusres.2022.03.059

EC (2023) LEAD Project: Low-Emission Adaptive last mile logistics supporting 'on Demand economy' through digital twins. https://cordis.europa.eu/project/id/861598

Elkington J (1998) Cannibals with forks: the triple bottom line of 21st century. New Soc. https://doi.org/10.1002/tqem.3310080106

Giret A, Julián V, Corchado JM, Fernández A, Salido MA, Tang D (2018) How to choose the greenest delivery plan: a framework to measure key performance indicators for sustainable urban logistics. In: Moon I, Lee G, Park J, Kiritsis D, von Cieminski G (eds) Advances in production management systems. Smart manufacturing for industry 4.0. APMS 2018. IFIP advances in information and communication technology, vol 536. Springer, Cham. https://doi.org/10.1007/978-3-319-99707-0_23

Gonzalez JN, Garrido L, Vassallo JM (2023a) Exploring stakeholders' perspectives to improve the sustainability of last mile logistics for e-commerce in urban areas. Res Transp Bus Manag 49:101005. https://doi.org/10.1016/j.rtbm.2023.101005

Gonzalez JN, Sobrino N, Vassallo JM (2023b) Considering the city context in weighting sustainability criteria for last-mile logistics solutions. Int J Log Res Appl 28(4):380–400. https://doi.org/10.1080/13675567.2023.2264788

Gonzalez-Feliu J (2016) Viability and potential demand capitation of urban freight tramway systems via demand supply modelling and cost benefit analysis. In: Proceedings of the 6th International conference on Information Systems, Logistics and Supply Chain (ILS 2016), Bordeaux, France, pp 1–4

Gonzalez-Feliu J (2017) Sustainability evaluation of green urban logistics systems. In: Advances in business strategy and competitive advantage book series, pp 103–134. https://doi.org/10.4018/978-1-5225-2662-9.ch005

Gonzalez-Feliu J, Basck P, Morganti E (2013) Urban logistics solutions and financing mechanisms: a scenario assessment analysis. Eur Transp/Trasporti Europei 54(11):1–16

Gudmundsson H, Hall RP, Marsden G, Zietsman J (2016) Sustainable transportation. Indicators, frameworks, and performance management. Springer Verlag, Heidelberg

Holguín-Veras J et al (2015). Improving freight system performance in metropolitan areas: a planning guide. En Transportation Research Board eBooks https://doi.org/10.17226/22159

Holmgren J (2018) Using cost–benefit analysis to evaluate city logistics initiatives: an application to freight consolidation in small- and mid-sized urban areas. In: Taniguchi E, Thompson RG (eds) City logistics, vol 2, pp 271–290

Jamshidi A, Jamshidi F, Ait-Kadi D, Ramudhin A (2019) A review of priority criteria and decision-making methods applied in selection of sustainable city logistics initiatives and collaboration partners. Int J Prod Res 57:5175–5193. https://doi.org/10.1080/00207543.2018.1540892

Janjevic M, Knoppen D, Winkenbach M (2019) Integrated decision-making framework for urban freight logistics policy-making. Transp Res Part D: Transp Environ 72(May):333–357. https://doi.org/10.1016/j.trd.2019.05.006

Jardas M, Hadžić AP, Ogrizović D (2024) Application of the MAMCA method in the evaluation of delivery flows within city centers: a case study of Rijeka. Urban Sci 8(3):149. https://doi.org/10.3390/urbansci8030149

Karbassi Yazdi A, Farshid A (2017) Designing robust model for banks benchmarking based on Rembrandt method and DEA. BIJ 24(2):431–444. https://doi.org/10.1108/BIJ-01-2015-0001

Katsela K, Pålsson H (2019) A multi-criteria decision model for stakeholder management in city logistics. Res Transp Bus Manag 33:100439. https://doi.org/10.1016/j.rtbm.2020.100439

Kin B, Macharis C (2015) Social cost-benefit analysis of a private urban consolidation centre in Antwerp. Steunpunt Goederenstromen

Krstić M, Tadić S, Kovač M, Roso V, Zečević S (2021) A novel hybrid MCDM model for the evaluation of sustainable last mile solutions. Math Probl Eng 2021:1–17. https://doi.org/10.1155/2021/5969788

Licata JW (1978) Consequence analysis: theory and practice in school problem-solving. Educ Technol 18(9):22–28. http://www.jstor.org/stable/44419211

Lyons T, McDonald NC (2022) Last-mile strategies for urban freight delivery: a systematic review. Transport Res Record J Transport Res Board 2677(1):1141–1156. https://doi.org/10.1177/03611981221103596

Macharis C, De Witte A, Turcksin L (2010) The Multi-Actor Multi-Criteria Analysis (MAMCA) application in the Flemish long-term decision making process on mobility and logistics. Transp Policy 17(5):303–311. https://doi.org/10.1016/j.tranpol.2010.02.004

Mangiaracina R, Perego A, Seghezzi A, Tumino A (2019) Innovative solutions to increase last-mile delivery efficiency in B2C e-commerce: a literature review. Int J Phys Distrib Logist Manag 49(9):901–920. https://doi.org/10.1108/ijpdlm-02-2019-0048
Mardani A, Jusoh A, Zavadskas EK (2015) Fuzzy multiple criteria decision-making techniques and applications - two decades review from 1994 to 2014. Expert Syst Appl 42:4126–4148. https://doi.org/10.1016/j.eswa.2015.01.003
Mouter N (2018) A critical assessment of discounting policies for transport cost-benefit analysis in five European practices. Eur J Transp Infrastruct Res 18(4):389–412
Olson DL, Fliedner G, Currie K (1995) Comparison of the REMBRANDT system with analytic hierarchy process. Eur J Oper Res 82(3):522–539. https://doi.org/10.1016/0377-2217(93)E0340-4
Olsson J, Hellström D, Pålsson H (2019) Framework of last mile logistics research: a systematic review of the literature. Sustainability 11(24):7131. https://doi.org/10.3390/su11247131
Porto (2023) Zonas de Acesso Automóvel Condicionado. https://mobilidade.cm-porto.pt/servicos/zonas-de-acesso-automovel-condicionado
Quak H, Nesterova N, Van Rooijen T, Dong Y (2016) Zero emission city logistics: current practices in freight electromobility and feasibility in the near future. Transport Res Proc 14:1506–1515. https://doi.org/10.1016/j.trpro.2016.05.115
Ranieri L, Digiesi S, Silvestri B, Roccotelli M (2018) A review of last mile logistics innovations in an externalities cost reduction vision. Sustainability 10(3):782. https://doi.org/10.3390/su10030782
Rao C, Goh M, Zhao Y, Zheng J (2015) Location selection of city logistics centres under sustainability. Transp Res Part D Transp Environ 36:29–44. https://doi.org/10.1016/j.trd.2015.02.008
Russo F, Comi A (2010) A classification of city logistics measures and connected impacts. Procedia Soc Behav Sci 2(3):6355–6365. https://doi.org/10.1016/j.sbspro.2010.04.044
Saha A, Simic V, Senapati T, Dabic-Miletic S, Ala A (2023) A dual hesitant fuzzy sets-based methodology for advantage prioritization of zero-emission last-mile delivery solutions for sustainable city logistics. IEEE Trans. Fuzzy Syst. 31(2):407–420. https://doi.org/10.1109/TFUZZ.2022.3164053
Silva V, Amaral A, Fontes T (2023) Towards sustainable last-mile logistics: a decision-making model for complex urban contexts. Sustain Cities Soc 96:104665. https://doi.org/10.1016/j.scs.2023.104665
Tadić S, Zečević S, Krstić M (2014) A novel hybrid MCDM model based on fuzzy DEMATEL, fuzzy ANP and fuzzy VIKOR for city logistics concept selection. Expert Syst Appl 41(18):8112–8128. https://doi.org/10.1016/j.eswa.2014.07.021
Tadić S, Zečević S, Krstić M (2018) Assessment of the political city logistics initiatives sustainability. EURO mini conference on "Advances in Freight Transportation and Logistics". Transp Res Proc 30:285–294. https://doi.org/10.1016/j.trpro.2018.09.031
The Hague (2022) Low-emission zones in The Hague. https://www.denhaag.nl/en/getting-there-and-around/low-emission-zones-in-the-hague/
Van Duin R, Quak H, Muñuzuri J (2008) Revival of cost benefit analysis for evaluating the city distribution centre concept. In: Taniguchi E, Thompson RG (eds) Innovations in city logistics. Nova Science, pp 97–114
Vassallo JM, Bueno PC (2021) Sustainability assessment of transport policies, plans and projects. Adv Transp Policy Plann 7:9–50. https://doi.org/10.1016/bs.atpp.2020.07.006
Vassallo JM, Gonzalez JN, Sobrino N (2023) Sustainable expert criteria weights. Database. https://doi.org/10.5281/ZENODO.7695814
Velickovic M, Stojanovic D, Ilin V, Mircetic D (2025) A combined multri-criteria decisión-making and social cost-benefit analysis approach for evaluating sustainable city logistics initiatives. Sustainability 17(3):884. https://doi.org/10.3390/su17030884
Wang Y, Li Y, Lu C (2023) Evaluating the effects of logistics center location: an analytical framework for sustainable urban logistics. Sustainability 15(4). https://doi.org/10.3390/su15043091

Watróbski J, Małecki K, Kijewska K, Iwan S, Karczmarczyk A, Thompson RG (2017) Multi-criteria analysis of electric vans for city logistics. Sustain 9. https://doi.org/10.3390/su9081453
Zhang X (2024) Road to sustainability: developing a sustainability performance measurement framework for City logistics from a multi- stakeholder perspective. Technological University, Dublin

Flanders' E-Commerce Living Lab

Julia Amaral, Carlos Granada, Yubing Lei, Jia Zhang, and Joris Beckers

1 Introduction

Flanders, home to approximately 6 million inhabitants in the northern region of Belgium, has been a slow adopter in terms of electronic commerce. The combination of a discerning and tradition-oriented consumer favoring in-person shopping with the sprawled urban landscape providing high overall accessibility to physical retail outlets resulted in a slow development of the demand for online shopping. On the supply side, six different governments and a tri-lingual country long put Belgium at the bottom of priority lists for online retailers, especially since the country could be served easily from major operations in The Netherlands, Germany, or France. With low investments, it lagged its fellow Western European countries in terms of internet users who bought or ordered goods or services for private use. A telling example is Amazon's notably late entry into the Belgian market, launching its *.be* website only in 2022, making it the tenth country in Europe to receive a dedicated website.

The pandemic was the real catalyst for online shopping in Belgium. By taking away the so prevalent physical accessibility, the traditional-minded consumer was forced to the online channel. Once the bridge was crossed, e-commerce was here to stay. With electronic sales accounting for just over a quarter of total business turnover in 2023, Belgium now ranks third among EU member states, just behind Denmark and Finland (Eurostat 2025). From March 2023 to March 2024, 90% of 18 to 70-year-old Belgians have made at least one online purchase (Comeos 2024). By now, a quarter of all retail spending is done online (becom 2025). The growth in

J. Amaral · C. Granada · Y. Lei · J. Zhang · J. Beckers (✉)
Department of Transport and Regional Economics, University of Antwerp, Antwerpen, Belgium
e-mail: Joris.Beckers@uantwerpen.be

A. Melkonyan-Gottschalk, Y. Li (eds.), *Urban Logistics Transformation*, Greening of Industry Networks Studies 15,
https://doi.org/10.1007/978-3-032-08387-6_8

e-commerce performance reflects the evolving consumer preferences in multiple retail channels.

Belgium's slow adoption of online shopping meant that, despite being ranked 7th globally in the 2023 Logistics Performance Index (LPI), the country largely missed the wave of e-commerce logistics growth. While Alibaba operates a highly debated major hub at Liège Airport (Holslag 2023), the majority of e-commerce fulfillment for the Belgian market happens in the well-developed logistics clusters in the southern Netherlands and other hubs in Germany and France. This reliance on cross-border infrastructure has sparked ongoing debate among stakeholders. At the same time, however, Belgium—particularly Brussels and the northern region of Flanders—has emerged as an international leader in the organization of urban logistics. The sprawled nature of the region, combined with its historic, often medieval city centers, has placed considerable pressure on logistics operators to cope with rising last-mile costs and urban planners to reduce logistics-related traffic while maintaining accessibility for residents. In response, several Flemish cities are piloting innovative alternatives to the traditional last-mile delivery model.

2 The Role of Logistics in the Online Shopping Behavior

Before diving into different e-commerce initiatives, it is worth noting the importance of shopping behavior as a fundamental driver for the e-commerce ecosystem (Beckers et al. 2026). The customer journey now takes place in an omnichannel environment, which means researching and testing (pre-purchase stage), purchasing and receiving (purchase stage), and returning (post-purchase stage) occur digitally and/or physically (Buldeo Rai et al. 2019; Lemon and Verhoef 2016). Hence, when intervening in any of these steps, one should consider the purchasing experience in its entirety; for example, the return options might influence the purchase decision.

Early studies of online shopping adoption have explored the factors affecting the heterogeneity of individual shopping channel preferences based on user characteristics and product categories. The results show that young and highly educated users are more willing to adopt new technologies and choose online shopping (Farag et al. 2006; Keng Kau et al. 2003), and there is a positive relationship between income and online shopping frequency (Clarke et al. 2015; Song 2022). Research shows that consumers' preferences toward online and offline channel choices also vary across products. Consumers behave differently when buying everyday products and products they buy less frequently and tend to use in the long term (Phau and Meng Poon 2000). Moreover, shopping attitudes, including subjective evaluations and feelings such as price sensitivity, risk aversion, and perceived convenience are important drivers of shopping channel choices (Clemes et al. 2014; Wieland 2021). Consumers with higher time and price sensitivity are more likely to shop online, while shoppers who are concerned about the risk of online shopping tend to choose in-store shopping (Etminani-Ghasrodashti and Hamidi 2020). With the development of the on-demand economy and multiple shopping channels, consumers expect a consistent retail offer and seamless shopping experience from one channel to another. As a result, consumer behavior cannot be fully understood by the traditional influencing factors.

Since online shopping relies heavily on reliable and efficient delivery methods, the last-mile delivery experience has a significant impact on consumers' online shopping perceptions and behaviors (Beckers and Cant 2024). Vrhovac et al. (2023) found that consumers' past experiences with last-mile delivery could be crucial in their decision to return to the same retailer or brand. A successful delivery experience encourages consumers to re-purchase in the same retailers, while a negative delivery outcome may cause a decrease in future re-purchase intentions. Moreover, last-mile delivery services also influence consumers' spending on online shopping. Xiao et al. (2018) examined the impacts of last-mile delivery solutions on e-shopping behavior and found that a positive assessment of delivery services would encourage more e-shopping spending. Given the significant impact of last-mile delivery on shopping behaviors, researchers have been focusing on how specific delivery factors influence consumer adoption of delivery methods. Delivery location, fee, speed, and time are relevant factors in this discussion (Molin et al. 2022; Nguyen et al. 2019).

In the Flemish context, Lei and Beckers (Forthcoming) examined the role of logistics in shopping behavior through an online survey in 2024, with a sample of 897 respondents. The authors queried respondents' socioeconomic characteristics, attitudes toward different delivery services (e.g., location, cost, time, etc.), and their online and in-store shopping frequencies across six product categories. The sample consisted of 56% female and 44% male (compared to 51% and 49% in the Flemish population, respectively, according to Statistics Flanders (2025)). Age distribution includes 18–29 years (13%), 30–39 years (15%), 40–49 years (17%), 50–59 years (20%), 60–69 years (19%), and 70+ years (16%). While the sample slightly overrepresented female and middle-aged groups (50–69 years), it maintained reasonable representativeness across all demographic categories. In addition, the survey employed structured questionnaires with 5-point Likert scale questions for respondents' delivery attitudes (from strongly disagree to strongly agree).

Analysis of consumer priorities regarding delivery service features (Fig. 1) shows that delivery location choices and lower delivery prices are consumers' primary concerns, with over 80% of the respondents preferring to have the option of choosing where their parcels arrive. Lower delivery costs encourage them to choose online shopping. Other appreciated services are having the flexibility to choose the delivery day or hour, an easy return process, having a sustainable delivery option, and a reliable delivery process. When it comes to the time the delivery is made, consumers find it important to be able to pick which day their parcels arrive, with less concern for the specific time slots. While consumers show awareness of the environmental impact of the delivery process and choose sustainable delivery options, they will not pay extra for a more sustainable delivery method.

Further investigation of the delivery preferences shows that consumers form an integrated perception of delivery service as a whole rather than evaluating each factor separately when making shopping decisions. Consumers' delivery preferences

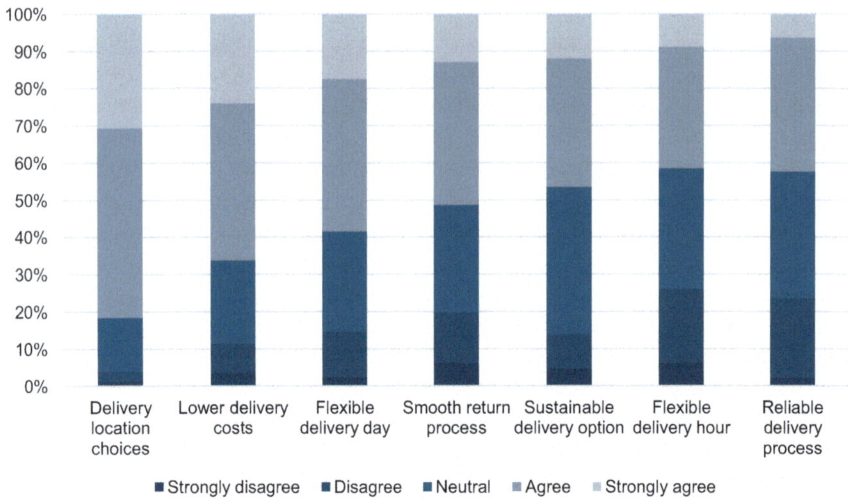

Fig. 1 Importance of different delivery service features to consumers

first shape their overall purchasing attitudes toward online shopping, which then determine their actual channel choices. The impact is strongest for groceries and clothing, while remaining significant across all product categories.

As e-commerce continues to reshape consumption patterns, logistics has become a critical interface between private behavior and public policy. While attention to last-mile impacts is growing among policymakers, operational changes remain limited. Yet, the survey results point to a rising consumer awareness of e-commerce logistics impacts, a necessary condition for enabling more sustainable last-mile practices. This shift in perception is crucial, as it opens the door to initiatives pursuing more sustainable practices, some of which are explored in the following sections.

3 Cargo Bike: Drivers, Dynamics, and Challenges

Partially fueled by consumers' expectations of the last-mile, Belgium, particularly the Flanders region, has seen a rapid acceleration in cargo-bike logistics since 2017, when the Belgian Cycle Logistics Federation (BCLF) was founded and later formalized as a national hub for advocacy, data exchange, and sectoral development. Leveraging public funding and triple-helix partnerships, the BCLF has positioned Belgian cycle logistics as a European reference for sustainable urban freight (Belgian Cycle Logistics Federation 2025).

Cargo bikes are increasingly regarded as a viable response to the environmental and operational challenges of last-mile logistics in dense urban areas. A wide body of research demonstrates their potential to reduce carbon emissions, noise, and road congestion while improving delivery speed and service reliability in compact cities

(Cowie and Fisken 2023; Ghazal et al. 2025). Compared to light commercial vehicles (LCVs), cargo bikes can offer lower operational and external costs, deliver parcels up to 60% faster than vans, with cost savings potentially reaching up to 30% compared with LCVs (Lebeau et al. 2023; Rudolph and Gruber 2017; Tipagornwong and Figliozzi 2014). However, the operational viability of cargo bikes is highly dependent on the context; the limitations in capacity and range make their operations rely on spatial conditions, such as short delivery radius, dense customer clusters, and cycling infrastructure (Dalla Chiara et al. 2023; Ghazal et al. 2025). Their effectiveness is enhanced when they operate through hubs, which could be shown in different layouts as micro hubs and mobile hubs, which extend their range and enable multimodality (Marujo et al. 2018; Rudolph et al. 2022). Thus, cargo bikes thrive when integrated into hub-based, multimodal networks. Considering the above, the Flemish context is very favorable to cargo-bike operations. The region consists of compact medieval cities, such as Antwerp, Ghent, Leuven, and Bruges, with active low-emission zones, good cycling infrastructure, and, in some places, a network of micro-hubs.

The local cargo-bike market can be divided into two distinct organizational models. Mixed-fleet carriers are large national and international freight logistics service providers (LSPs), often integrators or national postal operators, that add cargo bikes to their traditional fleets from a more tactical point of view. This can be to meet low-emission-zone (LEZ) requirements, preserve network resilience when vans are delayed or restricted, and/or signal environmental leadership. Dedicated cargo-bike operators, by contrast, are pioneering firms that rely almost exclusively on cargo bikes for last-mile delivery, with embedded environmental and social values in their operations from a strategic point of view. They capitalize on bikes' maneuverability to tap into niche markets, such as expedited deliveries. Although their scales and resource bases diverge, both groups rely on shared enablers but also constraints (see Table 1). The Belgian cycle logistics sector now comprises 42 professional firms: 29 dedicated cargo-bike operators, eight mixed-fleet carriers (e.g., Bpost, DHL), and five LSPs that subcontract rather than operate fleets directly (Belgian Cycle Logistics Federation 2024). In 2023, these companies executed 3.1 million urban deliveries, almost twice that of 2022, with mixed carriers accounting for roughly 80% of the volume, while dedicated cargo-bike operators accounted for the remaining 20% (Belgian Cycle Logistics Federation (BCLF) 2024).

Despite the optimism in the literature, the dedicated cargo-bike business model remains challenging. Although from our measurements, bikes tend to be 2–5 km/h faster than delivery vans in the city centers, the limited cargo-bike capacity constraints their competitiveness with a delivery van that carries 100 parcels or more. Hence, cargo-bike tours mostly transport B2B goods, usually serving (a couple of) retailers that want to pay extra for such high-service, emission-free deliveries.

In contrast to LCV-based last-mile deliveries, we observe different operational patterns within dedicated cargo-bike companies. More traditional rounds depart from micro-depots at full load, deliver clustered parcels, and return empty to reload. However, one competitive edge of the dedicated cargo bikes is the possibility to organize extremely flexible tours that mix last-minute pick-ups and drop-offs, a

Table 1 Organization models in the Flemish cargo-bike market (based on Lebeau et al. (2023))

Dimension	Mixed-fleet carriers	Dedicated cargo-bike operators
Organizational scale	National/international networks; high parcel volumes; low cargo-bike share	Local or regional SMEs, moderate volumes, bike-dominant fleets
Role of cargo bikes	Tactic	Strategic
Fleet composition	Trucks + vans + cargo bikes (selective urban use)	Cargo bikes (often with trailers); minimal motor vehicles
Geographic scope	Country-wide distribution, bikes confined to dense urban areas	Urban; last-mile focus
Typical flows	B2C e-commerce parcels, some B2B contracts	B2B and B2C micro-flows, on-demand services, hyper-local returns
Motives for adopting bikes	LEZ compliance, CSR targets, brand positioning, time savings in congestion	Mission-driven sustainability, local impact, access advantages
Core know-how	Advanced management and optimization tools (not specific to cargo bikes)	Rider expertise, locally developed dispatch tools
Key operational challenges	Integrating bikes into complex supply chains, software fit, and maintaining service levels	Achieving drop density, labor-cost dominance, capital constraints, and seasonality
Shared enablers	Micro-hubs for urban-core access; LEZ and access regulations, subsidies, curb space allocation, bicycle infrastructure	
Shared constraints	Payload and range, weather and topography, labor dependence, infrastructure gaps, specific routing software availability	
Example of companies	Bpost, PostNL, DHL, Amazon, UPS	Cargo Velo, Urbike, de Fietskoerier

pattern common in quick commerce (fresh or other time-constrained services). This operation allows the bike to limit the number of empty kilometers driven, for example, when returning to the hub. At the same time, however, accounting for deliveries and pickups makes routing more difficult.

Given the above observations, carrying capacity does not seem to be the only major factor affecting operations, as is sometimes assumed in the literature. Data in Flanders shows payload utilization reached only approximately 47% of the rated weight capacity and approximately 29% of their maximum volume. Typical tours span 4–12 km in contrast to the average of 40 km per tour for LCVs (World Economic Forum 2024). The low density of the operations implies that rider time, rather than weight or volume, is the binding operational constraint.

A detailed cost breakdown reveals that energy and other distance-based operational costs represent only 20–30% of total costs, whereas labor constitutes 70–80%. Therefore, under conditions of low drop density, per-parcel unit costs escalate rapidly, surpassing per-parcel costs of conventional LCVs. This effect is particularly visible in B2B flows where some operators have adopted cooled containers to serve markets such as food delivery. These adaptations reduce operational flexibility by locking carriers into a niche of clients, resulting in lower stop density and higher costs per delivery.

Besides low drop density, other factors constraint cost-effectiveness of cycle logistics across provider models. The lack of appropriate routing software is one of them. Standard optimization tools do not account for the cargo-bike-specific requirements, such as including topography, making their assumptions inaccurate and forcing operators to rely on riders' local knowledge (Taniguchi et al. 2024). Resultantly, it is not uncommon that dedicated cargo-bike operators use a self-developed routing software. In various Flemish cities, local administrations attempt to provide space for cargo-bike operations, recognizing that the investment is a significant cost for operators. However, these sites are often only temporarily available, which undermines the stability that dedicated cargo-bike companies need to grow, ultimately limiting expansion in the sector. Resultantly, despite Flanders' relatively favorable conditions—flat terrain, strong cycling infrastructure, and an embedded cycling culture—cargo bike logistics remain limited in scope. This indicates that cargo bikes are not a one-size-fits-all solution, but they can play a complementary role within a diversified urban logistics system. Their continued integration may benefit from supportive public policies and coordination with other urban freight strategies.

4 Receiver-Led Consolidation, Innovation Through EU Project Funding

In response to these barriers, and others observed in urban logistics innovations more broadly, Flemish stakeholders have initiated and participated in a range of collaborative projects to foster innovation and structural change in last-mile logistics. Many of these efforts were coordinated by the Flemish innovation platform for logistics (VIL). Locally funded projects such as R!SULT (VIL 2025) (potential of microhubs), IntelloCity (internet-of-things in the last mile), or the Green Deal Sustainable Urban Logistics (capacity building), together with EU-funded projects such as NOVELOG and SURFLOGH created a large stakeholder group and knowledge within the region. Led by a couple of active cities, the insights and networks from these projects resulted in the best practices such as the *convenant sustainable urban logistics* in Mechelen, a joint ambition document by policymakers and professionals, and the *GentLevert* initiative, a cross-department urban policy working group with professional stakeholders. Nonetheless, despite many years of triple helix collaboration between the cities, different research institutes and logistics actors yielded relatively few sustainable last-mile practices in the region. A key cause is that transport is a derived demand: The logistics operations merely depend on the sales of the online store. Resultantly, there is little room for the LSP to structurally change the e-commerce delivery chain. Even more, tight margins and strong competition limit overall innovation in the last mile.

While including the shipper would be the logical alternative, the online retail market is difficult to reach. These players are (a) too large to get around the table to organize the last mile in a Flemish city (e.g., Amazon, Shein, Zalando) or (b) too small to have a significant impact (e.g., 80% of Flemish online stores are micro-enterprises). Hence, within the GREEN-LOG project, three different Flemish cities together with the VIL bundled forces to try and break the last-mile dreadlock through the receiver. The GREEN-LOG Flemish living lab tests the potential of Logistics-as-a-Service (LaaS), as proposed by Beckers et al. (2023). Drawing inspiration from the concept of receiver-led consolidation (RLC), which is a group of initiatives that has proven potential to leverage the power of receivers to push for cargo consolidation (Aljohani and Thompson 2020; Holguín-Veras and Sánchez-Díaz 2016).

The LaaS proposes to connect receiver and carrier through a digitalized platform, enabling consumers to consolidate deliveries based on their own choices (Fig. 2). This way, through the LaaS, it is possible to steer consumers' choices by pricing and non-monetary nudging methods that raise consumers' awareness of last-mile external costs, encouraging more consolidated and sustainable choices (Beckers et al. 2023). This idea is supported by Rossolov and Susilo (2024), who suggest that digitalizing consumer integration is essential for guiding consumers toward more sustainable delivery choices through the case of e-grocery shopping, and Havrylenko et al. (2025), who also pointed out that innovation-driven last-mile solutions, such as tariff planning and pricing strategies, should place a strong emphasis on the consumer. Moreover, the adoption of the RLC by the receivers has the potential to influence the carriers to adopt more sustainable delivery practices, as receiver-driven

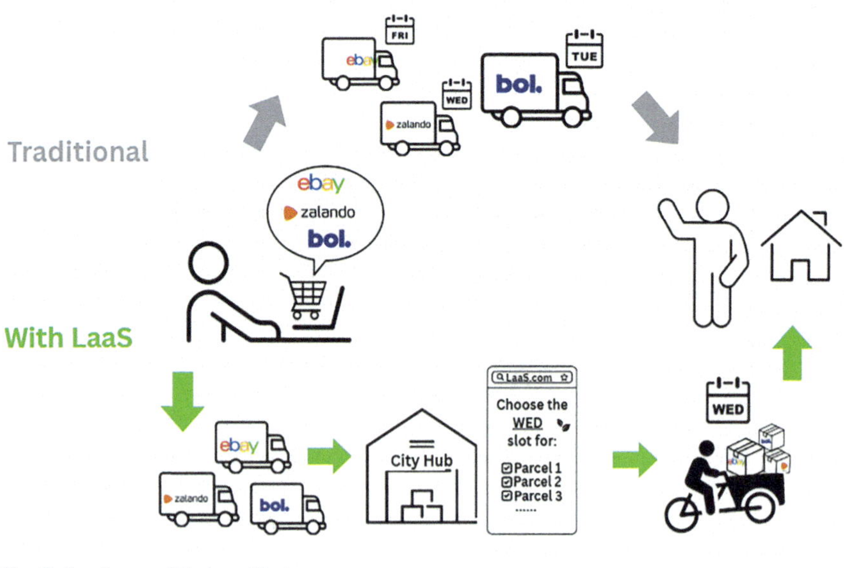

Fig. 2 LaaS consolidation effect

initiatives can prompt behavioral changes among carriers (Holguín-Veras and Sánchez-Díaz 2016).

The GREEN-LOG Flemish pilot has shown significant potential for last-mile RLC from end consumers. Pilot testing of the LaaS in Mechelen, for example, demonstrated the potential to reduce the number of door visits by 25% as consumers were able to consolidate shipments from different orders to their home. Sixty residents of the same neighborhood registered 500 shipments in a digital LaaS platform that enabled direct communications between receivers and carriers. Consumers were encouraged to ship their online orders to the city hub and then select differently-priced delivery time slots to their home addresses—fulfilled by cargo bikes—using the digital LaaS platform.

The pilot showed that consolidation to a desired delivery time slot can be achieved through the lowest prices and sustainability awareness nudging. From the perspective of individual consumers, a reduction in delivery trips to their home addresses was observed as they actively consolidated their orders from multiple shippers and different carriers into single trips.

The pilot results further showed that consumers' willingness to pay is strongly associated with the service level of the LaaS, especially the convenience of selecting the time slot of the delivery. Consumers reported a willingness to pay of at least €1 for using the digitalized LaaS services in real life, aligning with the findings from previous studies: 26% of Germans in major cities have a WTP of €1 or more for a conceptualized RLC delivery service (Hagen and Scheel-Kopeinig 2021) and an average WTP of €2.22 per delivery is found (Engelhardt 2023) for sustainable bundled last mile deliveries. It is also worth noting that LaaS users reported a significant increase in sustainability awareness by the end of the pilot testing period, highlighting the value of RLC in encouraging consumers to reflect on the impact of their shopping decisions.

The implementation of LaaS as an end-consumer-focused RLC initiative showed that digitalizing consumer integration in the last mile can enhance the effectiveness and potential of RLC, as end consumers can be effectively nudged toward behavioral changes—such as postponing and consolidating deliveries—through methods such as pricing and sustainability awareness nudging. An important implication for the effective application of RLC in real-life scenarios is that its digital integration into e-commerce platforms is essential, as the service level largely depends on user convenience. Carriers should optimize their capacities by leveraging RLC's consolidation potential, which can facilitate RLC operations and address inefficiencies resulting from low load utilization of delivery vehicles. Regulatory forces, on the other hand, should introduce policies that support RLC, such as providing incentives through sustainability-focused nudging methods—such as encouraging subscriptions to consolidated delivery services and implementing sustainability feedback mechanisms that give immediate insights into emission savings (Amiri et al. 2024). In addition, as with the cargo bikes, effective RLC implementation depends heavily on the efficient management of city hubs, where parcels from various shippers and carriers must be temporarily stored to enable consolidation for delivery on the consumer's preferred day.

5 Collection and Delivery Points

The previous two sections clearly underscore the need for robust urban logistics infrastructure. One of the most significant developments in this regard is the emergence of collection and delivery point (CDP) networks. These systems comprise both automated parcel systems (APS), such as parcel lockers, and attended pick-up points (PPs). APS units are typically located in publicly accessible areas and offer consumers 24/7 availability. In contrast, PPs are situated in venues such as convenience stores or gas stations, and their accessibility is restricted to the operating hours of the host establishment. However, the presence of staff at PPs allows for additional consumer services, including assistance with delivery issues. Initially intended as fallback locations for unsuccessful home deliveries, CDPs have since evolved into a promising and more sustainable alternative for last-mile delivery. By consolidating deliveries at the neighborhood level, CDPs can reduce the environmental impact of urban logistics, provided consumers retrieve their parcels through low-emission or active modes of transport, thereby offering benefits from both corporate efficiency and broader societal perspectives (Cárdenas et al. 2017).

In Flanders, the geographical distribution of CDPs follows that of the population, and related information is centralized by the Belgian postal regulator (BIPT) on www.postaalpunt.be. There are currently 961 lockers and 4681 pick-up points, including postal offices, in the region. This implies 1.6 lockers per 10,000 inhabitants, which is a sharp increase compared to 0.16 for the entire country in 2021 (Beckers and Verhetsel 2021). The sharp increase is mostly driven by bpost's (the national postal operator) *ecozone* expansion, which are zones in which the carrier promises CO_2-free deliveries with a zero-emission fleet in combination with a dense network of small parcel lockers and pick-up points.

This is a promising trend, however, because the carriers operating in the Flemish logistics market still look for exclusivity in CDPs, the expansion comprises two risks. On the one hand, it causes spatial cluttering; on the other, it would include multiple pick-up kilometers in case consumers need to pick up multiple shipments delivered to lockers, reducing the environmental and economic potential of the system compared to at-home deliveries. An interesting case is the City of Mechelen, which tendered the possibility to install a city-wide parcel locker network in 2021 on the public domain, with the intention to establish a uniform and open system, preventing the risks highlighted above. Bpost won this tender and installed 50 lockers in their first *ecozone*. While the reduction in emissions is undeniable, the national post asks other carriers to drop parcels destined for the lockers in their distribution center, instead of to the locker directly. As bpost delivers these parcels to the locker, the lockers are claimed to be "open." This practice, however, potentially delays the original shipment, and might provide the national post with useful insights on their competitors' volumes, as such limiting the appetite of these competitors to use the "open" system. The key lesson here is that cities need to be very careful when designing tenders and contracts with similar objectives, paying particular attention to the definition of an "open" system.

As the sustainability of the CDP implementation depends on the mode choice of the pick-up trip, the city of Mechelen has actively worked to integrate logistics infrastructure with urban mobility systems, encouraging consumers to use sustainable modes when accessing the parcel lockers. The goal is to embed parcel collection into people's existing travel routines, minimizing the need for additional trips. To support this, some parcel lockers have been placed within mobility hubs—key locations where public transport, shared mobility, and active travel options come together. While survey data shows that parcel lockers and mobility services are often used independently rather than as part of a combined trip chain, this may reflect the already dense network of parcel lockers in the city. The surveys indicate that most users live within 500 m of a parcel locker, making them easily accessible by foot or bike without necessarily requiring shared mobility options. This high level of proximity highlights the effectiveness of the locker network's spatial planning and suggests that accessibility by active modes is already well supported. These findings emphasize the value of continued strategic placement of parcel lockers to maximize their accessibility and sustainability benefits.

6 Conclusions

This chapter reveals several takeaways from Flanders' experience with urban logistics initiatives. First, surveys show that consumer awareness of sustainability in deliveries is growing, but convenience and price still dominate e-commerce choices. This means that simply promoting sustainability without addressing cost and service quality risks limited impact or even increased consumption through overly convenient and affordable delivery options. Policymakers and companies must recognize this trade-off to avoid unintended consequences. Nonetheless, in Flanders as in other study areas, there is a willingness to pay for logistics services.

Building on these behavioral insights, supply-side initiatives demonstrate promising potential for sustainable urban logistics. Although cargo bikes are not a one-size-fits-all solution for urban parcel delivery, they prove highly effective in specific niches, especially for business-to-business deliveries in dense urban areas. Evidence from Flanders highlights that last-mile delivery performance improvements do not come from simply swapping vans for cargo bikes. Rather, efficiency is achieved by addressing the challenges each mode faces within its optimal role. Cargo bikes excel in short-distance, time-sensitive deliveries where high parcel density exists, while vans and trucks remain indispensable for heavier loads and longer routes. Currently, the main limiting factor for cargo bikes is driver time, not vehicle capacity. Therefore, enhancing overall cost-efficiency and delivery volume depends on strategically integrating multiple transport modes. This, of course, applies to Flanders, which benefits from an excellent biking infrastructure. While there are many instances where cargo bikes operate without dedicated cycling lanes, such infrastructure is necessary to make the cargo-bike business model viable. However, looking at current Flemish and Dutch inner-city bike lanes, integrating cargo bikes into already busy lanes is not straightforward. Planning guidelines do exist to better accommodate the growing variety of green vehicles (Dutch Cycling Embassy 2023).

Complementing cargo bikes, the development of pick-up and delivery points, including parcel lockers, plays a crucial role in enhancing logistics efficiency. Strategic placement of these points within urban mobility hubs encourages sustainable travel modes such as walking and cycling for parcel collection. The rapid growth of parcel lockers and pick-up points across Flanders highlights their role in improving delivery efficiency and consumer convenience. However, funders and local authorities caution that fragmented locker networks, where carriers operate separate systems, lead to inefficiencies and spatial clutter. The public sector's role in promoting open, shared locker networks is critical to reducing overlap and ensuring equal access for carriers. Moreover, the public sector should also strategically place these points to encourage sustainable travel modes such as walking and cycling for parcel collection. In this regard, the former Belgian federal minister for postal affairs took a significant step by updating the Postal Law,[1] placing particular emphasis on "openness" as a requirement for collection and delivery networks operating in public spaces. This update is certainly worth examining, as it offers a notable example of a country attempting to curb the race to the bottom in the courier, express, and parcel market. However, its provisions aimed at addressing labor malpractices within the parcel sector remain more controversial.

In addition, urban consolidation hubs, loading bays, and real-time routing platforms serve as critical physical and digital connectors in this multimodal ecosystem. Policy tools such as low-emission zones, targeted subsidies for electric fleets, and standardized curb regulations should translate public benefits into private incentives, encouraging carriers, whether mixed-mode or dedicated cargo-bike operators, to collaborate effectively. The resilience and sustainability of urban freight hinge on such multimodal cooperation rather than simply substituting one mode for another. Freeing up space for logistics infrastructure in urban areas remains a complex challenge. Despite some successful examples, such as the *Hôtels Logistiques* in Paris, current approaches are often ad hoc, relying on shipping containers, vacant parking lots, or temporarily abandoned infrastructure. In Flanders, a critical reflection is needed on the coordination efforts across policy departments, as illustrated by the *GentLevert* initiative. Urban freight impacts fall under the jurisdiction of multiple domains, including mobility, economy, and sustainability. At the same time, its spatial requirements demand attention from urban planning departments or agencies. For meaningful progress, logistics must be addressed strategically rather than treated as a purely operational issue, as is too often the case today. Transport for London has made a commendable effort in this regard, developing a comprehensive urban freight plan that considers the full scope of the topic, including infrastructure planning and the integration of logistics requirements into building permits. The

[1] Wet houdende diverse bepalingen met het oog op het verbeteren van de werkomstandigheden van de postpakketbezorgers ("Act containing various provisions aimed at improving the working conditions of parcel delivery workers."), 17 December, 2023, https://etaamb.openjustice.be/nl/wet-van-17-december-2023_n2023048463.html

European Commission's promotion of Sustainable Urban Logistics Plans (SULPs), supported by EU project funding and clear guidelines, also marks an important step in the right direction.

Shifting focus to demand-side approaches, receiver-led consolidation initiatives emerge as an effective method to nudge consumers toward accepting grouped deliveries, reducing the total number of trips required. These systems rely on hubs and last-mile transport, often cargo bikes, to complete deliveries efficiently. A receiver-led consolidation pilot with 60 e-commerce consumers showed that consolidating deliveries can reduce trips and emissions. The pilot was enabled through an urban consolidation hub and last-mile cargo-bike deliveries, demonstrating the potential of synergy between multiple initiatives. Receiver surveys worldwide suggest that speed is not the primary logistics priority for most consumers; instead, the demand for ultra-fast delivery is largely driven by aggressive competition among major e-commerce players. This reveals significant potential for consolidation if delivery timelines were allowed to slow down. However, current e-commerce supply chains are tightly optimized for speed, resulting in reluctance to withhold goods, even for a short period, after a sale. Advancing receiver-led consolidation, therefore, requires a fundamental paradigm shift in how e-commerce supply chains are organized and managed.

Overall, the experience in Flanders highlights the critical role of the public sector in orchestrating a cohesive urban logistics strategy. Rather than focusing on isolated initiatives, success lies in aligning multiple initiatives—across infrastructure, operational modes, and consumer engagement—into a coordinated ecosystem. This includes enabling infrastructure sharing, incentivizing the use of sustainable transport modes, and fostering behavioral shifts through carefully designed incentives and communication. Crucially, these efforts must be grounded in ongoing evaluation to ensure they deliver the intended outcomes and remain adaptive to local needs. The practical lessons from Flanders offer valuable guidance for other cities seeking to build more efficient, resilient, and sustainable urban logistics systems.

Coordination, however, should extend beyond individual municipal borders. To date, nearly all last-mile logistics initiatives in the region have been driven by individual city efforts. This has resulted in a fragmented landscape of access restrictions, infrastructure provisions, and local freight regulations. A clear example is the uncoordinated implementation of delivery time windows across cities, which has led to inefficiencies: logistics providers have had to split delivery tours to comply with differing cut-off times, ultimately increasing the number of vehicles on the road. As a result, LSPs are calling for more harmonized urban logistics policies across municipalities. In response, cities are increasingly looking to the Flemish regional government for leadership. During the previous legislative term, the region, together with various stakeholders, developed a policy framework that acknowledged the need for integrated urban logistics planning. However, it failed to produce any binding decisions, and the current legislative agenda does not foresee further action. In the absence of a clear regional direction, cities are now reluctant to introduce new initiatives, fearing they will become isolated in their restrictions and risk alienating logistics partners. As such, the lack of commitment at higher levels of governance is actively hampering innovation in last-mile logistics across the region. Furthermore, the fragmented governance landscape in Flanders, with varying municipal policies, exemplifies a common challenge: without coordinated regional frameworks, urban logistics efforts risk inefficiencies and increased operational burdens for logistics providers. Other regions aiming to enhance urban logistics should prioritize early regional coordination to harmonize policies, delivery regulations, and infrastructure planning.

References

Aljohani K, Thompson RG (2020) Receivers-led delivery consolidation policy: estimating the characteristics of the most interested businesses to participate. Res Transp Econ 80:100808. https://doi.org/10.1016/j.retrec.2019.100808

Amiri B, Jafarian A, Abdi Z (2024) Nudging towards sustainability: a comprehensive review of behavioral approaches to eco-friendly choice. Discov Sustain 5(1):444. https://doi.org/10.1007/s43621-024-00618-3

Beckers J, Cant J (2024) Half a decade in two years: household freight after COVID-19. Transp Rev 44(2):544–565. https://doi.org/10.1080/01441647.2023.2266859

Beckers J, Verhetsel A (2021) The sustainability of the urban layer of e-commerce deliveries: the Belgian collection and delivery point networks. Eur Plan Stud 29(12):2300–2319. https://doi.org/10.1080/09654313.2021.1921118

Beckers J, Cardenas I, Le Pira M, Zhang J (2023) Exploring logistics-as-a-service to integrate the consumer into urban freight. Res Transp Econ 101:101354. https://doi.org/10.1016/j.retrec.2023.101354

Beckers J, Beziat A, Hagberg J, Jaller M (2026) Transport for retail in urban areas. In: The Routledge handbook of sustainable urban transport. Routledge

Becom (2025) Market monitor. Consulted online: https://wiki.becom.digital/nl/?post_type=product&s=market+monitor

Belgian Cycle Logistics Federation (2024) The yearly cycle logistics barometer. https://bclf.be/wp-content/uploads/2024/11/Barometer-BCLF-2024.pdf

Belgian Cycle Logistics Federation (2025) Transforming urban deliveries: data evidence from Belgium's cargo bike transition. https://bclf.be/wp-content/uploads/2025/03/Transforming-Urban-Deliveries-Data-Evidence-from-Belgiums-Cargo-Bike-Transition-FINAL-2.pdf

Buldeo Rai H, Mommens K, Verlinde S, Macharis C (2019) How does consumers' Omnichannel shopping behaviour translate into travel and transport impacts? Case-study of a footwear retailer in Belgium. Sustainability 11(9):2534. https://doi.org/10.3390/su11092534

Cárdenas I, Beckers J, Vanelslander T (2017) E-commerce last-mile in Belgium: developing an external cost delivery index. Res Transp Bus Manag 24:123–129. https://doi.org/10.1016/j.rtbm.2017.07.006

Clarke G, Thompson C, Birkin M (2015) The emerging geography of e-commerce in British retailing. Reg Stud Reg Sci 2(1):371–391. https://doi.org/10.1080/21681376.2015.1054420

Clemes MD, Gan C, Zhang J (2014) An empirical analysis of online shopping adoption in Beijing, China. J Retail Consum Serv 21(3):364–375. https://doi.org/10.1016/j.jretconser.2013.08.003

Comeos (2024) E-commerce survey 2024. https://static.comeos.be/Ipsos_Comeos_E--commerceSurvey2024_Results.pdf

Cowie J, Fisken K (2023) Cycle logistics. In: Monios J, Budd L, Ison S (eds) The Routledge handbook of urban logistics. Routledge. https://doi.org/10.4324/9781003241478

Dalla Chiara G, Donnelly G, Gunes S, Goodchild A (2023) How cargo cycle drivers use the urban transport infrastructure. Transp Res A Policy Pract 167:103562. https://doi.org/10.1016/j.tra.2022.103562

Dutch Cycling Embassy (2023) CROW updates bike lane width recommendations. DCE. https://dutch-cycling.nl/knowledge/blogs-by-experts/crow-updates-bike-lane-width-recommendations/

Engelhardt M (2023) Who is willing-to-pay for sustainable last mile innovations? Transport Res Proc 69:910–917. https://doi.org/10.1016/j.trpro.2023.02.252

Etminani-Ghasrodashti R, Hamidi S (2020) Online shopping as a substitute or complement to in-store shopping trips in Iran? Cities 103:102768. https://doi.org/10.1016/j.cities.2020.102768

Eurostat (2025) E-sales record a slight increase over recent years. https://ec.europa.eu/eurostat/statistics-explained/index.php?title=E-commerce_statistics

Farag S, Krizek, Kevin J, Dijst M (2006) E-shopping and its relationship with in-store shopping: empirical evidence from The Netherlands and the USA. Transp Rev 26(1):43–61. https://doi.org/10.1080/01441640500158496

Ghazal A, Narayanan S, Adeniran IO, Kehrt C, Antoniou C (2025) Analysis of logistics measures of CEP service providers for the last-mile delivery in small- and medium-sized cities: a case study for the Aachen city region. Eur Transp Res Rev 17(1):5. https://doi.org/10.1186/s12544-025-00706-z

Hagen T, Scheel-Kopeinig S (2021) Would customers be willing to use an alternative (chargeable) delivery concept for the last mile? Res Transp Bus Manag 39:100626. https://doi.org/10.1016/j.rtbm.2021.100626

Havrylenko N, Kasian S, Opalko V, Ptytsia N, Lytvynenko L (2025) Innovative marketing in intelligence urban logistics systems. E3S Web Conf 616:03014. https://doi.org/10.1051/e3sconf/202561603014

Holguín-Veras J, Sánchez-Díaz I (2016) Freight demand management and the potential of receiver-led consolidation programs. Transp Res A Policy Pract 84:109–130. https://doi.org/10.1016/j.tra.2015.06.013

Holslag J (2023) "Worse than cocaine" Assessing the Impact of Alibaba's Mega-Hub in Liege, Belgium. https://www.jonathanholslag.be/wp-content/uploads/2023/02/Alibaba-fin.pdf

Keng Kau A, Tang YE, Ghose S (2003) Typology of online shoppers. J Consum Mark 20(2):139–156. https://doi.org/10.1108/07363760310464604

Lebeau P, Cok B, Kees C, Macharis C (2023) Towards more sustainable vehicles for the last mile? Cycle logistics as a part of the solution. In: Marcucci E, Gatta V, Le Pira M (eds) Handbook on city logistics and urban freight. Edward Elgar Publishing, pp 178–189. https://doi.org/10.4337/9781800370173.00018

Lei Y, Beckers J (Forthcoming) The role of logistics factors in consumers' shopping channel choices. Manuscript in Preparation

Lemon KN, Verhoef PC (2016) Understanding customer experience throughout the customer journey. J Mark 80(6):69–96. https://doi.org/10.1509/jm.15.0420

Marujo LG, Goes GV, D'Agosto MA, Ferreira AF, Winkenbach M, Bandeira RAM (2018) Assessing the sustainability of mobile depots: the case of urban freight distribution in Rio de Janeiro. Transp Res Part D: Transp Environ 62:256–267. https://doi.org/10.1016/j.trd.2018.02.022

Molin E, Kosicki M, Van Duin R (2022) Consumer preferences for parcel delivery methods: the potential of parcel locker use in The Netherlands. Eur J Transp Infrastruct Res 22(2). https://doi.org/10.18757/ejtir.2022.22.2.6427

Nguyen DH, De Leeuw S, Dullaert W, Foubert BPJ (2019) What is the right delivery option for you? Consumer preferences for delivery attributes in online retailing. J Bus Logist 40(4):299–321. https://doi.org/10.1111/jbl.12210

Phau I, Meng Poon S (2000) Factors influencing the types of products and services purchased over the internet. Internet Res 10(2):102–113. https://doi.org/10.1108/10662240010322894

Rossolov O, Susilo YO (2024) Are consumers ready to pay extra for crowd-shipping e-groceries and why? A hybrid choice analysis for developing economies. Transp Res A Policy Pract 187:104177. https://doi.org/10.1016/j.tra.2024.104177

Rudolph C, Gruber J (2017) Cargo cycles in commercial transport: potentials, constraints, and recommendations. Res Transp Bus Manag 24:26–36. https://doi.org/10.1016/j.rtbm.2017.06.003

Rudolph C, Nsamzinshuti A, Bonsu S, Ndiaye AB, Rigo N (2022) Localization of relevant urban micro-consolidation centers for last-mile cargo bike delivery based on real demand data and city characteristics. Transp Res Record J Transport Res Board 2676(1):365–375. https://doi.org/10.1177/03611981211036351

Song Z (2022) The geography of online shopping in China and its key drivers. Environment and Planning B 49(1):259–274. https://doi.org/10.1177/23998083211002189

Statistiek Vlaanderen. (2025). Population by age and gender. Flanders.Be. https://www.vlaanderen.be/en/statistics-flanders/population/population-by-age-and-gender

Taniguchi E, Thompson RG, Qureshi AG (2024) Recent developments in urban freight analytics for collaborative city logistics. Transport Res Proc 79:3–12. https://doi.org/10.1016/j.trpro.2024.03.003

Tipagornwong C, Figliozzi M (2014) Analysis of competitiveness of freight tricycle delivery Services in Urban Areas. Transport Res Record J Transport Res Board 2410(1):76–84. https://doi.org/10.3141/2410-09

VIL (2025) R!sult (Responsive Sustainable Urban Logistics). https://vil.be/en/project/rsult-responsive-sustainable-urban-logistics/

Vrhovac V, Vasić S, Milisavljević S, Dudić B, Štarchoň P, Žižakov M (2023) Measuring E-commerce user experience in the last-mile delivery. Mathematics 11(6):1482. https://doi.org/10.3390/math11061482

Wieland T (2021) Identifying the determinants of store choice in a Multi-Channel environment: a hurdle model approach. Pap Appl Geogr 7(4):343–371. https://doi.org/10.1080/23754931.2021.1895875

World Economic Forum (2024) Transforming urban logistics: sustainable and efficient last-mile delivery in cities. https://reports.weforum.org/docs/WEF_Transforming_Urban_Logistics_2024.pdf?utm_source=chatgpt.com

Xiao Z, Wang JJ, Liu Q (2018) The impacts of final delivery solutions on e-shopping usage behaviour: the case of Shenzhen, China. Int J Retail Distrib Manag 46(1):2–20. https://doi.org/10.1108/IJRDM-03-2016-0036

Development and Assessment of Communications to Increase Consumer Choice of Sustainable Delivery

Ross O. Phillips and Delphine Pernot

1 Introduction

The EU and UN have set ambitious sustainability goals, and improving the sustainability of road transport is crucial to achieving them.[1] Freight transport is a particular challenge due to its inefficiency and disproportionate environmental and social impacts, with an increasing number of last-mile deliveries linked to e-commerce (Bosona 2020). This chapter concerns the role of consumer choice in the sustainability of e-commerce deliveries. It addresses how retailers can develop and assess communication interventions aimed at making the delivery choices of their customers more sustainable.

E-commerce has grown rapidly due to urbanization, technological advancements, and the expansion of direct-to-consumer markets (Kokkinou et al. 2024). To stay competitive, retailers offer a wide range of flexible delivery options, including home delivery with time slots, parcel lockers, service counters, and in-store pick-up (Amorim et al. 2024). Out-of-home delivery options are often more sustainable as they enable consolidation of deliveries and minimize failed delivery attempts. However, their effectiveness depends on consumer cooperation and transport choices.

Since consumer preferences strongly influence market supply, encouraging sustainable delivery choices can reduce environmental impacts substantially. One estimate suggests that if half of consumers opted for pick-up instead of home delivery,

[1] sdgd.un.org/goals

R. O. Phillips (✉) · D. Pernot
Institute of Transport Economics (TØI), Norwegian Centre for Transport Research, Oslo, Norway
e-mail: Ross.Phillips@toi.no

A. Melkonyan-Gottschalk, Y. Li (eds.), *Urban Logistics Transformation*, Greening of Industry Networks Studies 15,
https://doi.org/10.1007/978-3-032-08387-6_9

CO_2 emissions could be reduced by a third (Davydenko and Hopman 2020). Yet, consumers tend to favour delivery options that are cheap and convenient (Kokkinou et al. 2024; Pernot et al. 2025). Sustainable alternatives must therefore be developed that match traditional options on cost and convenience, *and* efforts should be made to influence consumer choices. Thus, there is a need to:

- Develop and test sustainable last-mile delivery and return solutions that appeal to consumers.
- Influence consumer choice of delivery option.

This chapter contributes to the latter by developing and describing advice for retailers about (i) how to influence consumer delivery choices through communications, and (ii) how to assess the mechanism and effectiveness of these communications as part of an iterative design process. The advice is presented in the form of a model of consumer choice of delivery option developed based on an understanding of what influences delivery choices by e-commerce consumers (Pernot and Phillips 2025). The model is then summarized as a set of user-friendly indicators from which retailers can select according to their situation and needs, as they consider how to communicate with their consumers and evaluate the effects of communication.

The model and indicators were developed using three sources of evidence:

1. A review of the most relevant theories on behaviour and decision-making.
2. A review of frameworks for evaluating sustainable consumer behaviour.
3. Interviews with online consumers about their delivery choices, carried out as part of the EU project CodeZERO.

The next section reviews this evidence. The model and indicators are presented in Sections 3 and 4, respectively. Section 5 presents an example of how a retailer might use the framework (= model + indicators) to communicate with its customers about sustainable delivery options. Finally, Section 6 addresses how the framework can bring value to the retailer and makes recommendations for the retailer.

2 Understanding Influences on Delivery Choice by E-Commerce Consumers

A model describing factors influencing delivery choice by online consumers was developed based on a review of theory and frameworks for evaluating sustainable consumer behaviour, along with interviews with online consumers conducted as part of the EU project CodeZERO (Pernot et al. 2025; Phillips and Pernot 2025). In the following section, factors that were ultimately included in our model are highlighted in bold.

2.1 Theory on Behavioural Change and Decision-Making

Behavioural change and decision-making theories suggest that the following psychological and cognitive dimensions should be considered when aiming to influence consumer delivery choices.

- Intention to choose sustainable delivery: A review of key behavioural change theories (Ajzen 1985; Schwartz 1977; Stern et al. 1999) suggests that a consumer's delivery choice can change if the consumer forms an intention to choose a sustainable delivery option. Such intentions can be formed if the consumer develops:
  - A positive attitude toward sustainable delivery (e.g. a consumer believing that choosing the most sustainable delivery option is good for society).
  - A belief that they are personally responsible for choosing a sustainable delivery option and that their choice will impact sustainability (which we consider as a component of attitude).
  - The belief that "valued others" choose sustainable delivery options or want them to choose sustainable delivery options (social norms).
  - The belief that they are in control of their choice of delivery option (e.g. a consumer perceives little control if they think "I cannot choose the most sustainable delivery option as it is always too far for me to collect") –formally conceived as perceived behavioural control (PBC).
- Planning, habit, and visibility: A review of the integrated behaviour model (Montaño and Kasprzyk 2008), habit theory (Klöckner and Verplanken 2018), and the Stage Model of self-regulated behavioural change (Bamberg 2013) informs that behaviour is shaped not only by intention, but also by:
  - The consumer's ability to plan and act on that intention, especially concerning know-how on how to choose a delivery option and collect after selecting a sustainable delivery option (e.g. parcel locker).
  - The degree to which the behaviour has become habitual (e.g. the consumer always chooses home delivery from Amazon).
  - Visibility of choice, or the degree to which sustainability or other goal-related attributes of delivery options are salient for the consumer (e.g. salient that an eco-labelled parcel locker delivery option is more sustainable than an unlabelled home delivery option).
- Goal trade-offs and intuitive choice: Theory in Naturalistic Decision Making (Klein 2008), The Decision Ladder (Vicente 1999), and Nudging Theory (Kahneman and Frederick 2005) suggest that:
  - Consumers will trade-off several goals, only one of which is sustainability, depending on the unique ordering situation, e.g. cost, convenience of delivery.
  - Ordering behaviour might often be guided by quick, intuitive thinking rather than deliberate reasoning.

- Design cues and subtle prompts could be used to influence more intuitive or habitual consumer choices.

In work carried out as part of CodeZERO, several of the above-mentioned factors have been conceptualized as a "motivation/habit/opportunity" framework describing main drivers of behavioural change (Cauwelier et al. 2025). This framework emphasizes the need to understand consumer motivation as central to being able to influence their sustainable delivery choices. Intrinsic motivation, stemming from personal values, may be more effective than extrinsic motivation, such as pressure from social norms or financial incentives. The framework emphasizes the importance of habits in overriding stated attitudes and intentions in choosing sustainable delivery. Opportunity describes the availability or accessibility of sustainable versus other delivery options available to the consumer.

2.2 Frameworks for Influencing and Evaluating Change in Sustainable Consumer Behaviour

A review of evidence-based frameworks used to effect and evaluate change in sustainable consumer behaviour (SCB) was conducted to inform how to model the communication intervention itself (characteristics, implementation, and reach). Frameworks reviewed included:

- The SHIFT framework (White et al. 2019). SHIFT can be used to assess the effect of communications on sustainable consumer behaviour. SHIFT is a mnemonic for five main drivers of behavioural change: social influences (including social norms), habits, individual self, feelings, cognitions, and message tangibility. "Individual self" describes the extent to which choice lets people maintain a positive view of themselves and be consistent with their values. It is important that the consumers believe they can choose and that doing so will lead to desirable effects in the real world. Thus, "Individual self" resembles a composite variable of perceived behavioural control and attitude as conceptualized earlier.
- The COM-B framework (Ran et al. 2022). COM-B can also be used to assess the effect of communications on sustainable consumer behaviour. It considers that consumers can make a choice if they have the capability to choose, the opportunity to choose, and the motivation to choose. Capability and motivation to choose can be assimilated into the concepts perceived behavioural control and know-how, which we have already considered. Opportunity to choose is linked to being given visibility of choice, considered above, and again knowing how to choose and collect.
- The RE-AIM framework (Gaglio et al. 2013, Holtrop et al. 2021). RE-AIM aims to improve reporting on the implementation of health communications. Components are: reach of the target audience; effect of communication on choice and factors that can influence choice; adoption; implementation (akin to process

evaluation); and maintenance, describing the extent to which the organization adopts communication in the longer term.
- McGuire's communication/persuasion model (Oyibo and Toyonaga 2024). This model addresses factors influencing how well the message works along its path from initial presentation to behavioural change in the targeted audience, e.g. presentation of the message, receiver's attention to the message, comprehension of the message, attitude to the message, and retention of the message.
- The EAST framework (BIT 2014). EAST aims to simplify guidance on achieving change in sustainability behaviour. It is based on the finding that four main factors make sustainable communications more appealing: social influence, tangibility of message, attractiveness and timeliness of message, and ease with which it can be followed by consumers, which can be linked to know-how and visibility of choice.
- B = MAP model (Shah and Agarwal 2020). B = MAP also distils much behavioural evidence into a mnemonic, finding that behavioural or choice change results from changing the extent to which consumers want to change (motivation, goals), they feel they can change (links to perceived behavioural control) and prompts or contextual cues, which refer to how nudging interventions are timed in relation to when the choice is made.
- MINDSPACE model (Hashemi et al. 2024). MINDSPACE assess the reception of communication by consumers. It is also a mnemonic and considers that choice is influenced by extent to which consumers trust the message and its sender, incentives (link to goals), social norms, attitude, visibility of choice and contextual cues, which a special emphasis on subconscious cues that influence consumer choice.

As can be seen, the above-mentioned frameworks incorporate several of the factors identified in Sect. 2.1, and especially goals, social norms, habits, attitudes, values, and the individual's capacity to make delivery choices. They also suggest, however, additional elements that can be critical for designing and assessing the effect of communications to consumers on choosing sustainable delivery:

- Message tangibility: A challenge unique to messages about sustainability is that they are perceived by consumers as lacking tangibility. Many sustainability communications ask consumers to forego short-term personal benefits in favour of distant, abstract collective gains—making the message difficult to be concerned about and act on. To be effective, sustainability messages should clearly articulate why individual choices matter.
- Timing of communication along consumer journey: communication can happen along the "consumer journey" from planning and online purchasing to the physical delivery and collection process; they should be designed and evaluated as such.
- Communication implementation and reach: evaluation should assess the quality of implementation and the degree to which consumers noticed, attended to, and processed the messages communicated.

- Interface and contextual cues: The characteristics of the user interface should also be designed and evaluated for effective communication of the need to choose sustainable delivery choices.

2.3 Interviews with Consumers About Their Online Delivery Choices

The analytical phase of the EU project CodeZERO included interviews with 43 online consumers from European countries, including Norway, Belgium, the Netherlands, and Italy, about their real-life delivery choices. The interviews included questions to consumers about their life contexts, sustainability perceptions, and influences on sustainability choices, and how they decided between online delivery options (Pernot et al. 2025). "Speak aloud" exercises were also conducted with consumers using online platforms (Phillips and Pernot 2025). The interviews revealed additional key influences on consumer choice of delivery option, which also need to be accounted for when advising retailers on how to influence and evaluate change in delivery choice by consumers:

- Delivery attributes—price of delivery; delivery time, punctuation and precision; vehicle type used—as consolidated by work elsewhere in CodeZERO assessing the current landscape of delivery and return options (Kin et al. 2025).
- Consumer goals and motivations— key goals are cost, convenience, delivery speed, safety/carrier quality, and sustainability; prioritization of these goals can vary depending on the type of product being ordered and the reasons for ordering.
- Consumer awareness of the potential effects on diverse aspects of sustainability that their delivery choices can have (e.g. emissions, working conditions, economic).
- Consumer trust in the sustainability message and its sender.
- Consumer belief that it is possible to choose and individual choices matter (links to perceived behavioural control and attitude).
- Physical situation of the consumer (e.g. accessible pick-up points and shops).
- Digital environment of the consumer (e.g. interface design, visibility of preferred choices).
- Previous experience with delivery options, which influences the likelihood that they will choose that option in the future.

3 Modelling the Influence of Communication on Consumer Delivery Choice

The knowledge summarized above was reviewed to develop a model of how communication can influence consumer choice of delivery option, with a view to increasing sustainable delivery choices (Fig. 1).

The model can be used to structure the assessment of:

- The characteristics of the communication intervention (pink box, Fig. 1).
- The mechanism by which communication influences consumer choice of delivery option (green box, Fig. 1).
- The effectiveness of the communication on delivery choice (the extent to which a change in delivery choice aligns with the hypothesized mechanism of communication).

The model was developed by considering how communications can be used to change each of the factors influencing consumer delivery choice outlined in Sect. 2. Factors in bold are included in the model based on their importance in the above-mentioned review. Based on their relation to each other in the review, the factors are structured along three communication routes or three ways in which the retailer's "messages" can influence delivery choice by their consumers. Each route is represented by an arrow labelled 1, 2, or 3, leading from the pink "retailer" to the green "consumer" box in Fig. 1.

The first route (1) is taken by retailer communications aimed at influencing consumer intentions to choose sustainable delivery (see Fig. 1). Here, tangible messages that the consumers trust can be aimed at changing the attitudes and social norms or behavioural control perceptions of consumers, as well as their practical

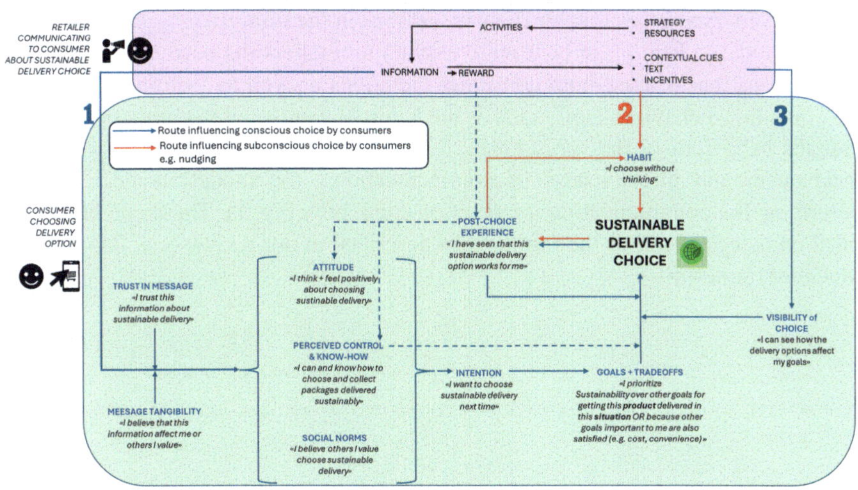

Fig. 1 Model of how communications influence choice of delivery option by consumer

know-how, such that these factors better align with (and therefore increase) intentions to choose sustainable delivery. The second route retailer communications can take is to influence the more intuitive, habitual delivery choices of consumers by changing or introducing contextual cues or messages on the online shopping interface (e.g. eco-labels) or elsewhere along the consumer journey [see route (2) in Fig. 1]. Route 2 can be considered as a route by which communications developed by the retailer to influence consumers subconsciously, e.g. via nudging

The third route retailer communications can take is to make visible how delivery options affect sustainability alongside other goals that the consumer values [see route (3) in Fig. 1]. This can also involve presenting information or contextual cues on the interface or elsewhere along the consumer journey, but incentives may also be used to increase the attractiveness of sustainable choices. Examples of incentives are extra points earned in a retailer's bonus scheme. Communications that take route 3 will be processed deliberately by the consumer.

An important finding from CodeZERO is that consumers, rather than comparing all presented delivery options, given a set of criteria, tend more to home in on a preferred or attractive delivery option and process information to see if it will work for them for that product in that ordering situation (singular rather than comparative evaluation; see Phillips and Pernot 2025). It may therefore be important to make sustainable delivery options presented on the interface immediately attractive to the consumer, such that they are not overlooked. This may involve designing the overall communication architecture using routes (2) and (3) together (see Fig. 1).

Finally, an additional route by which retailers can influence consumers is that of positive feedback or non-financial rewards using messages such as "thank you for contributing to green delivery". Such messages can give consumers a positive experience of a delivery solution and thereby help shape a positive attitude to sustainable delivery. This additional route can be considered as a special form of route 1 message, as it can be included as information that is designed to increase consumer intentions to choose sustainable delivery options in the future.

The model in Fig. 1 can be used to explain to retailers the need to design communications that address both consumer intentions as well as factors that address the "intention-behaviour gap", such as the habitual nature and visibility of the individual's behaviour or delivery choice. Since it was also designed to evaluate the mechanism and effectiveness of communications, the model includes factors describing the communication intervention (pink box, Fig. 1). These can be elaborated on using several of the frameworks described in our review (e.g. RE-AIM or McGuire framework).

4 Indicators

To present the above-mentioned model in terms that retailers can more easily understand and in a way that can assist the evaluation of attempted communications, a set of indicators was developed based on the model (Table 1). It comprises two subsets that:

(i) Describe the communication intervention.
(ii) Assess its impact using key performance indicators (KPIs).

Each indicator can act as a prompt for communication design or be converted to measures for the evaluation of communication on consumer choice of delivery option (Pernot and Phillips 2025).

The two above-mentioned indicator subsets can be supplemented with a third subset of indicators that do not necessarily directly link communication to consumer choice behaviour, but are needed to help assess *how* the communication affected consumer delivery choice. Examples are consumer demographics; consumer perception of information, cues, and symbols introduced onto the shopping platform; consumer sustainability values; determinants of delivery choice habits; or collection parameters such as location of pick-up points relative to home or reasons for choosing transport means. For more details on these indicators, see Pernot and Phillips (2025).

Changes in the KPIs in Table 1 can be compared with changes expected from a hypothesized communication intervention, to provide support as to whether the communication was responsible for any change in consumer choice. This is particularly useful in cases where it is difficult to carry out controlled or time-series evaluations. Thus, a "route 1" communication might be expected to bring about changes in "Attitude to target option", and consequently "Next-time intention" to choose the target option, which may then result in a change in actual "Delivery choice". Alternatively, "route 3" communications might be expected to change "Delivery choice" through changing "Choice salience".

The framework is designed to be flexible and adaptable. Practitioners can select evaluation measures from this framework according to the purpose of the evaluation, the specific communication intervention being evaluated, and the supposed mechanism of effect of communication on the behaviour of the consumers targeted. Sample size, experimental design, and data collection methods and tools can each influence the choice of KPIs and other components to evaluate when assessing the impact of communication interventions (for more detail, see Pernot and Phillips 2025).

Table 1 Indicators for the evaluation of retailer communications to online customers to encourage them to choose sustainable delivery

		Indicator	Description	Source
Subset	Communication	Strategy and resources	Theoretical basis for communication, list of resources used, description of communication content and channels.	Project manager
		Activities and execution	Description of activities executed with evaluation of process (what worked well, what didn't) and output, e.g. interface designed as intended.	Communication team
		Communication output	Tangible result of activities, e.g. brochures printed, number of radio spots, changed text on interface etc.	Project manager
	Impact on consumers and their delivery choice (KPIs)	Message retention	% consumers who recall message.	Survey
		Message comprehension	% consumers who understand message.	
		Attention to message	% consumers drawn to target delivery option.	Mouse tracking
		Trust in message	% consumers who trust the message	Survey
		Tangibility of message	% consumers who believe the message affects them	
		Choice awareness	% consumers aware of sustainability of delivery/return choice	
		Attitude to target option	% consumers with positive attitude to target delivery option; % consumers believing choosing target delivery option will have an impact on sustainability	
		Social norms	% consumers believing most others choose target delivery option; % others that consumers think would choose target delivery option	
		"Hassle" factor	% consumers thinking it would be difficult to choose or implement the target delivery option	
		Next-time intention	% consumers who say will choose target delivery option next time they shop online	
		Real-time intention	% consumers who say intended to choose target delivery option while shopping online	
		Goal prioritization	Average rank score for sustainability as goal for choice of delivery option – relative to cost, convenience, speed and safety	
		Platform satisfaction	% consumers who say they are satisfied with platform and usability	
		Implementation know-how	% consumers who say they know how to choose delivery option; % consumers who say they know how to execute (e.g. collect) option	
		Choice salience	% consumers who easily find goal-related information on options	
		Choice habit	% consumers who last time say they chose delivery option out of habit	
		Delivery choice	% of deliveries executed by each option over defined period; % consumers who chose target delivery option last time they ordered	Retailer data + survey
		Collection parameters	Av. distance (km) consumers travel to collect; % consumers using different transport means to collect; % consumers trip-chaining for parcel pick-up.	Survey

5 Illustrative Example of Evaluation Using Indicators

To illustrate the use of the indicators, the reader might consider the case of a book retailer who wanted to inform the development of a communication intervention to encourage consumers to choose delivery from pickup points over home delivery. An intervention was developed based on Fig. 1 and evaluated using the indicators in Table 1.

Communication
Strategy and resource. The retailer wanted to increase consumers' awareness about the lower emission impacts of pickup point deliveries compared to home deliveries. They used two different communication interventions:

- an information message about the pickup points when choosing the delivery option,
- a green label for the deliveries in pickup points.

The information message was: "By choosing delivery to a pickup point, you help us deliver more sustainably by reducing delivery vehicles' travel distances, resulting in fewer emissions." It was visible on the retailer's website and mobile app, where consumers were supposed to choose between the two delivery options. The green label shows a green leaf with the tagline "Eco-Friendly Pickup". When selecting pickup points, consumers got a map where they could visualize the locations of the pickup points. By clicking on them, they got the name of the pickup point and the opening hours. Information about the delivery date and carrier name was also provided. The retailer wanted to keep the same price policy, i.e. pickup points and home deliveries cost the same. To assist the evaluation of the communication, it was delivered to target consumers with four different versions of the platform, i.e. there were four consumer groups:

A. Control group (no change in online platform).
B. Consumers who were only exposed to the information message.
C. Consumers who were only exposed to the green label.
D. Consumers who were exposed to the information message and the green label.

Activities and efficacy of execution. The sustainability expert and web developer comprised a design team, who meet to draft alterations to the consumer interface and develop and get management approval for the message and green label used. Activities were also conducted to communicate or disseminate the changes among staff or consumers. The message increasing consumer awareness was implemented on the website for consumers with a delivery address within a city district from February 1st to February 28th. The green label was then implemented for the same targeted consumers from March 1 to 31. And then both the message and the label were made visible for these consumers from April 1 to 30. The implementation of the measures worked as planned, except that the information message had to be shortened on the mobile app.

Outputs. Outputs were documented by screenshots showing the concrete digital changes made to the mobile or PC consumer interface.

KPIs

The reach of the communication was assessed by measuring "*Message retention*" and "*Message comprehension*" by surveying a sample of consumers targeted with the communication before and after the platform changes were made (see Table 1 for examples of measurable KPIs). "*Platform satisfaction*" was assessed to ensure the changes had not been detrimental to the online consumer's user experience. Since the communication aimed at changing consumer attitude, it was important that consumers had "*Trust in message*" and believed that it impacted them ("*Tangibility of message*") (see Fig. 1). The purported mechanism of communication was to shape a positive attitude to the sustainable delivery option by increasing consumer awareness of the sustainability impact of pick-up options compared with home delivery. "*Choice awareness*" and "*Attitude to target option*" were also assessed using the consumer survey. The retailer also decided to measure consumers' "*Next-time intention*" to choose pick-up (through the survey) and to track changes in the relative share of delivery options consumers chose over time using back-end platform metrics. The retailer also wanted to assess changes in the sustainability impact of communications by measuring the average consumer distance travelled to collect parcels from pick-ups and the transport means used to do so.

On analysing the results of surveys and platform data, the retailer found that 15 per cent more of the consumers who had been exposed to the combined eco-label and information message chose to pick up. In support of communication as the cause of choice change, the retailer also found a higher share of next-time intentions, positive attitudes, and awareness of sustainable delivery choice among consumers targeted with both measures. This was assisted by a high share of these consumers trusting each message and finding the messages tangible. The retailer was concerned, however, by a high share of consumers using diesel cars to collect from pick-up points.

6 Retailer Value and Recommendations

The logistics of last-mile delivery have emerged as a pivotal arena for sustainable transformation. The challenge for e-commerce retailers is that they must satisfy consumers' demands for affordability and convenience while ensuring sustainable operations. The above-mentioned framework informs retailers about routes they can use to communicate sustainable delivery options—by influencing consumer intentions through persuasive messages and rewards, through framing/nudging approaches, or making clear how sustainable delivery options align with different consumer goals.

When used as part of a larger strategy, the framework has the potential to add tangible economic value. For example:

1. By helping retailers increase customer satisfaction with the checkout by reducing "cognitive friction" in the checkout process. The interviews we have conducted suggest that, when it is brought to their attention, consumers experience a feeling of guilt at having to prioritize cost and convenience at the expense of sustainability (Phillips and Pernot 2025). This may be reduced if sustainable delivery options that compete on cost and convenience are made visible. Any attitude- or intention-behaviour discrepancies might also be highlighted in communications and ways to reduce them provided.[2] A clearer decision path can contribute to less basket abandonment by the retailer's consumers.
2. By helping retailers move their consumers toward sustainable choices to reduce last-mile logistics costs, which often constitute a significant portion of fulfilment expenses (e.g. more sustainable out-of-home deliveries often afford lower fuel usage, more efficient route planning, and fewer failed delivery attempts).
3. Promotion and improved brand reputation. Effective communication of sustainable practices strengthens brand perception among ethically minded consumers.

7 Recommendations

Based on the above-mentioned analysis, we recommend that retailers:

- Develop sustainable delivery options that offer consumers just as much cost and convenience as less sustainable delivery options.
- Consider how to get consumers to choose the more sustainable delivery option by combining three routes of communication—aimed at aligning consumers' intentions, instinctive/habitual choices, and perception of differences between delivery options with sustainable delivery choices.
- Select from indicators presented to evaluate and iteratively develop consumer communications on sustainable delivery choice.
- Increase value by incorporating the framework for communications on delivery choice into the strategy for consumer marketing and influence.

8 Recap and Conclusion

This chapter has described the development of a model and indicators that can be used to inform the development and assessment of communication interventions aimed at changing consumer delivery choices in online retail settings. The model and indicators have been informed by supplementing models of individual/

[2] Less sustainably motivated consumers might be influenced with nudging measures, such as green color-coding, eco-labels or concise messages ("low-emission delivery arrives tomorrow"; see Route 2 on Figure 1).

consumer behaviour and decision-making with (i) research structuring the assessment of communications aimed at changing sustainable consumption behaviour and (ii) interviews conducted with online consumers. Together, the model and indicators form a framework that has been designed to be dynamic and adaptable. While it offers a structured approach, it can be refined to fit specific campaign objectives, target audiences, and strategies. We have illustrated its application in assessing communication strategies aimed at influencing delivery choice, but the framework can also be used to inform and evaluate communications to alter return choices. The evidence-based and context-sensitive framework should serve as a general tool for diverse communication campaigns targeting online consumers. It is intended to support the iterative design and evaluation of communication efforts by retailers. Finally, we should point out that the framework should be evolved and refined through use, and in particular through assessments of retailer usability and usefulness.

Acknowledgements

This work was carried out as part of CodeZERO (codezero-project.eu), which has received funding from the European Union's horizon Europe research and innovation programme under Grant agreement no. 101146909.

References

Ajzen I (1985) From intentions to actions: a theory of planned behavior. In: Kuhl J, Beckmann J (eds) Action control: from cognition to behavior. Springer, Berlin, Heidelberg, pp 11–39. https://doi.org/10.1007/978-3-642-69746-3_2

Amorim P, DeHoratius N, Eng-Larsson F, Martins S (2024) Customer Preferences for Delivery Service Attributes in Attended Home Delivery. Management Science 70(11):7559–7578. https://doi.org/10.1287/mnsc.2020.01274

Bamberg S (2013) Changing environmentally harmful behaviors: A stage model of self-regulated behavioral change. Journal of Environmental Psychology 34:151–159

BIT (2014) EAST: four simple ways to apply behavioural insights. Behavioural Insight Team, London

Bosona, T (2020) Urban Freight Last Mile Logistics—Challenges and Opportunities to Improve Sustainability: A Literature Review. Sustainability March 2020, 12, 8769. https://doi.org/10.3390/su12218769

Cauwelier K, Lebeau P, Rai HB (2025) Definition of consumer groups. CodeZERO Proj Deliv 2:1

Davydenko, IY, Hopman, WMM (2020) Effect of pick-up points and returns on CO2 emissions in last mile parcel delivery networks. TNO reports 2020 R10909, TNO The Hague, The Netherlands

Gaglio B, Shoup JA, Glasgow RE (2013) The RE-AIM framework: a systematic review of use over time. Am J Public Health 103(6):e38–e46. https://doi.org/10.2105/AJPH.2013.301299

Hashemi S, Bai L, Gao S, Burstein F, Renzenbrink K (2024) Sharpening clinical decision support alert and reminder designs with MINDSPACE: a systematic review. Int J Med Inform 181:105276. https://doi.org/10.1016/j.ijmedinf.2023.105276
Holtrop JS, Estabrooks GB, Gaglio B (2021) Understanding and applying the RE-AIM framework: clarifications and resources. J Clin Transl Sci 5(1):e126. https://doi.org/10.1017/cts.2021.789
Kahneman D, Frederick S (2005) A model of heuristic judgment. In: The Cambridge handbook of thinking and reasoning. Cambridge University Press, New York, pp 267–293
Kin B, Winkelmann F, Meijer L (2025) Impacts of existing delivery models and options. In: Deliverable 2.3 from EU project «CodeZERO». Institute of Transport Economics (TØI)
Klein G (2008) Naturalistic decision making. Hum Factors 50(3). https://doi.org/10.1518/001872008X288385
Klöckner CA, Verplanken B (2018) Yesterday's Habits Preventing Change for Tomorrow? About the Influence of Automaticity on Environmental Behaviour. In Environmental Psychology (eds L. Steg and J.I.M. Groot). https://doi.org/10.1002/9781119241072.ch24
Kokkinou A, Quak H, Mitas O, Mandemakers A (2024) Should I wait or should I go? Encouraging customers to make the more sustainable delivery choice. Research in Transportation Economics, vol. 103, 101388, ISSN 0739–8859. https://doi.org/10.1016/j.retrec.2023.101388
Montaño DE, Kasprzyk D (2008) Chapter 4: Theory of reasoned action, theory of planned behaviour and the integrated behaviour model. In: Health behavior and health education: theory, research, and practice. Jossey-Bass
Oyibo K, Toyonaga S (2024) Conceptual frameworks for designing and evaluating persuasive messages aimed at changing behavior: systematic review. Comput Human Behav Rep 15:100448. https://doi.org/10.1016/j.chbr.2024.100448
Pernot D, Phillips RO (2025) Assessment framework for consumer's behaviour change. EU deliverable. In: Deliverable 2.2 from EU project «CodeZERO». Institute of Transport Economics (TØI)
Pernot D, Phillips RO, Saghafian M (2025) Consumers' awareness and behavioural change. EU deliverable. In: Deliverable 2.2 from EU project «CodeZERO». Institute of Transport Economics (TØI)
Phillips R, Pernot D (2025) Consumers' real-time decision-making on delivery options. In: Deliverable 2.3 from EU project «CodeZERO». Institute of Transport Economics (TØI)
Ran Y, Nilsson Lewis A, Dawkins E, Grah R, Vanhuyse F, Engström E, Lambe F (2022) Information as an enabler of sustainable food choices: a behavioural approach to understanding consumer decision-making. Sustain Prod Consump 31:642–656. https://doi.org/10.1016/j.spc.2022.03.026
Schwartz SH (1977) Normative influences on Altruism1. In: Berkowitz L (ed) Advances in experimental social psychology, vol 10. Academic Press, pp 221–279. https://doi.org/10.1016/S0065-2601(08)60358-5
Shah PR, Agarwal A (2020) Cybersecurity behaviour of smartphone users through the lens of Fogg behaviour model. In: 2020 3rd international conference on communication system, computing and IT applications (CSCITA), pp 79–82. https://doi.org/10.1109/CSCITA47329.2020.9137773
Stern PC, Thomas D, Troy DA, Greg G, Linda K (1999) A value-belief-norm theory of support for social movements: the case of environmentalism. Hum Ecol Rev 6(2)
Vicente K (1999) Cognitive work analysis. Toward safe, productive and healthy computer-based work. LEA, New Jersey
White K., Habib R., Hardisty DJ (2019) How to SHIFT Consumer Behaviors to be More Sustainable: A Literature Review and Guiding Framework. Journal of Marketing 83(3):22–49. https://doi.org/10.1177/0022242919825649

Rethinking the E-Commerce Ecosystem: Returns, Packaging, and Circularity in Sweden

Josephine Darlington, Allan Salimi, and Yanying Li

1 Introduction

This chapter presents a series of Swedish case studies that offer practical insights into how e-commerce systems can transition toward more sustainable and circular models. As global e-commerce continues to expand rapidly, it brings with it mounting concerns over returns, packaging waste, and emissions. These issues are not only environmental in nature, they also intersect with social equity, regulatory compliance, and long-term business resilience. The Swedish context, marked by its strong innovation ecosystem, policy support, and collaborative culture, provides a valuable lens through which to explore potential pathways for sustainable transformation.

The added value of examining these Swedish initiatives lies in their systemic and integrative approach. Rather than addressing sustainability in isolated silos, these case studies demonstrate how digitalization, cross-sector partnerships, and circular economy principles can be effectively combined across the entire e-commerce value chain, from return management and packaging design to product reuse and last-mile delivery.

The case studies are particularly instructive because they provide real-world evidence that sustainable e-commerce is not only feasible but can also be economically and operationally beneficial. They challenge the prevailing notion that sustainability

J. Darlington · A. Salimi (✉)
Lindholmen Science Park, Göteborg, Sweden
e-mail: josephine.darlington@lindholmen.se; allan.salimi@lindholmen.se

Y. Li
Alliance for Logistics Innovation through Collaboration in Europe (ALICE), Brussels, Belgium
e-mail: yanying.li@etp-alice.eu

A. Melkonyan-Gottschalk, Y. Li (eds.), *Urban Logistics Transformation*, Greening of Industry Networks Studies 15,
https://doi.org/10.1007/978-3-032-08387-6_10

inevitably entails trade-offs and instead highlight how well-designed systems can generate cost savings, enhance consumer trust, and improve resource efficiency. For policymakers, researchers, and industry stakeholders beyond Sweden, these examples offer replicable models and critical lessons that can inform broader efforts to decarbonize and circularize the global e-commerce sector.

The expansion of e-commerce is reshaping shopping and goods movement, especially in urban areas, and is set to continue growing. This evolution drives innovations in delivery systems, warehouse efficiency, and return solutions. Simultaneously, increasing awareness of environmental limits is fostering a shift toward a resource-smart economy. In Sweden, the pre-loved market is projected to double from 16 billion SEK in 2024 to 30 billion SEK by 2030, with 70% of these purchases occurring online, highlighting e-commerce's role in sustainable consumption (Svensk Handel 2024).

Contrary to common belief, e-commerce often has a lower environmental impact than physical retail in last-mile delivery, with emissions savings of 81–88%. This is due to optimized routing, consolidated deliveries, and faster adoption of electric or biofuel vehicles. For a large country like Sweden, e-commerce supports rural accessibility while reducing car trips and overall emissions (Makrologik 2025).

From a systems perspective, e-commerce also promotes social equity. It provides critical access for individuals with disabilities and plays a key role for women, who account for 70% of purchases and manage most household buying, by easing unpaid domestic labor (Stanfors and Tärneberg 2020). Rural municipalities show the highest per capita usage, benefiting from broader product access.

Efforts to improve e-commerce sustainability are thus vital for equitable service access nationwide. E-commerce logistics—from deliveries to returns—offer opportunities to reduce environmental impact. Despite high return rates, especially in fashion, these challenges have sparked innovation, advancing a shift from linear to circular logistics systems that aim to reduce waste and recover product value (Hellström et al. 2022, Liu.se 2023a).

In addition, several key regulations are driving this transition, including:

- EU Packaging and Packaging Waste Regulation (PPWR): Sets reuse targets for transport packaging (40% by 2030, 70% by 2040) and requires recyclable packaging (European Commission 2025).
- EU Ecodesign for Sustainable Products Regulation (ESPR): Introduces Digital Product Passports (DPP) for better traceability and transparency (European Commission 2024).
- Extended Producer Responsibility: Policies for the textile industry to manage product lifecycles (Vinnova n.d.).

Despite these drivers, significant obstacles remain that hinder the widespread adoption of sustainable practices in the e-commerce sector. Key challenges include economic profitability, as many sustainability initiatives still struggle to demonstrate clear financial returns. Logistical complexity also poses a barrier, especially when integrating environmentally friendly solutions into existing supply chains and distribution networks.

Consumer behavior is another critical factor. While awareness of sustainability is growing, purchasing habits do not always align with environmentally conscious choices. Additionally, the lack of standardized data makes it difficult to measure and compare sustainability efforts across the industry, creating uncertainty and inefficiencies. Scalability presents yet another challenge, as successful small-scale solutions often face difficulties when expanded to broader markets or larger operations.

Addressing these challenges requires a system-wide perspective and extensive collaboration among various actors in the value chain. Recognizing this, the Swedish e-commerce ecosystem is currently developing a range of joint initiatives. These take the form of research and development projects that aim to generate innovative solutions, supported financially by the Swedish Government. Through these collaborative efforts, the sector is working to overcome existing barriers and pave the way toward a more sustainable future.

This chapter provides an overview of the Swedish landscape regarding research, policy, industry practice, and innovation initiatives related to e-commerce in relation to sustainability and the circular economy. It focuses on three central areas:

(i) **E-Commerce Returns:** Extent, drivers, consequences, and strategies for efficient and sustainable management.
(ii) **Sustainable E-Commerce Packaging:** Challenges with packaging waste, regulatory requirements, and innovation initiatives.
(iii) **Circular Logistics Flows and Business Models:** Potential and challenges of implementing circular business models, with a focus on textile reuse and the role of non-profit organizations.

2 Literature, Current Practices, Knowledge

2.1 Consumer E-Commerce Returns

Research into product returns and returns management has become a significant area within the broader fields of logistics and supply chain management (SCM). Despite a considerable number of academic contributions, the existing literature with an explicit strategic focus on returns management is noted as limited. Researchers have called for further investigation into strategies for managing product returns and a deeper exploration of the inherent trade-offs between handling customer returns in terms of operational costs versus considering returns as a crucial element of customer service (Karlsson et al. 2023).

More recent studies have begun to examine strategic choices related to aspects such as return insurance and the development of employee capabilities as foundational elements for effective returns management. Previous research also suggests that enhancing functional integration, particularly at the interface between marketing and operations, could enable companies to better address customer return demands (Liu.se 2023b). Critically, there is also a recognized lack of research concerning the decision-making processes related to return policies and how these policies specifically impact the efficiency and effectiveness of the return process itself. Despite the increasing academic interest, research focused on the strategic dimensions of returns management within e-commerce remains scarce, and strategies for managing returns are often less discussed or prioritized within e-commerce companies compared to outbound logistics (Karlsson et al. 2023; Pal and Sandberg 2023).

In response to the challenges presented by e-commerce returns, both e-retailers and researchers are actively working on and implementing a range of strategies. These strategies can generally be categorized into three main areas: preventive measures, policy adjustments, and optimized management. Preventive measures, often referred to as Return Prevention, aim to reduce the initial likelihood of a return occurring. A crucial aspect of this involves providing customers with the most accurate and detailed information possible before they make a purchase. This includes clear and comprehensive product descriptions, precise size guides, and high-quality visual content such as images and videos that accurately represent the product. A survey indicated that 31% of e-retailers are actively focusing on these types of preventive strategies (Petisme 2025).

Policy adjustments involve modifying return conditions. One potential strategy is to differentiate return conditions based on customer segments or their purchase and return history, although this approach does not appear to be widely adopted currently. Another policy adjustment can include implementing return fees. Once a return does occur, the objective shifts to handling it as efficiently and sustainably as possible through optimized returns management (Karlsson et al. 2023).

Digital return platforms are a key current practice in this area; they digitize the return process, making it smoother and more intuitive for the customer while simultaneously providing the retailer with improved control and access to real-time data. This digitalization can also facilitate faster processing of exchanges ("express exchanges"). A collective lesson from research and practice indicates that digitizing returns management should be an obvious initial step for e-retailers. This requires acquiring appropriate system support to effectively analyze return data and manage the entire process.

2.2 *E-Commerce Packaging*

Packaging is a critical element in e-commerce and logistics, serving to contain, protect, unitize, and inform. It functions as a multi-level system (primary, secondary, tertiary) and is categorized as consumer or industrial, with the latter vital for distribution. However, packaging significantly impacts the environment through emissions, resource consumption, and waste. Its weight, volume, and material choice are key factors, and its end-of-life management varies widely. Sustainable packaging strives for circularity, replacing single-use options and demanding holistic environmental assessment beyond just CO2 metrics. Strategic design is crucial, guided by principles like optimal protection, material use, fill rate, and user-friendliness, recognizing that design decisions impact the entire lifecycle (Silva and Nilsson 2024).

Integrating packaging with logistics is essential for efficiency and reducing environmental harm; the overall supply chain context heavily influences sustainable packaging choices. The circular economy framework emphasizes designing packaging for high-quality recycling or reuse. Life cycle assessment (LCA) helps compare alternatives. Reusable packaging systems are gaining traction in e-commerce, especially for last-mile deliveries where reverse logistics are more manageable (Hellström et al. 2016).

While challenges exist in handling, cost, and integration with linear systems, customer willingness to participate is evident. Success requires robust reverse logistics, IT support, and collaboration. Circular business models such as Product as a Service (PaaS), leasing, and sharing platforms further support packaging circularity by incentivizing durability and maximizing utilization. Addressing packaging sustainably requires an integrated approach across its design, logistical application, and role within circular economic models.

2.3 *Circular E-Commerce*

Current knowledge highlights the critical role of data in enabling circular processes and facilitating informed decision-making. Improved access to data regarding product content, material flows, lifespan, and repairability, potentially facilitated by tools like Digital Product Passports (DPP) mandated by regulations such as the EU Ecodesign for Sustainable Products Regulation (ESPR), is seen as crucial for optimizing circular processes. However, a significant knowledge gap and challenge currently exists due to a lack of standardized and easily accessible data on product content, lifespan, repairability, and environmental impact across the entire value chain.

This data deficiency hinders companies' ability to optimize their operations for circularity and limits consumers' ability to make truly informed purchasing choices. Furthermore, a persistent challenge in current practice is the difficulty in achieving and clearly demonstrating long-term profitability for circular business models, particularly when competing with deeply established and often more cost-effective linear models. Many e-tailers also lack a full overview of their existing packaging or return costs, which makes it difficult to build a strong business case for investments in new circular systems (Karlsson et al. 2023).

In the textile sector, there is an established practice of reuse that often operates outside the traditional commercial supply chains, frequently driven by non-profit organizations. The supply chain for textile-to-textile recycling involves core activities such as collection, sorting, and recycling. Non-profit organizations play a key and valuable role in the existing textile reuse system, drawing on their extensive knowledge and experience in handling large volumes of products with widely varying quality levels. Beyond their logistical role, these organizations also conduct activities that contribute to a broader spectrum of global goals, which is considered crucial for fostering a sustainable future. While logistics and SCM issues are inherently linked to the supply chains for used clothing, they have historically seldom been the primary focus of research in this specific domain (Pal and Sandberg 2023).

Current knowledge emphasizes that the successful implementation of both efficient returns management and circular economy models in e-commerce logistics requires significant digitalization. Digitalization transforms analog, often manual, processes into smart, data-driven systems. This provides real-time visibility into product flows, enabling efficient handling and facilitating the steering of goods toward appropriate destinations, whether for resale, repair, reuse, or recycling. Digitalization is seen as a crucial enabler for navigating the inherent paradox between efforts to prevent returns and the parallel need to promote and manage future circular flows. It is argued that mastering returns management, enabled by digitalization, is a fundamental prerequisite for successfully implementing circular economy models within e-commerce.

3 Case Studies

The added value of these case studies lies in their demonstration of how integrated, real-world innovations, combining digital tools, cross-sector collaboration, and circular economy strategies, can make sustainable e-commerce both practical and profitable. Rather than targeting isolated sustainability challenges, the Swedish model adopts a holistic strategy that embeds innovation across returns management, packaging systems, and circular logistics.

The following interrelated dimensions constitute the country's unique value proposition within the e-commerce sustainability domain.

(a) **Returns Management as a Strategic Function**

A defining characteristic of the Swedish system is its reconceptualization of e-commerce returns. Rather than viewing returns as a logistical burden, Sweden positions them as a strategic opportunity for value creation and environmental impact reduction. The Smart Fossil-Free Return Flows (SFR) project exemplifies this shift. Findings from this initiative demonstrated that digitalization can reduce physical handling costs by approximately 80% and expedite refund processing by 50%, highlighting the operational and customer experience gains. Moreover, data-driven approaches enable more sophisticated decision-making, such as determining whether returned products should be resold, refurbished, or recycled, thereby supporting the broader circular economy agenda.

(b) **Development of Circular and Reusable Packaging Systems**

Sweden is advancing the development of industry-wide infrastructure for reusable e-commerce packaging, as seen in initiatives such as CIRKLA and Returbar.

These projects integrate multiple dimensions—physical product design, reverse logistics, consumer interaction, and digital tracking—to facilitate the repeated use of packaging materials. The focus is placed on designing for circularity, ensuring that packaging is durable, easily cleaned, and efficiently reintegrated into supply chains. These systems are also tested in real-world conditions, reinforcing their practical viability and scalability.

(c) **Adoption of Hybrid Circular Economy Models**

The Swedish model also reflects a hybrid approach to circularity, combining technical and social innovation. Technical components include artificial intelligence, digital return platforms, and tools such as Digital Product Passports. Simultaneously, Sweden leverages the expertise and infrastructure of non-profit organizations in the collection, reuse, and redistribution of goods, particularly in the textile sector. This dual-track strategy enables the integration of community-based circularity with industrial-scale digital systems, enhancing both resilience and inclusivity.

(d) **Digitalization as a Foundational Enabler**

Digitalization is not treated as a secondary tool but as the foundation of Sweden's e-commerce sustainability transformation. It underpins real-time tracking of goods, optimized return flows, customer engagement, and lifecycle monitoring. Advanced digital infrastructure supports projects such as PassOnline, enabling consumers to donate used clothing via e-commerce packages, thereby promoting seamless integration of reuse into the shopping experience. The deployment of AI, automated decision systems, and CO_2 tracking tools further supports data-driven, environmentally optimized logistics.

(e) **Institutional Support and Collaborative Ecosystems**

A critical enabler of Sweden's e-commerce innovation is its strong public-private collaboration framework, supported by national agencies such as Vinnova and strategic programs like ASTER. These entities facilitate mission-driven research, development, and deployment projects that involve actors across the value chain—retailers, logistics providers, packaging firms, academia, civil society, and policy bodies. Furthermore, alignment with European regulatory frameworks, such as the EU Packaging and Packaging Waste Regulation, ensures long-term coherence and relevance.

This collection of case studies explores the diverse strategies and innovations driving Sweden's transition toward a more sustainable and circular e-commerce ecosystem. As e-commerce continues to grow rapidly, it has become a focal point for efforts to reduce carbon emissions, limit waste, and promote more responsible consumption. These case studies span four critical domains, consumer returns, sustainable packaging, circular business models, and last-mile deliveries, each illustrating how Swedish stakeholders are addressing environmental challenges with collaborative, data-driven, and system-wide approaches.

In the area of consumer returns, the studies demonstrate how digitalization can transform traditionally wasteful processes into streamlined, value-generating flows.

Projects such as Smart Fossil-Free Return Flows and E-commerce's Value-Creating Return Process provide strong evidence that return management, when strategically designed, can support both customer satisfaction and long-term profitability. Similarly, in sustainable packaging, the e-com ready, CIRKLA, and Returbar projects highlight the innovation potential in rethinking materials, logistics, and consumer participation—key components of shifting from linear to circular models.

Circularity is also explored in the context of product reuse and lifecycle extension. Projects like Sustainable Reconditioning and Pay It Forward emphasize the hybrid potential of combining technical solutions (e.g., AI, digital platforms) with social innovations and the involvement of non-profit actors, particularly within the textile sector. Finally, in the realm of sustainable deliveries, the Industry Agreement for Fossil-Free Deliveries and the Accessible and Efficient Omnichannels project illustrate how standardized practices, emissions tracking, and consumer transparency are reshaping the environmental footprint of last-mile logistics.

Collectively, these case studies underscore a few critical themes that define Sweden's approach to sustainable e-commerce. First, digitalization emerges as a foundational enabler across all areas. Whether through optimizing return flows, automating packaging processes, or tracking carbon emissions in deliveries, digital

Consumer returns

Project	Objectives	Statistics	Findings
Smart fossil-free return flows (SFR)	Assess how digital return systems can reduce climate impact and transform returns into a sustainable business asset.	80% reduction in handling costs; 50% faster refund process.	Digitalization enables smarter, more efficient return handling, reduces transport, and supports circular logistics.
E-commerce's value-creating return process	Explore the strategic role of return policies and their effect on consumer behavior.	Survey of 730 consumers; analysis of 12 companies.	Fair return policies build trust and loyalty. High-return customers are often the most profitable long-term.

Sustainable packaging

Project	Objectives	Statistics	Findings
E-com ready project	Optimize packaging for e-commerce to reduce material use and improve warehouse and transport efficiency.	Up to 83% efficiency and 90% cost savings in pilots.	Simple design changes significantly enhance logistics. Adoption depends on cross-actor collaboration.
CIRKLA project	Develop scalable systems for reusable e-commerce packaging aligned with EU PPWR targets.	Pilots with >1000 consumers testing return models.	Focus on integrating packaging reuse into logistics and ensuring consumer participation.
Returbar project	Create a national system for reusable grocery bags with deposit-return functionality.	CO_2 and cost reductions reported; customer interest confirmed via survey.	Demonstrated potential for early profitability. Requires further pilot scaling.

Circular e-commerce

Project	Objectives	Statistics	Findings
Sustainable reconditioning	Improve reconditioning processes to support circularity in e-commerce.	n/a	Identified key activities for extending product life. Scaling constrained by investment and infrastructure gaps.
Pay it forward	Assess the role of non-profits in textile reuse and identify innovation opportunities.	n/a	Non-profits are key but face digitalization challenges. Innovation lies in AI, shared platforms, and hybrid models.

Sustainable e-commerce deliveries

Project	Objectives	Statistics	Findings
Industry agreement for fossil-free deliveries	Establish common standards for fossil-free delivery and consumer transparency.	Apoteket: 42% → 84% fossil-free fuels; early bird: 165 g → 125 g CO_2/package.	Shared standards drive emissions reduction and cost efficiency. Digital tools boost consumer trust.
Accessible and efficient Omnichannels	Evaluate climate impact of different sales channels and delivery methods across Sweden.		E-commerce can have lower CO_2 emissions than physical retail. Packaging and transport are key emission sources.

tools are crucial for gathering data, improving efficiency, and enabling circularity at scale.

Second, the case studies reveal that collaboration across sectors and stakeholder groups is indispensable. Projects like CIRKLA and the fossil-free delivery agreement involve coordinated efforts among retailers, logistics providers, technology firms, municipalities, and research institutions. This collaborative model not only ensures broader impact but also accelerates innovation through shared learning and aligned incentives.

Third, there is a strong emphasis on systemic design thinking—embedding sustainability not as an add-on, but as a core principle in operational logistics, product handling, packaging, and customer interaction. This is particularly evident in the hybrid approaches seen in circular e-commerce projects, which recognize the interplay between technological capacity and social infrastructure, such as community-based textile reuse.

Finally, the results across these studies illustrate that sustainability and profitability are not mutually exclusive. Projects like Returbar and E-com Ready show that with careful planning and pilot testing, circular systems can deliver both environmental and economic value. However, many initiatives also highlight the need for continued investment, regulatory support, and larger-scale pilots to validate assumptions and refine models.

4 Analysis

The provided case studies offer valuable insights into the significant complexity and critical importance of transforming the e-commerce system, especially considering the expansion of online shopping and the clear need for a transition toward a circular economy. While the rapid growth of e-commerce, particularly its concentration in urban areas, results in notable environmental considerations related to emissions, congestion, and waste production, the sources equally highlight that these aspects present clear avenues for improvement. These can be effectively addressed through strategic planning and the dedicated implementation of circular economy principles across the entire logistics value chain.

4.1 Returns Management

Regarding returns management, it has an inherent complexity and the absolute necessity of adopting a strategic approach rather than treating it merely as a cost center or operational inefficiency. Research indicates that a fundamental prerequisite for developing truly effective returns management strategies is achieving a high degree of alignment between a company's stated return policy and its operational return process.

Despite this, the literature and case studies identify several multi-dimensional misalignments that can hinder organizations from developing adequate and coherent returns management strategies. These misalignments represent crucial strategic elements that managers must carefully consider when formulating objectives and strategies for handling returns.

4.2 Cross-Functional Activity

One key analytical insight is that returns management is fundamentally a cross-functional activity that requires close collaboration and coordination between various internal departments. Sales and marketing departments, logistics and operations teams, customer service, finance, and IT must work together effectively. Coordination between these departments can be fostered through active leadership and the establishment of clear, common goals that span organizational silos.

Strategies aimed at preventing unnecessary returns, which touch upon aspects like sales information, customer behavior, inventory management, product design, and packaging, necessitate careful coordination among all involved departments. It is also important to remember that not all returns are identical; therefore, the handling process should be adapted and differentiated based on factors such as the customer, the specific product, and the reason for the return.

4.3 *Digitalization*

Digitalization emerges from the literature and case studies as a crucial enabler and transformative force in returns management. It is highlighted as essential for navigating the apparent paradox between efforts to minimize returns in the first place and the need to build and manage future circular flows where returned items re-enter the value chain. Digitalization has the profound potential to fundamentally transform returns management from a burdensome operational necessity into a value-creating process that actively supports sustainability goals. The studies indicate that digitizing the return process, for instance, by implementing comprehensive digital systems for real-time tracking and management of returns, is a foundational step for establishing effective returns management. This involves seamless integration with existing IT systems and the automation of workflows.

The benefits are substantial: earlier availability of information about the return enables smarter decisions about how the returned item should be handled based on its category, condition, and customer data. This allows for guiding items toward the most appropriate destination early in the process, such as directing them back to the nearest physical store for quick resale, sending them to a repair center, or routing them directly for recycling. A digitized process facilitates standardized and faster handling for the majority of returns, enabling the system to quickly determine if an item can be returned to stock, sold through an outlet, or handled differently.

Beyond process efficiency, digitalization provides the crucial ability to collect the "right data points" necessary for data-driven management of the entire return flow. This data can be analyzed to understand return patterns by product or customer, providing invaluable insights for optimizing purchasing decisions, improving product design, and enhancing overall profitability. Improved transparency and data collection lead to more fact-based decision-making, further boosting profitability and the customer experience.

Furthermore, digitalization facilitates decentralized return management, which can lead to more resource-efficient logistics by enabling local handling and reducing the need for long-distance transport back to central warehouses. Paperless return processes, a direct outcome of digitalization, also contribute to reduced environmental impact. The analysis also points to the exciting potential of using AI for real-time analysis of return data to further streamline flows, potentially enabling returned items to be shipped directly from one customer to another without returning to a warehouse, representing a highly optimized circular flow.

4.4 Circular Packaging Systems

Circular packaging systems have a central role in reducing the environmental footprint of e-commerce, driven significantly by upcoming EU regulations like the PPWR. In the Sustainable Omnichannel project, it was discovered that about 20% of emission from the e-commerce products journey comes just from packaging and where the E-com Ready project shows several low-hanging fruits to reduce that emission post. Projects like CIRKLA and Returbar, as case studies, actively explore the practical implementation challenges and design requirements for stepping toward such circular systems.

A central principle is designing packaging for circularity, meaning it should be easily recyclable into high-quality secondary materials or, preferably, designed for multiple reuses. For reusable packaging systems, such as those involving bags or boxes, there is a need for the development of several interconnected components. These include the packaging itself (focusing on material properties, durability, and design for easy cleaning and end-of-life recycling), the establishment of functional deposit systems, the operation of efficient cleaning systems, and, crucially, the development of an effective logistics solution.

The logistics solution must prioritize ease and convenience for the customer to return the packaging to ensure high return rates and the viability of the system. It is critical that circulating products and packaging are seamlessly integrated into existing logistics flows. Failure to do so can render the circular business model counterproductive from an environmental perspective due to the creation of inefficient parallel logistics networks. Challenges are found in operationalizing these systems within existing warehouse environments and in actively engaging customers to participate consistently and take responsibility for returning packaging. However, consumers' initial positive responses and appreciation for avoiding single-use bags indicate a promising potential for successful behavioral change with well-designed systems.

4.5 Circular E-Commerce

Regarding circular e-commerce, the retail sector's shift toward circularity is evolving into a hybrid future, blending technical innovations like digitalization and AI-driven sorting with crucial social innovations. As mentioned earlier, the pre-loved market is growing by the day, where the current Swedish market is expected to double from 2024 and 16 billion SEK to 30 billion SEK in 2030. At the same time, 70% of pre-loved products are bought online—highlighting the crucial role e-commerce plays in the adoption of more sustainable consumption patterns (Svensk Handel 2024).

In the case studies, the role of non-profit organizations is seen as central to the push toward circular retail, leveraging their expertise in handling textiles and driving social sustainability. This hybrid model acknowledges that while technology offers automation, manual, local, and artisanal processes will remain essential alongside industrial systems. Integrating social aspects like work training and community building is also vital.

However, challenges persist, including the slow pace of new regulations, which can hinder sustainable product development and intensify competition for high-quality reused items. Economic viability can also differ between physical stores and digital channels, especially for lower-value textiles. Despite advances in digitalization, the need for physical and digital community spaces for repair, learning, and social engagement remains important for resilience.

4.6 Supporting Infrastructure

More generally, all areas highlight a pervasive need for robust supporting infrastructure and systems to enable the successful implementation of both circular systems and effective returns management. Digital tracking mechanisms, leveraging standards and upcoming technologies like Digital Product Passports (DPP), are seen as crucial facilitators for logistics, returns management, and ensuring compliance with regulatory requirements. For a system-wide transition toward sustainability and circularity to succeed, it is deemed essential to consider how multiple global goals interact and support each other. This holistic view involves simultaneously pursuing increased resource efficiency, waste reduction, improved traceability, the development of sustainable logistics networks, the creation of efficient digital services and working methods, and critically, considering the social implications and aspects of these changes.

4.7 Barriers to Deployment

Looking at all the case studies mentioned, it is possible to conclude that the widespread, large-scale implementation of circular business models within e-commerce is currently limited by a series of significant barriers. These barriers include the following.

- **Economic:** challenges in demonstrating profitability compared to linear models.
- **Logistical:** complexity of reverse flows, collection, cleaning, and redistribution.
- **Regulatory:** need for supportive policies and clear standards.
- **Behavioral:** changing consumer habits. Achieving and demonstrating long-term profitability for circular models, especially in direct competition with established linear alternatives, remains a central challenge.

Overcoming these deeply rooted obstacles necessitates not only technological innovation but also fundamental shifts in business models, consumer behavior, and extensive collaboration across the entire value chain.

5 Conclusion and Recommendations

5.1 Findings

The transition toward sustainable e-commerce is a complex and multifaceted process that requires concerted action on multiple fronts. While the continued growth of e-commerce offers convenience, economic opportunity, and increased access to pre-loved goods, it also places significant pressure on urban environments and logistics systems. Rising parcel volumes, high return rates, and escalating packaging waste underscore the need for effective, sustainable management strategies. Within this context, e-commerce's transformation must be guided by the integration of circular economy principles and a systems-based approach.

Circular packaging systems, particularly those involving reusable alternatives, represent a vital component of this transition. Projects such as CIRKLA and Returbar highlight the need for fully integrated systems, encompassing thoughtful packaging design, efficient cleaning operations, robust deposit management, and seamless return logistics. Without integration into existing transport flows, there is a risk of creating parallel systems that diminish environmental gains.

Finally, achieving meaningful transformation in urban logistics and circular e-commerce systems hinges on multi-stakeholder collaboration. This includes partnerships between retailers, logistics providers, packaging suppliers, policymakers, researchers, technology providers, and non-profits. However, collaboration is currently hindered by fragmented funding structures, which often target isolated aspects of the e-commerce system. A broader, more holistic funding approach is needed to support ecosystem-wide innovation.

5.2 Discussion

A central conclusion is the strategic importance of returns management. Rather than being treated merely as a cost or operational burden, returns must be seen as a value-generating component of the business. This reframing is essential to achieving sustainability and circularity goals in e-commerce.

Digitalization is identified as the key enabler of this transformation. It is not just a tool for efficiency but a foundational technology that enables smarter, faster, and more sustainable management of return flows. By leveraging detailed data on returns and customer behavior, businesses can optimize product handling, reduce emissions, enhance customer experience through quicker refunds, and even guide returned goods into reuse or recycling streams. Additionally, digitalization allows for decentralized, paperless return processes and the use of AI to support advanced logistics, such as direct consumer-to-consumer exchanges.

Research and pilot projects show that such digitalized returns systems can significantly reduce handling costs (up to 80%) and accelerate refund processing times (by 50%), highlighting the economic and operational benefits of these innovations.

5.3 Recommendations

To fully realize a circular and sustainable e-commerce system, the following actions are recommended:

(i) **Digitize Returns Management:** Systematic digitalization of return processes is the first step toward smart, fossil-free flows. This includes integrating digital platforms, automating workflows, and applying analytics and AI to optimize handling and routing of returned products.

(ii) **Develop Integrated Circular Packaging Systems:** Invest in reusable packaging solutions that are seamlessly embedded into current logistics networks. Focus on designing for durability, recyclability, and customer convenience to ensure high return rates and reduce waste.

(iii) **Promote Holistic Policy and Funding Support:** Advocate for funding models that support systems-based innovation across the entire value chain. Rather than siloed investments in logistics, materials, or urban planning, future funding calls should support integrated collaborations that span disciplines and sectors.

(iv) **Foster Multi-Stakeholder Collaboration:** Build platforms and partnerships across public, private, and civil sectors. Encourage knowledge-sharing and joint development of tools and infrastructure to overcome scalability and behavioral challenges.

(v) **Adopt Hybrid Innovation Models:** Combine technical innovations (like digital tracking and AI sorting) with social innovations (such as training, awareness campaigns, and community reuse hubs), especially in areas like textile circularity. This hybrid approach enhances both operational resilience and social inclusion.

(vi) **Continuously Monitor and Improve:** Implement a roadmap that includes phased digital transformation, policy alignment, stakeholder education, and ongoing evaluation through real-time data to adapt strategies effectively.

References

Aster (2023a) Hållbara omnikanaler. Available at: https://aster.lindholmen.se/sv/projekt/hallbara-omnikanaler [Accessed 19 May 2025]

Aster (2023b) E-com ready. Available at: https://aster.lindholmen.se/sv/projekt/e-com-ready [Accessed 19 May 2025]

Aster (2024) CIRKLA. Available at: https://aster.lindholmen.se/sv/projekt/cirkla [Accessed 19 May 2025]

Aster (2025) PassOn-line. Available at: https://aster.lindholmen.se/sv/projekt/passon-line [Accessed 19 May 2025]

Axfoundation (2025) Returbar—ett cirkulärt system för matkassar i e-handeln. Available at: https://www.axfoundation.se/projekt/returbar-aterbrukbara-kassar [Accessed 19 May 2025]

Dehghannejad M, Pal R, Dissanayake K, Sandberg E (2025) Exploring paradoxical tensions in scaling circular business models within the textile and clothing industry. Sustain Prod Consump. https://doi.org/10.1016/j.spc.2025.04.022

European Commission (2024) Ecodesign for sustainable products regulation. Available at: https://commission.europa.eu/energy-climate-change-environment/standards-tools-and-labels/products-labelling-rules-and-requirements/ecodesign-sustainable-products-regulation_en [Accessed 19 May 2025]

European Commission (2025) Packaging waste. Available at: https://environment.ec.europa.eu/topics/waste-and-recycling/packaging-waste_en [Accessed 19 May 2025]

Hellström D, Olsson A, Nilsson F (2016) Managing packaging design for sustainable development: a compass for strategic directions. Wiley. https://doi.org/10.1002/9781119151036

Hellström D, Olsson J, Darlington J, Westblom L, Hjort K (2021) Systemförändring för omställning till hållbar e-handel: Kraftsamling som accelererar innovation. Lunds universitet. Available at: https://portal.research.lu.se/sv/publications/systemf%C3%B6r%C3%A4ndring-f%C3%B6r-omst%C3%A4llning-till-h%C3%A5llbar-e-handel-kraftsamli [Accessed 19 May 2025]

Hellström D, Olsson J, Hjort K (2022) Transition towards sustainable E-commerce: a design thinking approach. Lunds universitet. Available at: https://portal.research.lu.se/sv/publications/transition-towards-sustainable-e-commerce-a-design-thinking-appro [Accessed 19 May 2025]

Hellström D, Hjort K, Karlsson S, Oghazi P (n.d.) Konsumentreturer i digital handel. Available at: https://handelsradet.se/app/uploads/2021/06/2017-5-Konsumentreturer-i-digital-handel.pdf [Accessed 19 May 2025]

Karlsson S, Hellström D, Hjort K, Oghazi P (2020) Så kan e-handeln fixa returerna. Lund University Available at: https://portal.research.lu.se/en/publications/s%C3%A5-kan-e-handeln-fixa-returerna [Accessed 19 May 2025]

Karlsson S, Oghazi P, Hellström D, Patel PC, Papadopoulou C, Hjort K (2023) Retail returns management strategy: an alignment perspective. J Innov Knowl 8(4):100420. https://doi.org/10.1016/j.jik.2023.100420

Klarna Holding AB (publ) (n.d.) Annual report Klarna Holding AB 2020. Available at: https://owp.klarna.com/legacy/assets/2021/03/30041745/Annual-Report-Klarna-Holding-AB-2020-210329.pdf [Accessed 19 May 2025]

Liu.se (2023a) Den snabba e-handeln—en motsats till den hållbara e-handeln? Available at: https://liu.se/forskning/hallbar-e-handel [Accessed 19 May 2025]

Liu.se (2023b) The role of logistics in circular retail business models. Available at: https://liu.se/en/research/circular-logistics [Accessed 19 May 2025]

Lunds universitet (2015) Fredrik Nilsson. [online] Available at: https://portal.research.lu.se/sv/persons/fredrik-nilsson-3/publications/ [Accessed 20 May 2025]

Makrologik (2025) Smartare handel och färre transporter. Available at: https://app.box.com/s/rm1h4w14bkiypbei4tnogeghaddckio9

Olsson J (2024) Effekter av branschöverenskommelsen för den fossilfria leveransen: nuläge och vägar framåt. Lunds universitet. Available at: https://portal.research.lu.se/sv/publications/effekter-av-bransch%C3%B6verenskommelsen-f%C3%B6r-den-fossilfria-leveransen [Accessed 20 May 2025]

Pal R, Sandberg E (2023) Circular supply chain valorisation through sustainable value mapping in the post-consumer used clothing sector. Int J Logist Manag. https://doi.org/10.1108/ijlm-01-2023-0023

Petisme A-M (2025) Conformity and contradictions: exploring return management and sustainability in fashion e-retail through neo-institutional theory. DIVA Portal Available at: https://hj.diva-portal.org/smash/record.jsf?pid=diva2%3A1931655&dswid=1920 [Accessed 19 May 2025]

RISE (2021) PayItForward - Digitala lösningar för ett ökat återbruk av textilier | RISE. Available at: https://www.ri.se/sv/expertisomraden/projekt/payitforward-digitala-losningar-for-ett-okat-aterbruk-av-textilier [Accessed 19 May 2025]

Silva N, Nilsson F (2024) Exploring the circular last mile: reusable packaging in home delivery of food and groceries using a design science approach. Int Rev Retail Distrib Consum Res:1–22. https://doi.org/10.1080/09593969.2024.2414082

Stanfors M, Tärneberg AW (2020) Det obetalda hem- och omsorgsarbetets fördelning - en fråga om jämställdhet med relevans för ekonomisk jämlikhet: Underlagsrapport till Jämlikhetskommissionen. Lunds universitet. Available at: https://portal.research.lu.se/sv/publications/det-obetalda-hem-och-omsorgsarbetets-f%C3%B6rdelning-en-fr%C3%A5ga-om-j%C3%A4mst [Accessed 20 May 2025]

Svensk Handel (2024) Pre-loved Indikatorn. Available at: https://www.svenskhandel.se/api/documents/pressmeddelanden/%C3%85rssummering%20Pre%20Loved-indikatorn%202024.pdf [Accessed 19 May 2025]

Svenskhandel.se (2025) Om Branschöverenskommelsen - Svensk Handel. Available at: https://www.svenskhandel.se/branschoverenskommelsen/om-branschoverenskommelsen [Accessed 19 May 2025]

Triple F (2022) Smarta Fossilfria Returer—SFR. Available at: https://triplef.lindholmen.se/projekt/smarta-fossilfria-returer-sfr [Accessed 19 May 2025]

Vinnova (n.d.) Klimatneutral e-handel. Available at: https://www.vinnova.se/globalassets/utlysningar/2023-00415/omgangar/klimatneutral-e-handel-utlysningstext_20230321.pdf1235194.pdf?cb=20230322125711 [Accessed 19 May 2025]

Mapping E-Commerce Consumer Personas: Behavioral Insights and Incentives for Sustainable Last-Mile Deliveries and Returns

Ioannis Karakikes, Amalia Polydoropoulou, and Efstathios Bouchouras

1 Introduction

The exponential growth of e-commerce has fundamentally reshaped how consumers purchase goods. E-commerce provides consumers with immediate access to products worldwide, offering unparalleled convenience and cost benefits. However, it also introduces challenges that necessitate efficient solutions for order fulfilment and returns (Nanda and Patnaik 2023).

Delivery processes are pivotal in determining customer satisfaction, as they directly impact perceptions of service quality (Uzir et al. 2021). Today's consumers demand fast, reliable, and flexible deliveries—a requirement that becomes increasingly complex as online order volumes continue to surge. Similarly, seamless and cost-free return processes are now considered a standard expectation, with any delays or inefficiencies risking customer dissatisfaction and brand damage (Karlsson et al. 2023). Managing these processes is also crucial from a financial perspective, as returns often incur substantial costs for transport and repackaging.

The environmental impact of delivery and return operations adds another layer of complexity. As online shopping grows, the increased transportation needs and packaging waste associated with returns amplify the carbon footprint of businesses (Kim et al. 2022). With sustainability gaining prominence, consumers are increasingly drawn to companies that adopt eco-friendly practices, such as recyclable packaging or parcel lockers for returns and pickups. By addressing these environmental concerns, businesses can attract sustainability-conscious customers while reducing their ecological footprint.

I. Karakikes (✉) · A. Polydoropoulou · E. Bouchouras
Department of Shipping, Trade and Transport, University of the Aegean, Chios, Greece
e-mail: ikarakikes@aegean.gr

A. Melkonyan-Gottschalk, Y. Li (eds.), *Urban Logistics Transformation*, Greening of Industry Networks Studies 15,
https://doi.org/10.1007/978-3-032-08387-6_11

The shift toward e-commerce and digital marketplaces necessitates targeted consumer engagement strategies to create tailored experiences for distinct audience segments. To achieve customer segmentation, marketers create personas which, through analyzing behavioral patterns, preferences, and motivations, offer comprehensive profiling (Abidar et al. 2024; An et al. 2018). The profiling can be constructed based on demographic information (age, gender, income and educational level, marital status, etc.), psychographic insights (values and beliefs, lifestyle preferences, interests and hobbies, personality traits, etc.), historical interactions with a brand, social media usage or a combination of the previous (Gomes and Meisen 2023; Jansen et al. 2021).

Effective customer segmentation is crucial for e-commerce businesses aiming to tailor operational aspects, such as website content, logistics options, and product offerings as well as their marketing strategies (Stevenson and Rieck 2024; Beckers and Cant 2023; Wu et al. 2022). Offering incentives is among the most effective strategies for managing deliveries and returns in e-commerce. Incentives can enhance customer experience, boost loyalty, and reduce costs associated with returns. When return policies are adapted to the distinct needs of consumer segments, they can support circular behaviors such as reuse, resale, and waste minimization. This eventually contributes to the broader shift from linear consumption models to circular logistics systems, where products remain in use for longer and resource loops are closed (EIB 2024).

A common delivery incentive is free shipping for orders above a specific amount (Polydoropoulou et al. 2022). This approach encourages consumers to increase their order value, thereby boosting sales while minimizing low-value orders that may inflate transportation costs. Moreover, free shipping enhances the shopping experience and fosters customer loyalty (Wang et al. 2023a, b). While offering free shipping encourages consumers, it is important to consider potential drawbacks. This strategy might inadvertently promote unnecessary consumption, leading to increased transportation demand and, consequently, greater environmental impacts.

For returns, many companies offer discounts or store credits for future purchases to consumers who avoid returning products or opt for returns through lockers (Shopify 2023). Such incentives reward consumers for responsible behavior while reducing operational costs for businesses. Additionally, companies may offer extra reward points to customers who return products to designated drop-off points instead of opting for home pick-ups. While this strategy can minimize transportation and support environmental sustainability from a logistics operations perspective, it is important to consider the potential environmental impact if consumers drive personal vehicles to reach these drop-off points. To mitigate this, businesses prioritize drop-off points in locations accessible by walking, biking, or public transportation. Moreover, some companies provide immediate discounts to consumers who commit not to return a product or who choose standard shipping over express delivery. This encourages customers to select cost-effective and sustainable options, especially when fast delivery is not essential. The latter benefits both parties;

customers enjoy a discount, and companies reduce the need for more expensive and energy-intensive deliveries.

Incentives can also strengthen consumer loyalty. Businesses may offer exclusive perks, such as free returns or express delivery, to members of their loyalty programs or customers who meet a specific purchase threshold. This fosters a sense of belonging to a valued community, increasing the likelihood of repeat purchases (Dietl et al. 2024).

The motivation for conducting this research is that there are only scarce attempts that examine the role of personas amidst the rise of e-commerce and digital analytics (Salminen et al. 2018; An et al. 2018). Another driving factor, is that, excluding research focused on the e-commerce consumer personas and behavior changes, there is no literature addressing the bridging of personas and market dynamics with last-mile delivery/return options. The existing literature predominantly addresses (i) factors influencing e-commerce adoption, such as technological infrastructure, trust, and socio-economic conditions (Hendricks and Mwapwele 2024); (ii) consumer behavior in online shopping contexts (Pundir et al. 2024; Xin et al. 2024; Bhukya and Paul 2023; Zhang and Benyousef 2016); and (iii) behavioral change related to e-commerce deliveries (Wang et al. 2023a, b; Campisi et al. 2023; Cruz-Cardenas et al. 2021). The topic of product returns remains largely underexplored.

To fill in these research gaps, the objective of this chapter is to investigate the pivotal role of e-commerce consumer personas in relation to goods' deliveries and returns. Specifically, the purpose is twofold: first, to identify underlying incentives and nudging mechanisms that could potentially influence a shift toward more sustainable options; and second, to understand how different personas behave with respect to delivery and return services or policies. The two arising research questions (RQ) are:

- RQ1: How do different e-commerce consumer personas behave in relation to delivery and return services or policies?
- RQ2: What incentives and nudging mechanisms can influence e-commerce consumer personas to adopt more sustainable delivery and return options?

The rest of the chapter is structured as follows: Section 2 presents the methodology as well as the details of the systematic literature review process, including the search strategy, screening criteria, and data analysis methods. Section 3 presents the key findings from the literature review. Section 4 synthesizes the information collected, while Sect. 5 discusses the results and concludes the chapter.

2 Methodology

The methodology applied is structured around four (4) discrete steps. The first step involves an analysis of the European e-commerce market, drawing on insights from the European E-Commerce Report (2024). This step provides the foundation for identifying emerging trends and sector-specific terminology. The second step entails conducting a systematic literature review. The third step distills and presents the key findings of the literature review, i.e., identified personas and their preferences, incentives, and nudging mechanisms. Finally, the fourth step synthesizes this information and interprets it in direct relation to the two research questions. Figure 1 illustrates the four-step methodological flow.

2.1 *Analysis of European E-Commerce Market*

The first phase of the methodology draws on key insights from the European E-Commerce Report 2024 to contextualize consumer behavior within the evolving digital retail environment. The report highlights that despite a modest B2C e-commerce turnover growth of 3% in 2023—rising from €864 billion to €887 billion—the European market is grappling with multiple structural and competitive challenges. These include rising inflation, increased competition from non-EU e-commerce players, and the pressures of implementing green and digital transitions. Consumers are increasingly demanding sustainability in products and logistics, yet many are financially constrained and unable or unwilling to pay premium prices for environmentally responsible options. This tension between sustainability and affordability, as well as the fragmented legislative landscape across EU member states, shapes the behavioral dynamics examined in the subsequent literature review.

Furthermore, the report identifies critical buzzwords and themes that guide the construction of consumer personas in later stages of the research. These include sustainability, cross-border logistics, e-payments innovation (notably mobile and digital wallets), digital intensity, and regulatory enforcement. Regional disparities in e-commerce penetration and digital maturity—such as Northern Europe's leadership in e-shopper penetration versus Eastern Europe's lagging digital infrastructure—underscore the need for segmentation in consumer profiling. Additionally, the report underscores the growing digital divide between large e-commerce players and SMEs, influencing not only adoption of technology but also the ability to deliver sustainable and personalized services.

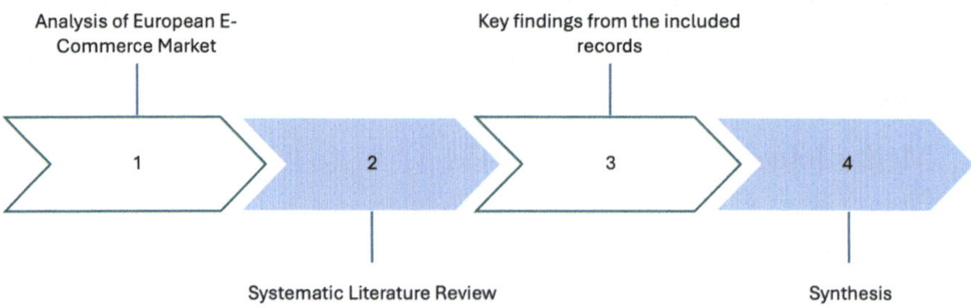

Fig. 1 Methodology flow

2.2 Systematic Literature Review Methodology

The performed systematic literature review follows the Preferred Reporting Items for Systematic Reviews and Meta-Analysis (PRISMA) guidelines (Page et al. 2021) and is executed in two steps; the first step defines the scope, and the second structures a protocol that outlines the objectives, criteria for inclusion and exclusion of records, and methods for data extraction and analysis. Analytically, as a first step, appropriate keywords were selected to identify records that provide answers to the RQs. The initial search terms were "e-commerce,*" "personas," "deliver*," and "return" "*", where "*" is a wildcard of the SCOPUS database when a word has multiple spelling variations. The search yielded a total of 313 articles.

Having the initial search results as a basis, in the second step, additional inclusion criteria were implemented to narrow down the number of retrieved documents and refine the search process. The inclusion criteria that were set are the following:

- Records should be written in the English language.
- Records should be published either as (i) articles in scientific journals, (ii) books, (iii) chapters in a book, or (iv) conference proceedings. In this way, a high-quality review process is ensured.
- Records should be contemporary—published after 2019.
- Records should be relevant to the research questions. This implies that retrieved records that belong to the following SCOPUS subject areas should be eliminated: Chemical Engineering, Biochemistry, Chemistry, Earth and Planetary Sciences, Health Professions, Immunology and Microbiology, Pharmacology, Toxicology and Pharmaceutics, Neuroscience, Nursing, Agricultural and Biological Sciences, Physics and Astronomy, Medicine.
- Records should be available and accessible online.

Integrating the criteria, a total of 139 records was finally yielded. The exact query with the exclusion criteria embedded can be found in Appendix A (Table A1).

The initial set of 139 records was further screened to exclude studies that did not address the research questions or were deemed only marginally relevant. This screening was conducted by reviewing the titles and abstracts, which reduced the pool to 12 eligible studies. An additional 47 records were manually identified through cross-referencing the bibliographies of these 12 studies. In total, 59 studies were selected for in-depth analysis. The full list of references is provided in Appendix B. The screening process is illustrated in Fig. 2.

2.3 Key Findings and Synthesis Methodology

The results of the systematic literature review were synthesized into actionable consumer personas, forming the basis for deriving behavioral insights related to sustainable delivery and return options in e-commerce. The synthesis was conducted

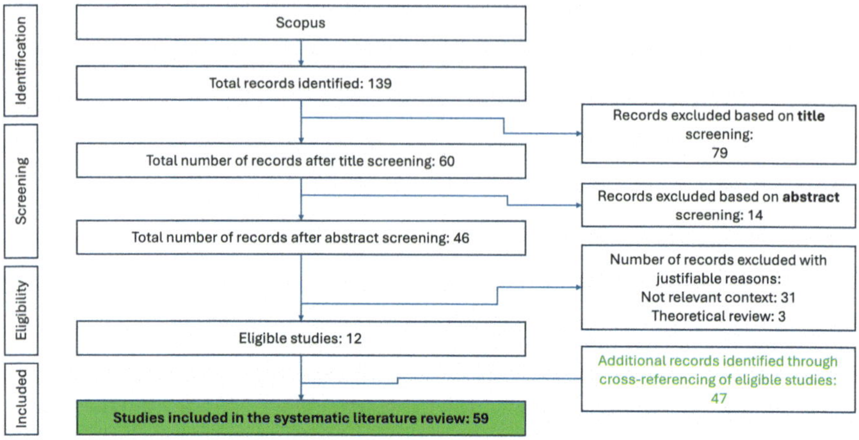

Fig. 2 Data collection and screening process

through manual coding and thematic clustering, following the procedure outlined as follows:

1. Identification of core dimensions: Key characteristics from the reviewed studies were extracted and mapped.
2. Clustering of persona types: Based on the identified characteristics, persona types were clustered and named after their most dominant motivation.
3. Persona journey mapping: For each persona type, an indicative "e-commerce journey" was developed, structured around five key stages: Discover, Engage, Decide, Purchase, and Feedback (Lemon and Verhoef 2016). Each stage was rated on a three-point scale (1 = low, 3 = high) to represent the level of engagement for each persona type, based on the authors' qualitative assessment.
4. Linking incentives and nudging strategies: Based on each persona's journey, the (i) incentives for sustainable delivery and return policies and (ii) nudging strategies for delivery return policies were matched with each persona type, as both emerged from the systematic literature review.
5. Persona table consolidation: All personas were compiled into a structured table, detailing their key characteristics, applicable delivery and return incentives, and corresponding nudging mechanisms.

3 Key Findings from the Included Records

3.1 Identified Personas and Preferences

A systematic analysis of the selected articles revealed two main thematic clusters relevant to RQ1. The first includes studies focused on the identification and classification of consumer personas. The second includes studies examining consumer behaviors and preferences that may influence the choice or acceptance of delivery and return service policies.

Personas

Xu and Lee (2020) identified four personas based on online behavior patterns, while Miva (2024), Ventura (2024) and Fenton et al. (2022) presented detailed persona classifications focusing on consumer motivations, engagement levels, and shopping habits. Companies like Delve.AI (2024) provided a streamlined approach with three core personas, emphasizing decision-making stages. Contentsquare (2024) introduced personas tailored to digital journey behaviors, reflecting users' browsing habits and purchasing confidence. Zhong et al. (2024) applied persona development to live-stream e-commerce platforms, categorizing consumers into clusters based on their interaction styles. Meanwhile, Romero Montero et al. (2024) used qualitative interviews to identify personas in the hospitality sector, showcasing distinct buyer profiles. Akre et al. (2019) focused on millennials' digital behaviors, highlighting personas driven by financial aspirations and consumer desires. Collectively, these studies emphasize the importance of understanding diverse personas to tailor engagement strategies, optimize user experiences, and align marketing efforts with evolving consumer behaviors. Table 1 summarizes the personas identified in a tabulated form.

Consumer Behaviors and Preferences

Consumer personas exhibit distinct behaviors and preferences regarding delivery and return services, shaped by demographic, behavioral, and psychographic factors. Alguacil et al. (2020) found that women demonstrate higher trust and loyalty toward services, while younger users excel in credibility and trustworthiness (as users of sports services). Frequent users exhibit the strongest loyalty, whereas unemployed individuals prioritize trust, and retirees show the highest loyalty levels. These findings highlight that different demographic groups may favor reliable and transparent delivery processes, while loyalty-driven personas value consistency in service offerings.

The DHL (2024) report underscores additional behavioral trends, particularly among social media, sustainable, and cross-border shoppers. Social media shoppers prioritize personalized, engaging shopping experiences and exclusive products, often influenced by trends on platforms like Instagram and TikTok. Sustainable shoppers, on the other hand, prefer environmentally friendly practices, such as transparent eco-impact information and sustainable delivery methods, even if it involves higher costs. Cross-border shoppers emphasize cost savings, clear customs processes, and flexible return policies to avoid unexpected issues like basket abandonment. These preferences suggest that delivery and return services must be tailored to address key motivators such as exclusivity, sustainability, and convenience.

Further studies, such as those by Hall and Towers (2017) and Appiah and Watson (2024), highlight the role of social validation and digital interactivity in influencing consumer decisions. For instance, Millennials rely on peer validation and social networks before finalizing purchases, while modern consumers increasingly seek

Table 1 Personas identified

Authors	Personas identified
Xu and Lee (2020)	Opportunists: visit online stores to find specific shopping info and leave once obtained. Contributors: actively engage in sharing information, write reviews, and answer questions. Explorers: browse extensively through product info, deals, and reviews, seeking recommendations. Followers: join online groups/communities to stay updated with trends and discussions.
Miva (2024)	Value hunter: hunts for best prices and deals. Researcher: takes time to learn about products. Brand devotee: loyal promoter of brands. Social butterfly: social media influencers sharing knowledge. Replenisher: repeat customers valuing products. Mobile shopper: prefers shopping on mobile devices. Gifter: enthusiastic buyers purchasing gifts for friends and families.
Delve.AI (2024)	Bouncers: only look and are unsure what they want. Passive browsers: haven't yet started shopping. Cart abandoners: know what they want but are undecided about the store to buy from.
Ventura (2024)	Impulse shopper: driven by emotions and instant gratification. Researcher: methodical shopping approach. Loyal customer: values personalized experiences and loyalty rewards. Tech-savvy shopper: seeks seamless online experiences and innovation. Ethical shopper: supports brands aligning with values. Luxury consumer: seeks exclusivity, quality, and superior experiences. Practical shopper: looks for functionality. Gift giver: values recommendations, packaging, and flexible return policies.
Contentsquare (2024)	Novice: spends 46% more time browsing than average new visitors. Demanding: avoids wasting time, spends35% less time oν pages. Thrifty: price-conscious shoppers. Anxious: uncertain buyers. Carefree: confident, relaxed shoppers.
Zhong et al. (2024)	Dedicated: enjoys searching but interacts minimally online. Dependent: needs time to figure out processes, alerted by family to risks. Active: ensures products are cost-effective before buying. Lurker: prefers observing over expressing but forwards interesting findings.
DHL (2024)	Social media shoppers: purchase frequently through Facebook, Instagram, and TikTok with focus on exclusivity and personalization. Sustainable shoppers: prefer eco-friendly, transparent deliveries despite cost. Cross-border shoppers: motivated by savings, unique offerings, and quality, requiring clear customs and returns.
Romero Montero et al. (2024)	Primary: main target customer in marketing strategies. Secondary: significant but less prioritized group. Tertiary: relevant but less significant group. Undesirable: customers accommodations prefer to avoid.

Table 1 (continued)

Authors	Personas identified
Fenton et al. (2022)	Jim smith: male, 52, avid sports fan, uses social media, motivated by health concerns and unsure about fitness app setup. Andrea Rogers: female, 33, football club marketing professional, explores apps, VR, and eSports to engage younger audiences.
Akre et al. (2019)	Millennial male: saving for a house and trades in stock market but struggles financially. Millennial female: struggles to save for shopping but values advertisements, desires comfort and happiness.

seamless, interactive experiences with brands throughout their digital journey. Together, these insights emphasize the importance of offering flexible, transparent, and sustainable delivery and return options to align with personas' unique expectations, enhancing overall satisfaction and loyalty.

Pasupuleti et al. (2024) highlight the values of customer segmentation in improving supply chain agility and sustainability. By categorizing customers based on purchasing behaviors and preferences, businesses can tailor their inventory management and logistic strategies more effectively. Asdecker (2021), by investigating the elements influencing consumer adoption of innovative last-mile delivery methods (parcel lockers and home delivery), identified that customer preferences are strongly influenced by perceived convenience, privacy, and security concerns. Iannaccone et al. (2021) concluded through their research that accessibility, security, and ease of use have a big impact on the delivery method selection; young e-consumers presented an increasing preference for parcel lockers, suggesting a shift toward alternative collection strategies in urban logistics. Sharda et al. (2025) highlighted through their research that policymakers should prioritize investments for digital literacy initiatives and Internet development to guarantee fair access to online purchasing and delivery services. To fulfill the growing demand for home deliveries, the authors suggest that urban regions should concentrate on enhancing their last-mile delivery infrastructure. Rangel et al. (2025) investigated the factors influencing e-commerce behavior, focusing on adoption, purchase frequency, and returns. Using data from a survey in Madrid, Spain, and a statistical modeling approach (GSEM), the study finds that mobility habits, gender, fear of COVID-19, and sociodemographic factors all play a role in shaping online shopping behavior. Castaño et al. (2025) investigated how e-commerce affects people's travel habits, using data from Madrid, Spain. The authors explored the connections between socio-demographics, mobility patterns, and online shopping behavior, with a focus on the sustainability implications of these changes. Fardin et al. (2025) examined the relationship between telework and online shopping/delivery activity in post-pandemic New York City, U.S.A. The findings suggest that telework is associated with increased online shopping and deliveries, especially for groceries, and that this relationship is influenced by factors like income, household size, and car ownership. Patwary and Khattak (2025) through their research identified how online shopping is affecting in-person shopping activities, with a specific focus on how these changes impact

Table 2 Consumer characteristics and their influence on e-commerce experiences based on the literature review

Factors	Impact on online shopping and delivery	Source
Demographics		
Age	Younger users prioritize credibility and trustworthiness. Young e-consumers prefer parcel lockers.	Alguacil et al. (2020); Iannaccone et al. (2021)
Gender	Women demonstrate higher trust and loyalty toward services.	Alguacil et al. (2020)
Employment status	Unemployed individuals prioritize trust. Retirees show the highest loyalty levels.	Alguacil et al. (2020)
Psychographics		
Shopping behavior	Frequent users exhibit the strongest loyalty.	Alguacil et al. (2020)
Values & attitudes	Sustainable shoppers prioritize eco-friendly practices.	DHL (2024)
Socioeconomic factors		
Income	Influences online shopping and delivery activity (higher income = more online shopping).	Fardin et al. (2025)
Mobility	Mobility habits influence e-commerce behavior.	Rangel et al. (2025)
Residential location	Access to internet and security concerns in informal settlements impact delivery access.	de Oliveira et al. (2025)
Technological factors		
Internet access	Essential for online shopping and delivery.	Sharda et al. (2025); de Oliveira et al. (2025)
Digital literacy	Crucial for fair access to online purchasing and delivery services.	Sharda et al. (2025)

disadvantaged communities. de Oliveira et al. (2025) investigated how residential location, especially living in informal settlements, affects access to parcel delivery in Belo Horizonte, Brazil. The authors concluded that income, Internet access, and age are important factors, and that security concerns in informal settlements create additional barriers to delivery access.

Table 2 presents the key findings of the literature review in relation to consumer characteristics and their influence on e-commerce experiences.

3.2 *Incentives and Nudging Mechanisms*

To address RQ2—*what incentives and nudging mechanisms can influence e-commerce consumer personas to adopt more sustainable delivery and return options*—several strategies have been identified. These strategies aim at reducing the environmental impact of deliveries and returns while encouraging responsible consumer behavior.

Incentives for Sustainable Delivery and Return Policies
One effective approach to influencing consumer behavior is through incentives that reward sustainable choices. Offering discounts on future purchases can motivate customers to forgo returns altogether or opt for eco-friendly return methods (Kiba-Janiak et al. 2024). Similarly, free return shipping can be provided for customers who consolidate returns or use slower, greener logistics options, reducing unnecessary emissions and costs associated with expedited shipping.

Cashback offers also encourage customers to avoid returning low-value items, thereby minimizing handling and transportation demands (Rodrigue 2022). To reinforce long-term sustainable habits, businesses may introduce loyalty points programs. Extra loyalty points can be awarded to customers who minimize returns or adopt environmentally friendly methods, with points redeemable for discounts, free shipping, or special offers (Kiba-Janiak et al. 2024). For customers already committed to sustainability, additional rewards such as priority processing of refunds or exchanges for eco-friendly returns can serve as tangible benefits.

Innovative incentives like eco-friendly packaging discounts provide rewards for customers who reuse original packaging or select recyclable materials (Hadaś et al. 2024). Extending the return window, i.e., from 30 to 60 days, for customers choosing green options can further alleviate time pressures, promoting consolidated returns and reducing the environmental burden of rushed logistics (Rodrigue 2022). Companies may also leverage charity donations, allowing customers to contribute to environmental causes when opting for more sustainable return options or deciding not to return an item.

For consumers seeking long-term convenience, subscription-based models offer value while aligning with sustainability goals. Subscription-based return services allow unlimited returns for a flat fee, provided customers consolidate returns or use eco-friendly shipping methods (Sułkowski et al. 2022). Bundling returns into a single shipment can also be encouraged by offering discounts or rewards, significantly cutting down emissions generated by multiple deliveries. Green labeling—highlighting products with sustainable return policies—can help customers make informed decisions while offering small incentives, such as discounts, for choosing these options (Kiba-Janiak et al. 2024) (Fig. 3).

Nudging Strategies for Delivery Return Policies
Beyond incentives, nudging strategies can play a critical role in promoting sustainable choices by subtly guiding consumer behavior. For instance, integrating a visible carbon footprint calculator into the return process allows customers to see the environmental impact of their decisions in a visual, easy-to-understand format (Abdullahi et al. 2024). This transparency helps consumers better appreciate the consequences of their choices.

Nudges and incentives are complementary strategies aimed at influencing consumer behavior, but they differ in their approach and mechanisms. Incentives typically involve tangible rewards, such as discounts, cashback, or loyalty points, to encourage specific actions. They appeal directly to consumers' financial interests, offering clear, measurable benefits for adopting sustainable delivery and return

Fig. 3 Categorization of incentive policies for sustainable delivery and return behavior

practices. In contrast, nudges subtly guide decision-making by leveraging behavioral psychology principles without restricting choices or offering material rewards. Nudging strategies, such as emphasizing environmental benefits, framing eco-friendly options as the default, or using social proof, rely on influencing consumers' attitudes and perceptions to promote sustainable behavior. While incentives provide external motivation, nudges aim to internalize the desired behavior, fostering long-term change.

Businesses can further influence behavior through emotional appeals, emphasizing the positive impact of eco-friendly returns on the environment and future generations (Jiang et al. 2023). Such messaging creates an emotional connection, increasing the likelihood of sustainable behavior. In addition, companies can make default green options the standard choice in the return process, requiring customers to actively opt out if they prefer less sustainable alternatives. By framing eco-friendly options as the default, businesses nudge customers toward more responsible decisions.

Implementing a pre-return check system can also reduce unnecessary returns. Before initiating a return, customers can be prompted with questions about whether the return is truly necessary, offering alternatives like troubleshooting support or exchanges (Abdullahi et al. 2024). Once a sustainable return is completed, impact feedback can reinforce positive behavior by stating, "By choosing this return method, you helped reduce carbon emissions by X amount.

To encourage informed decision-making, businesses can use information nudges, such as presenting concise and clear details about the environmental impacts of various return methods at the point of decision. Gamification elements, like earning badges or levels for choosing sustainable options, can further enhance engagement (Cauwelier et al. 2024). Tangible rewards for reaching milestones (e.g., discounts or exclusive offers) provide an added incentive for customers to continue adopting green practices.

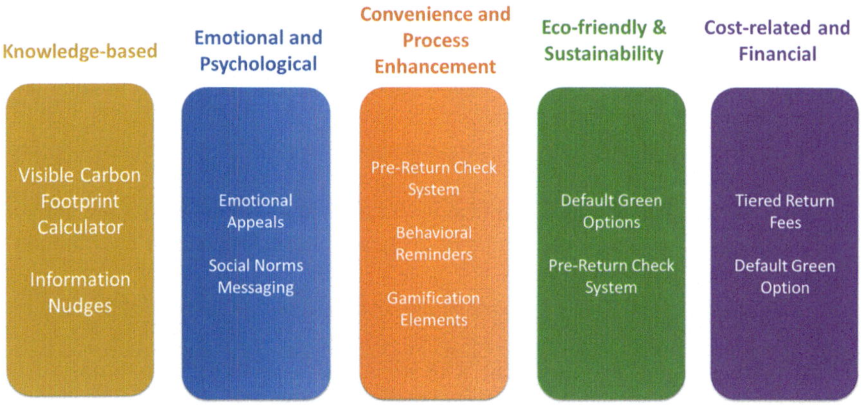

Fig. 4 Nudging strategies clustering

Behavioral reminders are another effective tool, particularly during the return window. Sending notifications that highlight the benefits of slower or consolidated returns can influence customer choices. Similarly, introducing tiered return fees—where greener return methods incur lower or no fees, while less sustainable options come at a higher cost—can make eco-friendly options more attractive (Moroz and Polkowski 2016).

Lastly, businesses can leverage social norms messaging by showcasing how other customers are making sustainable choices. For instance, statements like "80% of our customers choose green returns" create a sense of social proof, encouraging individuals to align their behavior with the majority (Kiba-Janiak et al. 2024) (Fig. 4).

4 Synthesis

The key findings related to personas, and the corresponding incentives and nudging mechanisms, are synthesized using the five-step methodological process (Sect. 2.3): (i) identification of core dimensions, (ii) clustering of persona types, (iii) persona journey mapping, (iv) linking incentives and nudging strategies, and (v) persona table consolidation. Based on the identified dimensions (Step 1), Step 2 resulted in the classification of 12 persona types, each defined by a dominant motivational trait. The 12 identified persona types and their key characteristics are presented as follows[1]:

[1] It is acknowledged that a regular e-customer could fall under several persona types as traits of his/her behavior.

1. *Time-savers*: Prioritize convenience, value time over price, and avoid physical stores.
2. *Brand seekers*: Prefer trusted brands, willing to pay more for quality and reputation.
3. *Review enthusiasts*: Check reviews, ratings, and product authenticity before purchasing.
4. *Family-centric*: Make decisions heavily influenced by family needs and values.
5. *Health-conscious*: Focused on buying organic, sustainable, and environmentally friendly products.
6. *Cultural guardians*: Value products and brands that are respectful of cultural traditions.
7. *Tech-savvy shoppers*: Confident with apps and e-commerce platforms, they enjoy personalized, tech-enabled shopping experiences.
8. *Social shoppers*: Heavily influenced by social media trends and recommendations.
9. *Cost-conscious*: Seek value through deals and discounts, focus on cost-effectiveness.
10. *Impulsive buyers*: Make quick, emotional decisions for instant gratification.
11. *Global explorers*: Shop internationally for unique items, balancing cost and exclusivity.
12. *Premium shoppers*: Value exclusivity, quality, and premium service experiences.

4.1 Mapping the Engagement Journey of E-Commerce Persona Types

For each persona type, an indicative "e-commerce journey" was developed to map engagement level around five key stages: discover, engage, decide, purchase, and feedback (Lemon and Verhoef 2016). Mapping the engagement journey of e-commerce persona types is a critical intermediate step that enables a more accurate and targeted alignment of incentives and nudging mechanisms. Rather than assuming uniform behavior across the customer lifecycle, this step reveals when and how each persona is most engaged across key stages. Understanding these engagement patterns is essential for timing and positioning behavioral interventions. For instance, a persona highly active during the feedback stage may respond more effectively to post-purchase nudges, while another with peak activity in the decision stage may benefit from incentives tied to eco-friendly delivery options at the point of purchase. Skipping this step would risk applying strategies that do not correspond to the persona's most influential touchpoints, thus reducing the effectiveness of the proposed sustainable behavior interventions.

The first stage, Discover, is where consumers initially become aware of a product, service, or e-commerce platform. This awareness can arise through various channels, such as online advertisements, social media platforms, recommendations from others, search engines, or random browsing. Different personas display unique discovery patterns; for instance, tech-savvy individuals may rely heavily on social media and influencer endorsements, whereas others might prefer more traditional approaches, like email marketing or word-of-mouth recommendations.

The second stage, Engage, involves the consumer's initial interaction with the e-commerce platform or product. During this stage, consumers might browse websites or mobile apps, read product reviews, compare options, or add items to wish lists or shopping carts. Personas differ significantly in their engagement behaviors. For example, some personas are meticulous researchers who spend considerable time analyzing reviews and exploring options, while others, such as impulse buyers, may move quickly to the next stage with minimal interaction.

The third stage, Decide, is where consumers deliberate and make the critical choice of whether to proceed with a purchase. Several factors influence this decision-making process, including price, brand reputation, convenience, trust in the platform, and the availability of deals or discounts. Personas vary in their approach to this stage. Cost-conscious buyers might wait for sales or discount coupons to finalize their decision, whereas premium shoppers might prioritize product quality and brand reputation over price, leading to quicker decision-making.

The fourth stage, Purchase, marks the point at which the transaction is completed. This stage includes selecting a payment method, choosing delivery preferences, and finalizing the order. The purchase experience itself—such as the ease and user-friendliness of the checkout process—plays a significant role in shaping consumer behavior at this stage. Personas may prefer different payment options, such as mobile wallets, credit cards, or cash on delivery, and their delivery preferences may range from same-day service to eco-friendly options. Challenges during this stage, such as a complex checkout process, might lead some personas to abandon their carts altogether.

The final stage, Feedback, occurs post-purchase when the consumer evaluates both the product and the overall shopping experience. This can involve leaving reviews, reaching out to customer support to address any issues, or sharing their experience with others through social media or other platforms. Personas behave differently at this stage as well; for instance, those who value community and social sharing may actively post reviews or engage online, while others may only provide feedback when incentivized, such as receiving discounts for leaving a review. In some cases, feedback may only be provided when the experience was particularly positive or negative.

Based on the five key stages, the authors scored on a scale from 1 (low) to 3 (high) each persona's type to reflect on the engagement level, based on the authors' judgment. The scores attributed can be seen in Fig. 5.

The figure shows that time-savers display increasing engagement through decision-making and maintain high involvement in feedback. Brand seekers begin and end strongly, with moderate engagement in between, reflecting loyalty and post-purchase reflection. Review enthusiasts are most active during early research, sustaining moderate involvement thereafter. Family-centric shoppers show consistently high engagement in the first three stages, with a slight decline post-purchase.

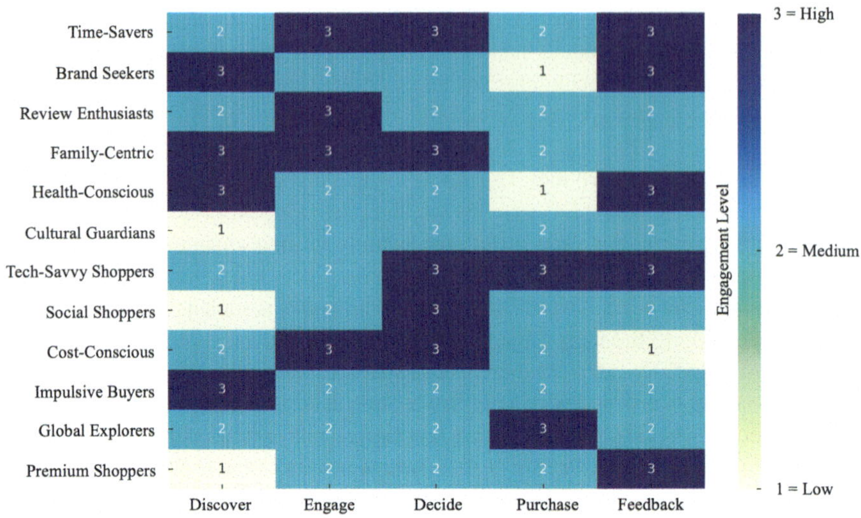

Fig. 5 Persona journey map—engagement patterns across key stages

Health-conscious consumers start and end with high engagement, dipping in the middle stages. Cultural guardians show moderate engagement throughout after a low-discovery start. Tech-savvy shoppers intensify mid-journey, peaking from decision to feedback stages. Social shoppers begin with low engagement but peak during decision-making, maintaining moderate levels afterward. Cost-conscious users are most engaged during the evaluation phases, with minimal post-purchase activity. Impulsive buyers are highly engaged at discovery and maintain medium levels across the journey. Global explorers are steady throughout but peak during purchase. Finally, premium shoppers show a back-loaded pattern, with low initial engagement that rises to a peak at the feedback stage.

4.2 *Consolidating Personas with Incentives and Nudging Strategies*

Based on each persona's journey, the (i) incentives for sustainable delivery and return policies and (ii) nudging strategies for delivery return policies were matched manually with each persona type, following the narrative of personas' journeys, as both emerged from the systematic literature review. Table 3 shows the persona types compiled into a structured table, detailing their key characteristics, applicable delivery and return incentives, and corresponding nudging mechanisms.

Table 3 E-commerce personas

Persona type	Key characteristics	Incentives for sustainable delivery and return policies	Nudging strategies for delivery return policies
Time-savers	Prioritize convenience, value time over price, avoid physical stores	Customer convenience and experience; Eco-friendly and sustainability	Convenience and process enhancement; Eco-friendly and sustainability
		Incentives focused on convenience and eco-friendly options (e.g., fast checkouts, bundled deliveries) are suitable given their priority for time efficiency, while nudges enhancing process simplicity and default green options align with their preference for low-effort experiences.	
Brand seekers	Prefer trusted brands, willing to pay more for quality and reputation	Loyalty and engagement; Eco-friendly and sustainability	Emotional and psychological; Eco-friendly and sustainability
		They respond well to loyalty programs and green-labeled trusted brands, hence incentives from engagement and sustainability categories fit; emotional nudges and green defaults reinforce their values of brand integrity and reputation.	
Review enthusiasts	Check reviews, ratings, and product authenticity before purchasing	Loyalty and engagement; Eco-friendly and sustainability	Knowledge-based; Eco-friendly and sustainability
		Since they seek transparency and credibility, recognition-based and eco-aware incentives match the loyalty and sustainability categories, while knowledge-based and green nudges help reinforce trust and informed decisions.	
Family-centric	Make decisions heavily influenced by family needs and values	Cost-related; Customer convenience and experience	Convenience and process enhancement; Emotional and psychological
		This group values affordable and practical solutions, thus cost-related and extended return benefits appeal to their need for flexible planning; convenience-focused nudges and emotional appeals resonate with family-centered thinking.	

(continued)

Table 3 (continued)

Persona type	Key characteristics	Incentives for sustainable delivery and return policies	Nudging strategies for delivery return policies
Health-conscious	Focused on buying organic, sustainable, and environmentally friendly products	Eco-friendly and sustainability; Customer convenience and experience	Eco-friendly and sustainability; Emotional and psychological
		Driven by values, eco donations and sustainable packaging incentives match sustainability goals, and emotional and green-default nudges reinforce their alignment with environmental and wellness principles.	
Cultural guardians	Value products and brands respectful of cultural traditions	Cost-related; Eco-friendly and sustainability	Knowledge-based; Eco-friendly and sustainability
		They value authenticity and respectful branding, making eco-sensitive product discounts fitting under cost and sustainability incentives; information nudges and default green return flows respect their need for clarity and trust.	
Tech-savvy shoppers	Confident with apps and e-commerce platforms, enjoy personalized, tech-enabled shopping experiences	Loyalty and engagement; Eco-friendly and sustainability	Convenience and process enhancement; Eco-friendly and sustainability
		They are drawn to gamification and app-based experiences, making loyalty engagement and green innovation incentives relevant, while process enhancement nudges like app-based returns and green defaults align with their tech-first habits.	
Social shoppers	Heavily influenced by social media trends and recommendations	Loyalty and engagement; Eco-friendly and sustainability	Emotional and psychological; Eco-friendly and sustainability
		Influenced by visibility and peer norms, they are responsive to eco-branded social campaigns under loyalty and sustainability, while emotional and social norm nudges amplify sustainable behavior adoption.	
Cost-conscious	Seek value through deals and discounts, focus on cost-effectiveness	Cost-related; Customer convenience and experience	Knowledge-based; Cost-related and financial
		Motivated by savings, free shipping, and return offers represent cost-related and convenience incentives, while financial nudges (like tiered fees) and information tools highlight value in green choices.	

Table 3 (continued)

Persona type	Key characteristics	Incentives for sustainable delivery and return policies	Nudging strategies for delivery return policies
Impulsive buyers	Make quick, emotional decisions for instant gratification	Cost-related; Loyalty and engagement	Emotional and psychological; Convenience and process enhancement
		Incentives like immediate eco-reward points cater to their desire for instant gratification under cost and loyalty themes, while emotional triggers and gamified reminders support spontaneous yet sustainable actions.	
Global explorers	Shop internationally for unique items, balancing cost and exclusivity.	Cost-related; Eco-friendly and sustainability	Knowledge-based; Cost-related and financial
		They need clear eco-labels and flexible global shipping options, fitting cost-efficient and sustainability incentives; knowledge-based nudges and financially framed green defaults aid in decision-making for cross-border returns.	
Premium shoppers	Value exclusivity, quality, and premium service experiences.	Loyalty and engagement; Eco-friendly and sustainability	Convenience and process enhancement; Eco-friendly and sustainability
		This group values quality and service, making exclusive green perks and sustainable packaging discounts ideal under loyalty and sustainability incentives; convenience-enhancing and eco-default nudges suit their expectation for seamless, premium experiences.	

5 Concluding Discussion

This chapter explores how e-commerce consumer personas influence and respond to delivery and return services, providing answers to the two research questions set out.

In response to RQ1, the behavioral segmentation of online shoppers reveals distinct delivery and return preferences that are not only shaped by demographic and psychographic factors but also by digital engagement and environmental awareness. Time-savers prioritize speed and simplicity, favoring services that reduce friction, such as doorstep pickups or app-enabled returns. Brand seekers and review enthusiasts, driven by trust and transparency, exhibit low return rates due to informed and loyal purchasing behavior. Family-centric and cultural guardians seek flexibility and reliability, especially in return windows and product appropriateness. Meanwhile, tech-savvy and social shoppers demand seamless digital experiences and peer validation, which influence their return decisions and adoption of green logistics practices.

In addressing RQ2, the study demonstrates that a strategic mix of incentives and nudging mechanisms, tailored to each persona's core motivation, can foster more sustainable behavior. For instance, gamified eco-reward schemes resonate with tech-savvy users, while emotional appeals and immediate benefits effectively guide Impulsive Buyers toward greener delivery choices. Cost-related incentives are essential for cost-conscious personas, whereas premium shoppers respond best to exclusive, high-quality green services. Across all personas, combining default eco-friendly options, transparent return policies, and cause-driven incentives shows potential to shift behaviors without compromising customer satisfaction.

These insights directly support the transformation of urban logistics within a circular economy paradigm. Profiling consumer behavior enables sustainable delivery and return strategies that are persona-sensitive. When embedded into urban freight planning, such insights help align last-mile models with real consumer needs. For example, micro-hubs can serve time-savers and impulsive buyers seeking immediacy, while route optimization for cost-conscious or health-conscious users supports consolidated deliveries. About returns, tailored choices help reduce unnecessary trips and keep products in use through reuse, resale, or donation, contributing to reducing waste and emissions, making urban logistics a stronger enabler of circular cities.

Moreover, while this chapter provides a structured synthesis of e-commerce personas and maps them to targeted incentives and nudging strategies, a key limitation is that the proposed persona/incentive/nudge mappings are conceptually derived and not empirically tested. Future research should validate these relationships through behavioral experiments or stated preference surveys to assess their effectiveness in real-world logistics settings.

Appendix A

Appendix A consists of the Table A1, which represents the query used in SCOPUS database.

Table A1 Query used in SCOPUS database

Database	Search query	Records	Search date
Scopus	e-commerce AND persona AND (deliver* OR return) AND PUBYEAR > 2018 AND PUBYEAR < 2026 AND (LIMIT-TO (SRCTYPE , "j") OR LIMIT-TO (SRCTYPE , "p") OR LIMIT-TO (SRCTYPE , "b") OR LIMIT-TO (SRCTYPE , "k")) AND (LIMIT-TO (PUBSTAGE , "final")) AND (EXCLUDE (DOCTYPE , "sh") OR EXCLUDE (DOCTYPE , "no") OR EXCLUDE (DOCTYPE , "ed") OR EXCLUDE (DOCTYPE , "cr") OR EXCLUDE (DOCTYPE , "re")) AND (EXCLUDE (SUBJAREA , "MEDI") OR EXCLUDE (SUBJAREA , "PHYS") OR EXCLUDE (SUBJAREA , "AGRI") OR EXCLUDE (SUBJAREA , "NURS") OR EXCLUDE (SUBJAREA , "NEUR") OR EXCLUDE (SUBJAREA , "PHAR") OR EXCLUDE (SUBJAREA , "IMMU") OR EXCLUDE (SUBJAREA , "HEAL") OR EXCLUDE (SUBJAREA , "EART") OR EXCLUDE (SUBJAREA , "CHEM") OR EXCLUDE (SUBJAREA , "BIOC") OR EXCLUDE (SUBJAREA , "CENG")) AND (LIMIT-TO (LANGUAGE , "English")) AND (LIMIT-TO (OA , "all"))	139	18.02.2025

Appendix B: List of Records (N = 59) that Were Included in the Systematic Literature Review

1. Alves Gomes, M., & Meisen, T. (2023). A review on customer segmentation methods for personalized customer targeting in e-commerce use cases. Information Systems and e-Business Management, 21(3), 527–570. https://doi.org/10.1007/s10257-023-00640-4
2. Kocaoğlu, B., & Bulut, M. (2024). Circular supply chain network design for e-commerce. Gazi University Journal of Science, 37(2), 840–852. https://doi.org/10.35378/gujs.1234548
3. Tsai, C.-A., & Chang, C.-W. (2022). Development of a partial shipping fees pricing model to influence consumers' purchase intention under the COVID-19 pandemic. Energies, 15(5), 1846. https://doi.org/10.3390/en15051846
4. Zennaro, I., Finco, S., Calzavara, M., & Persona, A. (2022). Implementing e-commerce from logistic perspective: Literature review and methodological framework. Sustainability, 14(2), 911. https://doi.org/10.3390/su14020911
5. Sułkowski, Ł., Kolasińska-Morawska, K., Brzozowska, M., Morawski, P., & Schroeder, T. (2022). Last mile logistics innovations in the courier-express-parcel sector due to the COVID-19 pandemic. Sustainability, 14(13), 8207. https://doi.org/10.3390/su14138207

6. Bellio, E., & Buccoliero, L. (2021). Mapping the research landscape of reverse logistics in e-commerce: A bibliometric analysis from 2003 to 2023. Journal Européen des Systèmes Automatisés, 57(3), 345–356.
7. Klumpp, M., & Loske, D. (2021). Persona perception scale: Development and exploratory validation of an instrument for evaluating individuals' perceptions of personas. Proceedings of the 2021 CHI Conference on Human Factors in Computing Systems, 1–12.
8. Dol, A., van Strien, T., Velthuijsen, H., van Gemert-Pijnen, L., & Bode, C. (2023). Reducing retail supply chain costs of product returns using digital product fitting. International Journal of Physical Distribution & Logistics Management, 53(4), 456–472.
9. Díaz-Meneses, G., Amador-Marrero, M., & Spinelli Guedes, C. (2023). The criteria of inbound marketing to segment and explain the domain authority of the cellars' e-commerce in the Canary Islands. Systems, 11(11), 527.
10. Vrhovac, V., Dakić, D., Milisavljević, S., Ćelić, Đ., Stefanović, D., & Janković, M. (2024). The factors influencing user satisfaction in last-mile delivery: The structural equation modeling approach. Mathematics, 12(12), 1857.
11. Li, Z., Gu, W., & Meng, Q. (2023). The impact of COVID-19 on logistics and coping strategies: A literature review. Regional Science Policy & Practice, 15(8), 1768–1794. https://doi.org/10.1111/rsp3.12665
12. Hadaś, Ł., Domański, R., Wojciechowski, H., Majewski, A., & Lewandowicz, J. (2024). The role of packaging in sustainable omnichannel returns—The perspective of young consumers in Poland. Sustainability, 16(6), 2231. https://doi.org/10.3390/su16062231
13. Abbasimehr, H., & Baghery, F. S. (2022). A novel time series clustering method with fine-tuned support vector regression for customer behavior analysis. Expert Systems with Applications, 200, 117004. https://doi.org/10.1016/j.eswa.2022.117004
14. Chen, S. S., Li, T. L., Wu, Y. C., & Singh, V. (2023). An algorithm-based approach for mapping customer journeys by identifying customer browsing behaviors on e-commerce clickstream data. In Proceedings of the 2023 5th International Congress on Human-Computer Interaction, Optimization and Robotic Applications (HORA) (pp. 1–8). IEEE. https://doi.org/10.1109/HORA58378.2023.10156808
15. Hamidi, H., & Haghi, B. (2024). An approach based on data mining and genetic algorithm to optimizing time series clustering for efficient segmentation of customer behavior. Computers in Human Behavior Reports, 16, 100520. https://doi.org/10.1016/j.chbr.2024.100520
16. Neto, L. G., & Vieira, J. G. V. (2023). An investigation of consumer intention to use pick-up point services for last-mile distribution in a developing country. Journal of Retailing and Consumer Services, 74, 103425.
17. Hsu, P.-F., Lu, Y.-H., Chen, S.-C., & Kuo, P. P.-Y. (2024). Creating and validating predictive personas for target marketing. International Journal of Human-Computer Studies, 181, 103147.

18. Hu, Y., Chen, Y., Zhou, S., Yang, S., Jiang, H., & Ma, Y. (2024). Determinants of consumer intention to use smart food lockers during COVID-19: A multi-method approach. International Journal of Hospitality Management, 117, 103858.
19. Lai, P.-L., Jang, H., Fang, M., & Peng, K. (2022). Determinants of customer satisfaction with parcel locker services in last-mile logistics. Asian Journal of Shipping and Logistics, 38(1), 25–30.
20. Kim, C., & Lee, J. (2024). Discovering patterns and trends in customer service technologies patents using large language model. Heliyon, 10(3), e34701.
21. Vakulenko, Y., Arsenovic, J., Hellström, D., & Shams, P. (2022). Does delivery service differentiation matter? Comparing rural to urban e-consumer satisfaction and retention. Journal of Business Research, 142, 476–484. https://doi.org/10.1016/j.jbusres.2021.12.079
22. Kiba-Janiak, M., Cheba, K., de Oliveira, L. K., & van Duin, R. (2024). Drivers and barriers for sustainable last-mile delivery based on e-customers' preferences: Examples from Poland and Brazil. Sustainability, 16(3), 1234. https://doi.org/10.3390/su16031234
23. Jiang, Y., Lai, P. L., Yang, C. C., & Wang, X. (2023). Exploring the factors that drive consumers to use contactless delivery services in the context of the continued COVID-19 pandemic. Journal of Retailing and Consumer Services, 72, 103276.
24. Rodrigue, J. P. (2022). Home-based parcel deliveries: Consumer and logistics patterns from retail digitalization. Case Studies on Transport Policy, 10(4), 2413–2419. https://doi.org/10.1016/j.cstp.2022.11.004
25. Kiba-Janiak, M., Cheba, K., Mucowska, M., de Oliveira, L. K., Piecyk, M., Evangelista, P., Prockl, G., & Rzesny-Cieplińska, J. (2024). How to design a sustainable last-mile delivery and returns business model from e-customers' expectations perspective? Sustainability, 16(5), 2345. https://doi.org/10.3390/su16052345
26. Ferreira, M. S., Antão, J., Pereira, R., Bianchi, I. S., Tovma, N., & Shurenov, N. (2023). Improving real estate CRM user experience and satisfaction: A user-centered design approach. Journal of Open Innovation: Technology, Market, and Complexity, 9(2), 100076.
27. Chen, L., Wu, P., Dou, Y., & Wu, Y. (2023). Investigating senders' switching intention to smart lockers: An extension of push-pull-mooring model. Journal of Retailing and Consumer Services, 74, 103414. https://doi.org/10.1016/j.jretconser.2023.103414
28. Barrera, F., Segura, M., & Maroto, C. (2024). Multiple criteria decision support system for customer segmentation using a sorting outranking method. Expert Systems with Applications, 232, 122310.
29. Ma, B., Wong, Y. D., & Teo, C. C. (2022). Parcel self-collection for urban last-mile deliveries: A review and research agenda with a dual operations-consumer perspective. Transport Policy, 116, 71–85. https://doi.org/10.1016/j.tranpol.2022.06.009

30. Inverardi, P., Migliarini, P., & Palmiero, M. (2023). Systematic review on privacy categorisation. Computer Science Review, 49, 100574. https://doi.org/10.1016/j.cosrev.2023.100574
31. Moroz, M., & Polkowski, Z. (2016). The last mile issue and urban logistics: Choosing parcel machines in the context of the ecological attitudes of the Y generation consumers purchasing online. Transportation Research Procedia, 16, 378–393.
32. Cauwelier, K., Buldeo Rai, H., Puttemans, K., Macharis, C., & Mommens, K. (2024). From cart to door: Unravelling consumer behaviour through attitudinal sustainability profiles. Transportation Research Part D: Transport and Environment, 126, 104168.
33. Gabryel, M., Kocić, E., Kocić, M., Patora-Wysocka, Z., Min, X., & Miroslaw, P. (2024). Accelerating user profiling in e-commerce using conditional GAN networks for synthetic data generation. Journal of Artificial Intelligence and Soft Computing Research, 14(4), 309–319.
34. Fedorko, R., Bačík, R., Rigelský, M., Horváth, J., & Král', Š. (2024). The importance of social networks for e-commerce: An analysis of young users' perspectives in the European Union. In Proceedings of the 2023 International Conference on Business and Technology (ICBT 2023) (pp. 221–234). Springer.
35. Rai, H. B., Verlinde, S., & Macharis, C. (2021). Who is interested in a crowdsourced last mile? A segmentation of attitudinal profiles. Travel Behaviour and Society, 22, 22–31.
36. Akre, V., Rajan, A., Ahamed, J., Al Amri, A., & Al Daisi, S. (2019). Smart digital marketing of financial services to millennial generation using emerging technological tools and buyer persona. In 2019 Sixth HCT Information Technology Trends (ITT) (pp. 120–125). IEEE.
37. Alguacil, M., Crespo-Hervás, J., & Pérez-Campos, C. (2020). Sociodemographic analysis of brand perception in a public sports service: From target to person buyer. Retos: Nuevas Tendencias en Educación Física, Deporte y Recreación, 37, 139–146.
38. Appiah, D., & Watson, A. (2024). Smartphones and digital customers' journey. In W. Ozuem, M. Willis, & S. Ranfagni (Eds.), Digital Transformation for Fashion and Luxury Brands: Theory and Practice (pp. 299–314). Palgrave Macmillan.
39. Cruz, A., & Karatzas, S. (2020). Understanding your buyer persona. In A. Heinze, G. Fletcher, A. Cruz, & S. Karatzas (Eds.), Digital and Social Media Marketing: A Results-Driven Approach (pp. 69–97). Routledge. https://doi.org/10.4324/9780429280689-6
40. Delve AI. (2024). Online shopper personas for e-commerce stores. Delve AI Blog. https://www.delve.ai/blog/ecommerce-personas
41. DHL eCommerce. (2024). Beyond the basket: 2024 online shopper trends. DHL eCommerce Insights. https://www.dhl.com/global-en/microsites/ec/ecommerce-insights/insights/reports/2024-beyond-the-basket.html

42. Fenton, A., Heinze, A., Osborne, M., & Ahmed, W. (2022). How to use the six-step digital ethnography framework to develop buyer personas: The case of Fan Fit. JMIR Formative Research, 6(11), e41489
43. Hall, A., & Towers, N. (2017). Understanding how Millennial shoppers decide what to buy: Digitally connected unseen journeys. International Journal of Retail & Distribution Management, 45(5), 498–517
44. Miva Blog. (2024). 7 types of ecommerce consumer personas and how to influence them. Miva Blog.
45. Romero Montero, A., Sellers Rubio, R., & Cely Álvarez, A. M. (2024). ¿Conoces a tu Buyer Persona? Identifica a tu cliente para mejorar tu estrategia de Inbound Marketing. Investigaciones Turísticas, (27), 53–76.
46. Samadi, S., & Akhtar, I. (2023). A review of social media marketing on digital savvy brand shoppers. In C. Iwendi, Z. Boulouard, & N. Kryvinska (Eds.), Proceedings of ICACTCE'23 — The International Conference on Advances in Communication Technology and Computer Engineering (Vol. 735, pp. 641–655). Springer.
47. Ventura Web Design & Marketing. (2024). Common buyer personas in e-commerce.
48. Xu, Y., & Lee, M. J. (2020). Identifying personas in online shopping communities. Multimodal Technologies and Interaction, 4(2), 19.
49. Zhong, R., Han, S., & Wang, Z. (2024). Developing personas for live streaming commerce platforms with user survey data. Universal Access in the Information Society, 23, 1705–1721.
50. Lax, A., & Mau, N. (2013). Customer value and value for customer – Retail sector's need for transformation. Journal of Marketing Research and Case Studies, 2013, Article ID 812332.
51. Asdecker, B. (2021). Building the e-commerce supply chain of the future: What influences consumer acceptance of alternative places of delivery on the last-mile. International Journal of Physical Distribution & Logistics Management, 51(2), 123–138.
52. Iannaccone, G., Marcucci, E., & Gatta, V. (2021). What young e-consumers want? Forecasting parcel lockers choice in Rome. Research in Transportation Business & Management, 39, 100591. https://doi.org/10.1016/j.rtbm.2021.100591
53. Pasupuleti, V., Thuraka, B., Kodete, C. S., & Malisetty, S. (2024). Enhancing supply chain agility and sustainability through machine learning: Optimization techniques for logistics and inventory management. Sustainable Computing: Informatics and Systems, 38, 100862.
54. Sharda, S., Alonso, P., Garikapati, V., Laguarda, I., Goulias, K., & Viz, N. (2025). New norms or old habits: Evaluating interlinked trajectories of online shopping and work commute post-pandemic. Transportation Research Part A: Policy and Practice, 195, 104461.
55. Rangel, T., Gomes, J., Alves, G., & Vassallo, J. M. (2025). Analyzing e-commerce last-mile: Key variables in adoption, frequency, and returns.

Transportation Research Part E: Logistics and Transportation Review, 178, 103046.
56. Castaño, D., Gomes, J., Garrido, L., Tapiador, L., & Vassallo, J. M. (2025). Connecting e-purchases and mobility: How does e-commerce impact people's mobility habits? Transportation Research Part A: Policy and Practice, 190, 104295. https://doi.org/10.1016/j.tra.2024.104295
57. Fardin, F. R., Le Morvan-Chevestre, M., & Conway, A. (2025). How does telework influence online shopping and home delivery activity? Findings from post-pandemic New York City. Transportation Research Part A: Policy and Practice, 195, 104460.
58. Patwary, A. L., & Khattak, A. (2025). Interaction between the emerging components of online shopping and in-person activities: Insights from behavioral survey and the Justice40 initiative. Transportation Research Part A: Policy and Practice, 195, 104462.
59. de Oliveira, L. K., Colaco, R., Araújo, G. G. F., & de Abreu e Silva, J. (2025). Analyzing the parcel delivery pattern in the Global South: The case of the Brazilian favelas. Transportation Research Part E: Logistics and Transportation Review, 179, 103047.

References

Abdullahi M, Ibrahim MH, Mohamud FAS (2024) Mapping the research landscape of reverse logistics in E-commerce: a bibliometric analysis from 2003 to 2023. J Eur des Syst Autom 57:671–679. https://doi.org/10.18280/jesa.570304

Abidar L, El Asri I, Zaidouni D, En-Nouaary A (2024) Evaluating customer segmentation efficiency via sentiment analysis: An e-commerce case study. In: Modern artificial intelligence and data science, pp 223–234. https://doi.org/10.1007/978-3-031-65038-3_18

Akre V, Rajan A, Ahamed J, Al Amri A, Al Daishi S (2019) Smart digital marketing of financial services to millennial generation using emerging technological tools and buyer personas. In: Information technology trends: Emerging technologies blockchain and IoT ITT2019, 120–125, 6th HCT Information Technology Trends, ITT 2019, Ras Al Khaima, UAE. https://doi.org/10.1109/ITT48889.2019.9075106

Alguacil M, Crespo-Hervás J, Pérez-Campos C (2020) Sociodemographic analysis of brand perception in a public sports service: from target to person buyer. Retos 37:139–146. ISSN: 15791726

An J, Kwak H, Jung S (2018) Customer segmentation using online platforms: isolating behavioral and demographic segments for persona creation via aggregated user data. Soc New Anal Min 8:54. https://doi.org/10.1007/s13278-018-0531-0

Appiah D, Watson A (2024) Smartphones and digital customers' journey. In: Digital transformation for fashion and luxury brands: theory and practice. Palgrave Macmillan, Cham, pp 407–421. https://doi.org/10.1007/978-3-031-35589-9_19

Asdecker B (2021) Building the E-commerce supply chain of the future: what influences consumer acceptance of alternative places of delivery on the last-mile. Logistics 5:90. https://doi.org/10.3390/logistics5040090

Beckers J, Cant J (2023) Half a decade in two years: household freight after COVID-19. Transp Rev 44(2):544–565. https://doi.org/10.1080/01441647.2023.2266859

Bhukya R, Paul J (2023) Social influence research in consumer behavior: What we learned and what we need to learn? – a hybrid systematic literature review. J Bus Res 162:113870., ISSN 0148-2963. https://doi.org/10.1016/j.jbusres.2023.113870

Campisi T, Russo A, Basbas S, Bouhouras E, Tesoriere G (2023) A literature review of the main factors influencing the e-commerce and last-mile delivery projects during COVID-19 pandemic. Transport Res Proc 69:552–559. https://doi.org/10.1016/j.trpro.2023.02.207

Castaño D, Gomes J, Garrido L, Tapiador L, Vassallo JM (2025) Connecting E-purchases and mobility: how does E-commerce impact people's mobility habits? Poster TRBAM-25-02003 presented at the TRB annual meeting 2025, Washington, U.S.A

Cauwelier, K., Buldeo Rai, H., Puttemans, K., Macharis, C., Mommens, K. From cart to door: Unravelling consumer behaviour through attitudinal sustainability profiles, Transport Res D Trans Environ, 130, 2024, 104168., ISSN 1361-9209, https://doi.org/10.1016/j.trd.2024.104168

Contentsquare (2024) Personas: a guide to understanding your customers. Available at: Contentsquare Personas Guide

Cruz-Cardenas J, Zaabelina E, Guadalupe-Lanas J, Palacio-Fierro A, Ramos-Galarxa C (2021) COVID-19, consumer behavior, technology, and society: a literature review and bibliometric analysis. Technol Forecast Soc Change 173. https://doi.org/10.1016/j.techfore.2021.121179

de Oliveira LK, Colaco R, Araújo GGF, de Abreu e Silva J (2025) Analyzing the parcel delivery pattern in the global south: the case of the Brazilian favelas. Poster TRBAM-25-00390 presented at the TRB annual meeting 2025, Washington, U.S.A

Delve.AI (2024) Online shopper personas for e-commerce stores. Delve.AI. Retrieved from https://www.delve.ai/blog/ecommerce-personas

DHL eCommerce (2024) Beyond the basket: 2024 online shopper trends. Retrieved from http://www.dhl.com/global-en/microsites/ecs/ecommerce-insights/insights/reports/2024-country-reports.html

Dietl M, Voigt S, Kuhn H (2024) From rush to responsibility: Evaluating incentives on online fashion customers' willingness to wait. Transport Res D Trans Environ 133:104280., ISSN 1361-9209. https://doi.org/10.1016/j.trd.2024.104280

European E-Commerce Report (2024) Published by Ecommerce Europe and the Centre for Market Insights, Amsterdam University of Applied Sciences. Available as a PDF document: CMI2024_Report_Complete_v3.pdf

European Investment Bank (2024) A catalogue of circular city actions and solutions. Circular City Centre—C3. Prepared in cooperation with Circle Economy. Available at: https://advisory.eib.org/c3

Fardin FR, Le Morvan-Chevestre M, Conway A (2025) How does telework influence online shopping and home delivery activity? Findings from post-pandemic new York City. Poster TRBAM-25-04841 presented at the TRB annual meeting 2025, Washington, U.S.A

Fenton A, Heinze A, Osborne M, Ahmed W (2022) How to use the six-step digital ethnography framework to develop buyer personas: the case of fan fit. JMIR Format Res 6(11). https://doi.org/10.2196/41489

Gomes AM, Meisen T (2023) A review on customer segmentation methods for personalized customer targeting in e-commerce use cases. Inf Syst E-Bus Manag 21:527–570. https://doi.org/10.1007/s10257-023-00640-4

Hadaś Ł, Domański R, Wojciechowski H, Majewski A, Lewandowicz J (2024) The role of packaging in sustainable Omnichannel returns—the perspective of young consumers in Poland. Sustainability 16(6):2231. https://doi.org/10.3390/su16062231

Hall A, Towers N (2017) Understanding how millennial shoppers decide what to buy: digitally connected unseen journeys. Int J Retail Distrib Manag 45(5):498–517. https://doi.org/10.1108/IJRDM-11-2016-0206

Hendricks S, Mwapwele SD (2024) A systematic literature review of the factors influencing e-commerce adoption in developing countries. Data Inform Manag 8(1). https://doi.org/10.1016/j.dim.2023.100045

Iannaccone G, Marcucci E, Gatta V (2021) What young e-consumers want? Forecasting parcel lockers choice in Rome. Logistics 5:57. https://doi.org/10.3390/logistics5030057

Jansen, B. J., Salminen, J., Jung, S.-G., & Guan, K. (2021). Data-driven personas. Synthesis lectures on human-centered informatics. Springer. https://doi.org/10.1007/978-3-031-02231-9

Jiang Y, Lai PL, Yang CC, Wang X (2023) Exploring the factors that drive consumers to use contactless delivery services in the context of the continued COVID-19 pandemic. J Retailing Consumer Serv 72:103276. https://doi.org/10.1016/j.jretconser.2023.103276. Epub 2023 Jan 27. PMCID: PMC9883787

Karlsson S, Oghazi P, Hellstrom D, Patel P, Papadopoulou C, Hjort K (2023) Retail returns management strategy: An alignment perspective. J Innov Knowl 8(4):100420, ISSN 2444-569X. https://doi.org/10.1016/j.jik.2023.100420

Kiba-Janiak M, Cheba K, Mucowska M, Oliveira LK, Piecyk M, Evangelista P, Prockl G, Rześny-Cieplińska J (2024) How to design a sustainable last-mile delivery and returns business model from E-customers' expectations perspective? Res Transp Bus Manag 56:Article 101194. https://doi.org/10.1016/j.rtbm.2024.101194

Kim Y, Kang J, Chun H (2022) Is online shopping packaging waste a threat to the environment? Econ Lett 214:110398., ISSN 0165-1765. https://doi.org/10.1016/j.econlet.2022.110398

Lemon KN, Verhoef PC (2016) Understanding customer experience throughout the customer journey. J Mark 80(6):69–96. https://doi.org/10.1509/jm.15.0420

Miva (2024) 7 types of ecommerce consumer personas and how to influence them. Miva Blog. Retrieved from https://blog.miva.com/7-types-of-consumer-personas1

Moroz M, Polkowski Z (2016) The last mile issue and urban logistics: choosing parcel Machines in the Context of the ecological attitudes of the Y generation consumers purchasing online. Transport Res Proc 16:378–393, ISSN 2352-1465. https://doi.org/10.1016/j.trpro.2016.11.036

Nanda P, Patnaik S (2023) A multi-agent coalition-based approach for order fulfilment in e-commerce. Decis Analy J 7:100227., ISSN 2772-6622. https://doi.org/10.1016/j.dajour.2023.100227

Page MJ, McKenzie JE, Bossuyt PM, Boutron I, Hoffmann TC, Mulrow CD et al (2021) The PRISMA 2020 statement: an updated guideline for reporting systematic reviews. BMJ 372:n71. https://doi.org/10.1136/bmj.n71

Pasupuleti V, Thuraka B, Kodete CS, Malisetty S (2024) Enhancing supply chain agility and sustainability through machine learning: optimization techniques for logistics and inventory management. Logistics 8:73. https://doi.org/10.3390/logistics8030073

Patwary AL, Khattak A (2025) Interaction between the emerging components of online shopping and in-person activities: insights from behavioral survey and the Justice40 initiative. Poster TRBAM-25-04971 presented at the TRB annual meeting 2025, Washington, U.S.A

Polydoropoulou A, Tsirimpa A, Karakikes I, Tsouros I, Pagoni I (2022) Mode choice modeling for sustainable last-mile delivery: the Greek perspective. Sustainability 14(15):8976. https://doi.org/10.3390/su14158976

Pundir S, Garg H, Singh D, Rana PS (2024) A systematic review of supply chain analytics for targeted ads in E-commerce. Supp Chain Analy 8. https://doi.org/10.1016/j.sca.2024.100085

Rangel T, Gomes J, Alves G, Vassallo JM (2025) Analyzing e-commerce last-mile: key variables in adoption, frequency, and returns. Poster TRBAM-25-00328 presented at the TRB annual meeting 2025, Washington, U.S.A

Rodrigue J-P (2022) Home-based parcel deliveries: consumer and logistics patterns from retail digitalization. Case Stud Transp Policy 10(4):2413–2419, ISSN 2213-624X. https://doi.org/10.1016/j.cstp.2022.11.004

Romero Montero A, Sellers Rubio R, Cely Alvarez AM (2024) Do you know your buyer persona? Identify your customer to improve your inbound marketing strategy. Invest Turist 27:53–76. https://doi.org/10.14198/INTURI.23961

Salminen J, Jansen BJ, Hamad JA, Kwak H, Jung SJ (2018) Are personas done? Evaluating their usefulness in the age of digital analytics. Persona Stud 4(2). https://doi.org/10.21153/psj2018vol4no2art737

Sharda S, Alonso P, Garikapati V, Laguarda I, Goulias K, Viz N (2025) New norms or old habits: Evaluating interlinked trajectories of online shopping and work commute post-pandemic. Poster TRBAM-25-05107 presented at the TRB annual meeting 2025, Washington, U.S.A

Shopify (2023) E-commerce returns: expert guide to best practices (2024). Available here: https://www.shopify.com/enterprise/blog/ecommerce-returns

Stevenson AB, Rieck J (2024) Investigating returns management across e-commerce sectors and countries: trends, perspectives, and future research. Logistics 8(3):82. https://doi.org/10.3390/logistics8030082

Sułkowski Ł, Kolasińska-Morawska K, Brzozowska M, Morawski P, Schroeder T (2022) Last mile logistics innovations in the courier-express-parcel sector due to the COVID-19 pandemic. Sustainability 14(13):8207. https://doi.org/10.3390/su14138207

Uzir H, Al Halbusi H, Thurasamy R, Hock RLT, Aljaberi MA, Hasan N, Hamid M (2021) The effects of service quality, perceived value and trust in home delivery service personnel on customer satisfaction: evidence from a developing country. J Retailing Consum Serv 63:102721., ISSN 0969-6989. https://doi.org/10.1016/j.jretconser.2021.102721

Ventura Web Design (2024) Common buyer personas in e-commerce. Ventura Web Design. Retrieved from https://www.venturawebdesign.com/learn/common-buyer-personas-in-e-commerce

Wang H, Li D, Jiang C, Zhang Y (2023a) Exploring the interactive relationship between retailers' free shipping decisions and manufacturers' product sales in digital retailing. Sustainability 15(17):12762. https://doi.org/10.3390/su151712762

Wang Z, Wong YD, Kim TY, Yen KF (2023b) Does consumers' involvement in e-commerce last mile delivery change after COVID-19? An investigation on behavioural change, maintenance and habit formation. Electron Commerce Res Appl 60. https://doi.org/10.1016/j.elerap.2023.101273

Wu Z, Jin L, Zhao J, Jing L, Chen L (2022) Research on segmenting E-commerce customer through an improved K-Medoids clustering algorithm. J Appl Math Comput 2022:9930613. https://doi.org/10.1155/2022/9930613

Xin M, Liu W, Jian L (2024) Live streaming product display or social interaction: how do they influence consumer intention and behavior? A heuristic-systematic perspective. Electron Commer Res Appl 67. https://doi.org/10.1016/j.elerap.2024.101437

Xu Y, Lee M (2020) Identifying personas in online shopping communities. Multimodal Technol Interact 4:19

Zhang. KZK, Benyousef M (2016) Consumer behavior in social commerce: a literature review. Decis Support Syst 86. https://doi.org/10.1016/j.dss.2016.04.001

Zhong R, Han S, Wang Z (2024) Developing personas for live streaming commerce platforms with user survey data. Univ Access Inf Soc 23:1705–1721. https://doi.org/10.1007/s10209-023-00996-x

Sustainable Cities and Circular Economy

Sustainable Urban Logistics and the Suitability of Freight Demand: A Comprehensive Analysis in Paris Region, France

Jesus Gonzalez Feliu

1 Introduction

Delivering cities is confronted by a major paradox (Crainic 2008): freight is necessary to ensure urban activities' operations and developing urban economies, but delivering that freight leads to major nuisances such as congestion, noise, global warming and pollution. However, the trivial solutions are not real ones (Benjelloun and Crainic 2008), since taking out all trucks and polluting vehicles from the cities will not solve the problem but create another one (Ackoff 1970), since the needs of urban activities and inhabitants, and their locations, do not change. Moving those locations only is not a solution, since the problem is moved geographically, but it remains, and sometimes is increased (Gonzalez-Feliu et al. 2012).

Urban logistics needs to be examined as a complex social system which is characterized by various elements, such as a set of involved stakeholders with their interactions, a demand structure, a supply structure (or a logistics network with its capacities and layout) and a system of constraints that need to be addressed and characterized to identify the main urban logistics problems and the way to solve them.

Concerning stakeholders (Lagorio et al. 2023), shippers, receivers, transporters and planners are the main ones. They interact and need coordination and collaboration tools to ensure the feasibility and deployment of suitable solutions for a more sustainable urban logistics planning and management. Planners and policy makers organize the access and global management of public space hosting those operations. Shippers and receivers exchange freight, and are demand generators, having a high impact on transport flows since they orient the freight transport needs.

J. G. Feliu (✉)
Excelia Business School, La Rochelle, France
e-mail: gonzalezfeliuj@excelia-group.com

A. Melkonyan-Gottschalk, Y. Li (eds.), *Urban Logistics Transformation*, Greening of Industry Networks Studies 15,
https://doi.org/10.1007/978-3-032-08387-6_12

Transporters represent the supply, and ensure the execution of those flows. Other stakeholders (residents, associations, real estate actors, control entities, etc.) have supporting or indirect roles but also need to be considered in the system.

Two main approaches to the development of urban logistics plans and solutions can be observed (Holguín-Veras et al. 2020). Traffic (or supply) approaches aim at proposing solutions based on regulating traffic or changing the capacities and conditions of the transport system. They are traditionally developed, and a wide variety of actions can be found nowadays. Demand approaches aim at changing the demand generation patterns and needs, in order to orient transport actions and solutions. According to Holguín-Veras et al. (2020), demand approaches allow us to propose more continuous and durable changes than traffic approaches.

But to develop those approaches, a good knowledge of freight transport demand is necessary. Although there is a plethora of frameworks and methods to analyze and assess freight transport demand (Comi et al. 2012; Gonzalez-Feliu and Routhier 2012; Marcucci et al. 2023), those methods are mainly based on general, aggregated trip generation, for which the categorial FTG framework starts to be a standard (Holguín-Veras et al. 2021). However, various elements that influence the demand, such as the vehicles used and the mode of management, have a high impact on the freight generation patterns.

This book chapter aims to address the issue of demand segmentation and the possibilities of deploying freight demand generation models, which take into account determinants that have a direct impact on sustainability transformations, such as the vehicle use or the mode of management. First, an overview of urban freight demand characterization and modelling is proposed. Then, a methodological framework for analyzing freight demand in a metropolitan area and constructing suitable freight generation models is proposed. After that, the framework is applied to the metropolitan area of Paris. Finally, research and practice implications are addressed.

2 Urban Logistics and Freight Transport Demand

Urban logistics is a complex system where various stakeholders exchange goods, services, information and financial flows. Those flows constitute freight transport demand, and need a supply network (with its capacities, needs and layout) to be transported. This demand plays an essential role in sustainability transitions, since it conditions the transport means, organizations and needs, and can be crucial to ensure the continuity and viability of sustainable urban logistics systems (Nimtrakool et al. 2018). We present in this section the main notions of freight transport demand proper to urban logistics.

2.1 The Main Physical Flows of Urban Goods Transport and Logistics

According to (Gonzalez-Feliu 2018), urban logistics systems are mainly composed of a demand, a supply and their interactions. This leads to the generation and execution of freight transport flows of different natures and characteristics. Urban logistics flows themselves are often assimilated to physical flows of goods transport, and can be divided into four macro-categories of flows:

- **Inter-establishment flows** include all movements of goods between two economic activities, and are mainly assimilated to urban logistics. These flows can be divided into several categories, which are in general associated with the type of activity generating those flows or with their mode of management (own account or third-party transport).
- Flows to supply the **end consumer** include all movements of the goods between the points of sale or the warehouses where it is located (in the case of online purchases) and the location of the final consumer. These flows include household shopping trips (whether motorized or not), logistics flows for home delivery and those for delivery to a pick-up point (relay point, concierge or locker).
- **Flows for urban management** include all travel related to urban space management activities, as well as the construction and maintenance of this space. This includes flows for the collection of household waste, travel related to construction activities (buildings or infrastructure, flows for the maintenance of public networks, or removals.
- To these three categories, we can add **internal or intra-establishment flows**, which do not have a direct impact on the occupation of the road because they take place inside a private space. Although difficult to estimate due to their nature, it is important to take into account their existence and their influence on the generation of inter-establishment flows.

It is important to remember that some of these flows are sometimes pooled. For example, courier services or parcel distribution (*small* parcel courier service, express or not) carry out deliveries with the same rounds (and therefore the same vehicles) for B2B and B2C flows. Moreover, the explosion of B2B flows contrasts with the needs of rationalization for B2B ones, which pose challenges to sustainable urban logistics planning, since both flows coexist. This takes particular importance if we consider also shopping trips, in a broader way (Gonzalez-Feliu and Peris-Pla 2017).

The estimation of these physical flows is complex, because these flows are heterogeneous and the level of information available to estimate them is not always sufficient to have a similar precision for each category. Nevertheless, a general methodology for estimating these flows exists and is currently used in several countries of the world with different applications (Amaral et al. 2025). A first estimation of such flows for the urban area of Lyon is made in Gonzalez-Feliu (2018). Table 1 presents an update of this estimation.

Table 1 Overall estimation of freight flows in the urban area of Lyon, from Gonzalez-Feliu's (2018) methodology, adding e-commerce flow estimation consideration as in Holguín-Veras et al. (2018a, b)

Feed Categories		Trips (%)		Distances (%)		Road occupancy (%)	
Year		2014	2018	2012	2018	2012	2018
Inter-establishment travel	Shops	16%	14%	6%	6%	10%	9%
	Services	29%	24%	9%	8%	17%	15%
	Other generators	16%	13%	10%	8%	13%	11%
Travel to supply the end consumer	Shopping trips	27%	22%	63%	55%	41%	34%
	Home deliveries*	5%	17%	5%	15%	6%	17%
	Relay point deliveries	3%	5%	1%	3%	2%	5%
Flow for urban management		3%**	4%**	5%	5%	10%	8%
Intra-establishment flows		No direct contribution to urban traffic					
Flows not falling into previous categories		1%	1%	1%	<1%	1%	1%
Total		100%	100%	100%	100%	100%	100%

*Excluding meal deliveries (telephone and online orders)

**In flows for urban management, the collection of household waste cannot be counted in the number of trips but the distances and corresponding road occupancy rates are well estimated

2.2 *Freight Transport Demand Definitions and Characteristics*

Freight transport demand is defined as the set of needs in freight transport an actor has. In urban logistics, two main types of transport demand can be defined (Bastida and Holgúín-Veras 2009): freight trips, which represent the number of deliveries, or vehicles, that an actor expedites or receives, and commodity, being the quantity that this actor expedites or receives. This demand is characterized by various sets of characteristics, which are defined, in terms of information needed, by Gonzalez-Feliu (2019):

1. The first the set characteristics of what is received or expedited, i.e. the *shipment*. It is characterized by the following elements:

   1.1. Its origin (and then the nature, location and size of its shipper) and its destination (and the related information regarding the receiver).
   1.2. The type of commodity: nature, specific characteristics (fragile, fresh, etc.).
   1.3. The characteristics of the shipment: type and number of shipping units, quantity (weight, volume), specific needs to carry the shipment (specific material needed).

2. The second is more related to the logistics operation at the vehicle stop, i.e. the pickup or delivery operation. It is characterized by the following elements:

   2.1. The carrying material used for pickup or delivery operations, like chariots, mechanized engines, or no carrying material, among others.

2.2. The possible consolidation-deconsolidation schemes or more precisely the layout of the transport system used. Indeed, in urban logistics, consolidation or multi-stage transport is often required and needs to be considered when characterizing freight demand.
2.3. Temporal information of the operation, like date, hour or frequency of delivery, among others. This information is often obtained after executing the operation.
2.4. Parking conditions: double line, in legal space, in specific zones.

3. The third group of information is related to describe the vehicle, its characteristics and organizational issues, including the path it follows. Two main subgroups of information can then be identified:

   3.1. Vehicle characteristics: type (size, weight, engine characteristics), freight capacity (surface, weight, volume), vehicle autonomy (when applicable), special equipment.
   3.2. Delivery path (number of stops, duration of stops, travel path, speed, travel time), i.e. what characterizes the transport journey.

4. The last, but not least, group is that of external characteristics, mainly linked to the infrastructure or to the environment, mainly composed of the following subsets of elements:

   4.1. Road network: characteristics, vehicle capacity, peak hours, traffic, and incidents.
   4.2. Other external factors: weather conditions, strikes.

To obtain those data, it is important to define suitable data production methods. Data production englobes tools, methods and approaches that are used to generate data, in different manners. It goes beyond surveying or observing and includes a wide variety of methods. Those methods are categorized by the ways they contribute to generating data:

- By observation, which can be simple or complex, and need the implementation of observation guides (Gonzalez-Feliu 2019).
- By measurement, which can be manual or automatic and needs the deployment of specific devices, as captors, counters or energy measurement tools, among others (Pluvinet et al. 2012).
- By declaration, which can be made by reporting (company data, mainly confirming delivery execution), by quantitative surveys (which can be of different types, Holguín-Veras and Jaller 2013; Gonzalez-Feliu 2019) or by qualitative methods (individual or collective interviews, focus groups or Delphi analyses, among others.
- By estimation (using statistical modelling, analytical models, combinatorial optimization or simulation, Gonzalez-Feliu 2018).
- Randomly (mainly via probability analyses and generation; Gonzalez-Feliu et al. 2014; Lagorio et al. 2018).

Moreover, data can be of a different nature:

- Quantitative, i.e. data which is represented by numbers, that can be aggregated and combined using mathematics.
- Qualitative, of textual nature, which are transcriptions of writings and are grouped in corpuses and not databases, or included in databases in columns as pure qualitative and processed via text analysis methods.
- Categorial, which are of three types:
  - Non-ordinal, which defines categories that cannot be compared among themselves but represent different elements of an answer.
  - Ordinal from qualitative appreciations, which define a level of conformity/agreement from low to high, and can be quantified using scales of measure.
  - Ordinal from quantitative aggregation, when processing information where the difference between various values is considered as negligible, or when, for processing purposes, those data are aggregated into a number of categories within which data have similar values. Main examples of those data are age or frequency, among others.

2.3 Freight Transport Demand Modelling

In modelling freight demand, different stages need to be considered (Gonzalez-Feliu 2018):

1. Generation, in which demand needs are associated to urban logistics actors, i.e. generators. It can be done for only shippers (production), receivers (attraction) or to both jointly (generation).
2. Commodity O/D matrix construction, i.e. the definition of the delivery needs in terms of matching origins and destinations. More precisely, it includes the assignment of destinations to origin and construction of customer sets per origin.
3. Trip chain construction, i.e. the definition of trips, by assigning freight to a vehicle use type (FTL or LTL), to a vehicle (and its capacity) and the definition of direct trips or transport routes.
4. Distance estimation, which is derived from the trip chains defined. Those trip chains can also be included in traffic models to estimate traffic impacts.
5. Environmental issues forecasting, via pollution emission models.

Figure 1 presents the main variables used in such models and their interactions/relationships.

In this work, we deal with the generation phase. In this phase, two main logics of generation can be observed (Holguín-Veras et al. 2011): the generation of freight trips, which are related to vehicles and are examined in terms of transport needs (frequencies, vehicles and eventually contribution to congestion) and commodity, which are in general weekly quantities (in weight or volume) more related to the

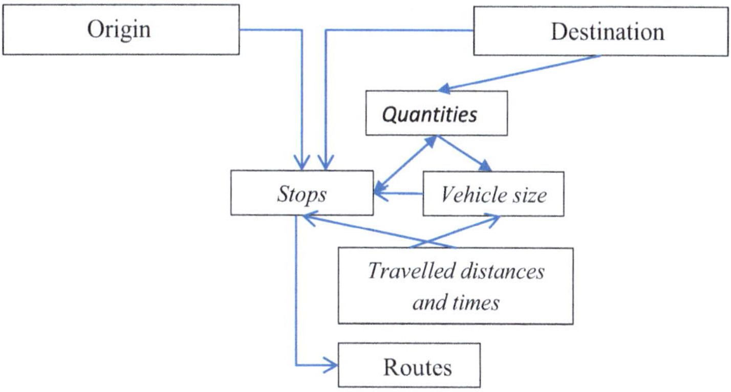

Fig. 1 Main variables and parameters to take into account when modelling urban goods flows (adapted from Gonzalez-Feliu 2019, chapter 2)

Table 2 Functional forms of FTG and FG (Gonzalez-Feliu 2024a, b)

Functional form	Formulation at the level of a single establishment
Constant	$F_i^a = k^a$
Linear	$F_i^a = K^a + a^a.x_i^a$
Logarithmic	$F_i^a = K^a + a^a.\log\left(x_i^a\right)$
Potential	$F_i^a = a^a.x_i^{ab^a}$
Exponential	$F_Z^a = a^a.e^{b^a.x_i^a}$

logistics needs of shippers and receivers. Those two logics are different and lead to different models and sometimes survey methods, but are related.

In both cases, the demand model needs the definition of the analysis of the functional forms that determine the generation logic. A functional form can be defined as the definition of the algebraic form of the function $f(X;a)$ that explains a phenomenon (or explained variable) Y where $Y = f(X;a)$. Those functional forms relate an explained variable (freight demand) to one or more explanatory variables, which account for the size of the establishment generating the demand, and are mainly the employment, the area, the revenue or the frontline, although the first one is the most widely used. The main functional forms used in freight demand are the following (Gonzalez-Feliu 2024a, b) (Table 2).

All those forms can be linearized and then calibrated using linear regression techniques (Gonzalez-Feliu and Sánchez-Díaz 2019). The main tests and validation procedures of linear regression are often used to assess the quality of a model (Holguín-Veras et al. 2011), although error measures, like MAPE and RMSE, are adopted for analyzing and comparing them (Gonzalez-Feliu and Sánchez-Díaz 2019).

3 Methodological Framework to Characterize and Model Freight Demand for a Metropolitan Area in Terms of Sustainable Urban Logistics

In this section, a general framework to both characterize and model freight demand in terms of sustainable urban logistics is proposed. This framework starts from the foundational works of Ambrosini and Routhier (2004), Holguín-Veras et al. (2011), Holguín-Veras and Jaller (2013) and Gonzalez-Feliu (2018) and is generalized for use in urban areas and applied to the metropolitan area of Paris (France).

The main purpose of the framework is to both quantify and qualify urban freight demand in a comprehensive way. Indeed, most surveys and models focus mainly on aggregated freight trip and commodity generation rates without an in-depth characterization of other aspects beyond the category of activities and the generation nature (production/attraction), although other parameters are important in freight trip and commodity generation, like the vehicles used, vehicle use mode (full truckload/less-than truckload), the mode of management (own account/third party), the parking place of the time slot of the operation, among others (Gonzalez-Feliu 2013). Previous works showed the importance of those issues (Ambrosini and Routhier, Objectives, methods and results of surveys carried out in the field of urban freight transport: an international comparison, 2004), but when presenting urban freight surveys and proposing demand models, methodologies and analysis are aimed at generating (at origin, destination or both) the transport flows in an aggregated way, and the distinction between vehicles or management modes appears after the generation phase, mainly in the form of a fixed, percentage distribution (González-Feliu et al. 2014).

The proposed analysis and modelling framework, which needs the deployment or the analysis of an establishment survey, can be structured as follows:

1. Data processing and preparation.
2. Data analysis, descriptive statistics.
3. Correlation and functional form suitability analysis.
4. Freight trip generation and trip generation modelling (Fig. 2).

To implement the framework, an establishment database and a generation survey are necessary. Different types of surveys, with different granularities, can be implemented (Holguín-Veras and Jaller 2013). However, a recent study showed that if the suitable generation patterns (functional form, categories, etc.) are identified, a limited set of information is enough to produce accurate models and analyses (Gonzalez-Feliu and Sánchez-Díaz 2019). In this work, the data used are those of the French Urban Goods survey of Paris (Enquête Marchandises en ville Régione

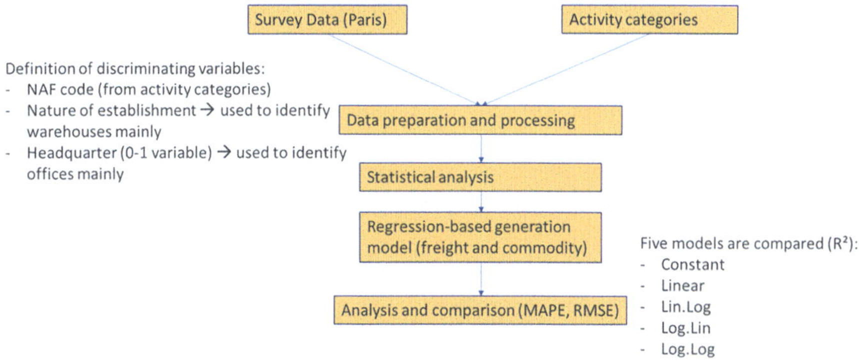

Fig. 2 Chart of the proposed methodology

Ile-de-France, ETMV-RIF). The survey, commanded by the Ile-de-France Region (Paris Metropolitan Area) and co-funded by the region, the French Ministry of the Environment and the French Agency of the Energy and the Environment, was carried out between 2011 and 2013. A first release, open to researchers and practitioners, was produced in 2014, for which permissions and a first access[1] were obtained in 2017.

The methodology of this survey is an update of that of the well-known surveys carried out in France in the mid-1990s (Ambrosini and Routhier 2004), and the survey led to the development of various freight demand models (Ambrosini et al. 2008, 2013; Gonzalez-Feliu et al. 2012; Gonzalez-Feliu and Sánchez-Díaz 2019; Holguín-Veras et al. 2021). The survey combines three connected data collection procedures, one for each of the following target populations (Ambrosini et al. 2010):

- The establishments.
- The drivers serving these establishments.
- The transporters operating in the study area.

The establishment surveys contain various tables, from which we used the two main ones: the establishment survey, which contains 1188 adjusted and verified establishments and the operations table, having a total of 5089 validated operations and characterizing them. The establishment survey includes, among others, the main characteristics of the establishment (category of activity, location, employment and area mainly) and the operations table contains, for each establishment, the characteristics of each type of operation associated with it:

- Frequency (number of operations per week).

[1] The survey access is managed by Région Ile-de-France and was carried out by a consortium following the methodology of Laboratoire d'Economie des Transports (Toilier et al. 2016). Authors got the necessary permissions of Région Ile-de-France to analyze the data, as released by Laboratoire d'Economie des Transports in 2014, and guarantee that all analyses and transformations respect the statistical secrent and the French RPGD protocol.

- Mode of management (third-party transport TPT, shipper's own account SOA and receiver's own account ROA).
- Vehicle type (including bikes/cargo bikes, light vehicles LV, light goods vehicles LGV, single trucks and articulated trucks, other types of vehicles having a small number of answers and being grouped by the survey operator with the non-responses).
- Vehicle use (full truckload FTL or direct transport, less than truckload LTL or routes).
- Quantities delivered (freight in Kg and volume in m^3).
- Place of parking of the delivery vehicle.
- Time period of the delivery.

To process the data, an SQL query was used to extract the necessary information. For frequencies, they were aggregated to define for each establishment a set of freight trip rates per week. In the same way, weights and volumes were weighted by the frequency to define for each establishment a weekly quantity delivered (in Kg and in m^3).

Data are processed and analyzed using descriptive statistics. Monovariate and bivariate statistics are proposed to make a comprehensive analysis of the demand, and a dispersion analysis is made to assess the suitability of demand generation models. Finally, a correlation analysis of employment and area is also made.

After that, three models are assessed, one for Freight Trip Attraction (FTA), one for Freight Trip Production (FTP), and a third one for Freight Trip Generation (FTG) without precising attraction and production. These models are built and calibrated using the same database used for the clustering analysis, on the obtained categories. The results of this model are assessed using MAPE and RMSE and discussed, relating them to previous work.

The proposed methodology is adapted from the works of Gonzalez-Feliu and Sánchez-Díaz (2019) and follows a similar sequencing, although the categories being defined in the previous step, the main focus is on building the FTA, FTP and FTG models. For each category and type of model (attraction, production and generation), the procedure is the following:

1. Assessment of four functional forms (as on Sanchez-Diaz et al. 2016; Gonzalez-Feliu and Sánchez-Díaz 2019): constant estimation (using the method in Holguín-Veras et al. 2011).
2. Assessment of R^2 and linear regression metrics for each model.
3. Choice of the model with higher R^2 verifying all latent hypotheses (Gonzalez-Feliu et al. 2020).
4. Analysis and comparison of the different models.

Concerning the data preparation and processing, the categories of establishments presented in Gonzalez-Feliu et al. (2024) are selected since they constitute the minimum number of categories that characterize urban freight transport while ensuring both exhaustivity and representativity. For each of those categories, operations are grouped in receptions, expeditions and joint operations (a joint operation is a reception that is immediately followed by an expedition, so an operation including a joint reception and expedition). Data also show modes of management, which are of three types: third-party transport (TPT) includes deliveries made by professional transport carriers, whereas own account can be of two types: by the receiver (receiver's own account or ROA), or by the shipper (shipper's own account or SOA). As shown by Cattaruzza et al. (2017), transport paths made by professionals (TPT) and by shippers' in own account (SOA) share many characteristics, since companies shipping in own account with trucks need to have a transport license and deploy freight transport processes by law. As for vehicles, the survey includes various categories of vehicles, including trucks, light vehicles, but also cargo cycles or barges, among others. An overview of the data and a first descriptive analysis of the database extracted and processed are presented in the next section.

4 Analysis of Freight Demand in Paris Metropolitan Area from a Sustainability Viewpoint

In this section, we present the main results of the application of the proposed framework to the metropolitan area of Paris. For the first time, a descriptive analysis of results is proposed. We can see in Table 3 the main characteristics of the economic activities of the area considered. We consider the five categories of activities as proposed in (Gonzalez-Feliu et al. 2024) as the minimum number of categories which present enough homogeneity to propose high-accuracy freight trip and freight generation models. We observe that retailing/stores and craftsmen/services represent respectively the two highest categories in both number of establishments and of employees, accounting to about two-thirds of the total number of establishments in Paris and 83% of the employment, which confirms the estimations shown in Holguín-Veras et al.

Table 3 Mean workforce, area and trip generation rates for the 5-category activity classification

Category	No. of Premises in the area	No. of Employees in the area	Mean workforce	Mean area (m^2)	Mean Trips/p	Mean Exp.	Mean Rec.	Mean joint
Agriculture	10,293	10,520	6.3	1578.5	3.6	0.6	1.8	1.2
Craftsmen/ services/ tertiary	341,167	259,232	87.4	2377.7	14.1	6.2	6.6	1.4
Industry/ wholesaling	111,726	114,898	46.1	3231.3	39.6	12.3	23.8	3.5
Retailing/ stores	394,154	392,579	36.2	1225.3	15.8	10.7	3.0	2.2
Warehousing/ transport	3394	2517	46.8	48,209.5	222.0	60.8	151.4	9.8
Overall	860,734	779,746	**47.8**	**3638.7**	**30.3**	**12.2**	**15.4**	**2.7**

(2018a, b). Agriculture and warehousing/transport represent a small number of establishments (less than 2%), this last category being very interesting since it represents only 0.3% of the establishments and workforce but due to its high generation rates (more than 200 trips per week and establishment), it accounts for 5% of the trips, those being mainly made by single trucks, which result into a non-negligible impact to road occupancy. Retailers and services contribute to near 70% of the overal trip rates per week, which justifies the high interest that public planners make to those activities, but in all the urban area (not only in the city centre). Moreover, industry and wholesaling represent 13% of the establishments (mainly in periphery), 15% of the employment, and 27% of the trip generation rates with the double of expeditions than receptions. This is mainly due to the fact that those establishments receive freight in big trucks and semi-articulated vehicles and send them to retailers in smaller vehicles.

To address the issue of vehicles, Tables 2 and 3 compile respectively the total number of trips (both receptions and expeditions) and their percentages in Paris metropolitan area for the same five categories as in Table 1. About 32% of the trips fall in the category others/non-indicated (which is mainly made of trips not characterized in terms of vehicle, and includes a small amount of trips made by barges, trains and other vehicles). This categorization includes cyclo-cargos, bikes and motorbikes (less than 3§ of the weekly trips in the area, and mainly in the city centre), Light vehicle (LV), i.e. vehicles with less than 2.5 Tons of total authorized weight (19%), light goods vehicles (LGV), i.e. vehicle up to 3.5–4 Tons (21%), single trucks (about 21%) and articulated trucks (about 5%). That means that the majority of freight transport is made by small vehicles (about 40%), some of them electric. Unfortunately, the survey did not include this information to state on the pollution impacts but some simulation works are dealing with this issue based on the same survey (Coulombel et al. 2018) (Table 4).

We observe that industry and wholesaling are those sectors that generate a higher number of trips in trucks (both single and articulated, about 11% of total trips together), and account for about 10% of trips non-identified or with other modes, being the main sector having fluvial or railway transport in urban areas. Moreover, craftsmen and service present about 13% of trips uncharacterized in terms of vehicles. Cargo-cycles are mainly used by craftsmen and postal services, but also by wholesalers, with a small application (mainly receptions) to retail distribution. Moreover, light vehicles (of both categories) are mainly used by craftsmen/services and retailers, but industry/wholesalers also see a non-negligible utilization (Table 5).

Concerning the mode of management, only 2% of the trips do not indicate it, which can be included as part of the statistical uncertainty, so we can take out them from the estimation. For that reason, we adjust the estimation of weekly trips an present them in Table 6.

Third-party transport represents about 50% of the total weekly trips, with industry/wholesaling and craftsmen/services being the main trip generators in this mode (20% and 17% of the total weekly trips, respectively). Concerning their own account, retailers, industry and craftsmen/services have similar percentages (about 5% of the total for each of them concerning receiver's own account, ROA and around 11% each regarding shippers' own account, SOA). Overall, the shipper's own account, which presents similar characteristics to third-party transport,

Table 4 Total number of trips per week for the main types of vehicles, per category in the 5-type classification

Category	Cyclo-cargo	LV	LGV	Articulated	Truck	Others/n.i.	Overall
Agriculture		7440.8	11,522.5	18.4	1701.4	1059.1	21,742.2
Craftsmen/services/tertiary	56,517.5	265,220.9	281,007.9	41,530.9	168,278.6	561,196.4	1,373,752.2
Industry/wholesaling	50,309.4	254,568.3	309,429.7	91,501.3	412,005.7	455,393.6	1,573,208.1
Retailing/stores	8417.7	273,792.6	294,684.4	38,387.9	262,104.9	208,682.3	1,086,069.8
Warehousing/transport		2866.1	5871.1	23,376.5	34,590.2	137,285.2	203,989.1
Overall	**115,244.7**	**803,888.7**	**902,515.5**	**194,814.9**	**878,680.9**	**1,363,616.7**	**4,258,761.3**

Table 5 Percentage of trips per week for the main types of vehicles, per category in the 5-type classification

Category	Cyclo-cargo	LV	LGV	Articulated	Truck	Others/ n.i.	Overall
Agriculture		0.2%	0.3%	0.0%	0.0%	0.0%	0.5%
Craftsmen/services/ tertiary	1.3%	6.2%	6.6%	1.0%	4.0%	13.2%	32.3%
Industry/wholesaling	1.2%	6.0%	7.3%	2.1%	9.7%	10.7%	36.9%
Retailing/stores	0.2%	6.4%	6.9%	0.9%	6.2%	4.9%	25.5%
Warehousing/transport		0.1%	0.1%	0.5%	0.8%	3.2%	4.8%
Overall	2.7%	18.9%	21.2%	4.6%	20.6%	32.0%	100%

Table 6 Number of trips per week per mode and mode of management

Category	TPT	ROA	SOA	Overall
Agriculture	2037.6	9503.7	10,737.4	22,278.7
Craftsmen/services/tertiary	721,321.8	211,136.7	454,828.8	1,407,650.5
Industry/wholesaling	881,865.4	204,895.3	480,263.9	1,612,028.1
Retailing/stores	425,242.7	199,251.9	459,085.6	1,112,869.3
Warehousing/transport	127,530.9	20,709.6	50,349.9	209,022.6
Overall	**2,157,998.4**	**645,497.3**	**1,455,265.6**	**4,363,849.2**

represents about 35% of the total weekly trips, and the receiver's own account (the most difficult to optimize and the least sustainable), 15%.

5 Modelling Results

To address potential demand for sustainable transport solutions, various models are tested and compared. First, a model for freight transport demand on retailers susceptible to be rendered more sustainable is constructed and calibrated. Thus, demand being able to be transported by cargo cycles (i.e. that being nowadays transported by small vehicles) is analyzed and tested in terms of demand generation modelling.

After testing the possible models, we obtained a freight trip attraction model combining both SOA and TPT flows, and for freight currently transported in small vehicles (up to 3.5–4.5 T, i.e. LV and LGV). Two models, for attraction (only receptions), were tested: one for freight trips and the second for commodity, in weight. From the five categories, only three presented significant results: industry/wholesalers, retailing/services and craftsmen/services. Agriculture did not present a high quantity of data; thus, the results were not significant. Concerning Warehousing and Transport, they are made by trucks or big vehicles, and data concerning small vehicles were not enough to provide statistically significant models. The resulting models on the three categories retained are presented in Let us start with demand attraction. Table 7 presents the best models obtained for demand attraction. For each category, two models are presented, one for trip attraction (i.e. relating the weekly number of trips received by an establishment to its employment rates) and one for commodity production, also known as freight production (i.e. relating total weight received weekly by an establishment to its employment). For each model, its functional form, the R^2 value (stating about the quality of the linear relationship, being also seen as a proxy of the model's accuracy) as well as the parameters obtained in the regression (first value in the corresponding cells is the parameter value, and the second value inside a parenthesis is the t-value of the test of Student carried out) and the F-value of the Fischer's test. The whole set of values allow to state about the quality, accuracy and significance of each model.

Tables 7 and 8 present the proposed models for demand attraction (i.e. reception or deliveries at destination) and demand production (i.e. expeditions or deliveries at origin), respectively. Joint deliveries did not result in significant models because of the small set of data available (24–26 individuals in terms of trips and 10–12 in terms of weights, which led to Fisher tests being unsatisfactory). Those flows should be analyzed more in-depth in further works with an aim of integrating them into current models.

We observe low values of R^2 in all models, which remain acceptable according to Gonzalez-Feliu and Sánchez-Díaz (2019) and Gonzalez-Feliu et al. (2024). Indeed, models when segmenting more precisely the demand remain on the same order of magnitude than aggregated models (presented in Gonzalez-Feliu et al. 2024). Indeed, if we compare the accuracy of the proposed models to that of models

Table 7 Proposed models for demand attraction in relevant activity categories regarding flows in third-party transport or shipper's own account (together) transported in LVs and LGVs

Category	Model	Function	R^2	Constant	Coefficient	F-value
Industry/wholesaling	Trips	Logarithmic	0.25	–	1.572 (7.275)	52.929
	Commodity	Potential	0.42	–	1.004 (8.004)	64.060
Retailing/stores	Trips	Logarithmic	0.35	–	1.823 (9.498)	90.216
	Commodity	Logarithmic	0.43	–	1.284 (8.841)	78.161
Craftsmen/services	Trips	Logarithmic	0.17	–	1.386 (6.021)	36.251
	Commodity	Potential	0.27	–	0.674 (5.952)	35.423

*In parentheses, the t-value for the Student's test

Table 8 Proposed models for demand production in relevant activity categories, same demand segmentation (vehicles, mode of management)

Category	Model	Function	R^2	Constant	Coefficient*	F-value
Industry/wholesaling	Trips	Potential	0.44	–	0.447 (7.322)	53.613
	Commodity	Potential	0.63	–	1.476 (8.653)	74.873
Retailing/stores	Trips	Logarithmic	0.27	–	1.757 (3.630)	13.179
	Commodity	Potential	0.23	–	1.203 (2.393)	5.727
Craftsmen/services	Trips	Logarithmic	0.15	–	1.967 (5.205)	27.088
	Commodity	Potential	0.44	2.135 (9.229)	0.994 (7.785)	60.607

*In parentheses, the t-value for the Student's test

without segmenting demand, we observe that for retailing/stores R^2 is higher (the non-segmented model's R^2 is 0.30) but remain of same order of magnitude. The model for industry/wholesaling with non-segmented demand presents an R^2 of 0.61, which is quite higher than that of the segmented demand model. Finally, the model for craftsmen and services is less accurate (0.17) but remains in the orders of magnitude as presented in Gonzalez-Feliu et al. (2024). Indeed, trip attraction model R^2 vary from 0.15 to 0.61 in the non-segmented demand approach, and here we observe, for trip attraction, R^2 between 0.14 and 0.6, knowing that the quantity of data used here is lower than that in Gonzalez-Feliu et al. (2014) since some lines had to be eliminated during data processing, and this data preparation process derives on approximations and aggregations that do not add significant bias. Thus, we can state that the segmented demand models remain at similar accuracies globally. In all cases, the logarithmic functional form was the one giving the most accurate models.

Concerning freight attraction, an estimation of non-segmented demand models leads to R^2 between 0.27 and 0.74. Segmented demand models in this work obtain R^2 of 0.41 and 0.43, respectively, which remain in the lower range but within the order of magnitude of such models. In the case of freight attraction (in weight), two models are potential (industry/wholesaling and craftsmen/services) and one is logarithmic (retailing).

If we apply the same logic to demand production (i.e. the origin of trips), we obtain the following relevant models for the same categories and same flows (Table 8). Retailing/stores, which have a smaller number of declared and complete expeditions, present tests which are at the limit of acceptance. For trip production, the F-value is low but remains over the limit that allow to validate it, but for commodity, both tests (Fischer and Student) are closer but lower to the limit value. The functional forms of production models remain exactly the same as for the attraction ones, showing the stability of segmented demand models. Only in craftsmen/services freight production (i.e. commodity), a constant term is found; all other models do not have a constant term.

Concerning the accuracy of the models, freight trip production presents R^2 between 0.15 and 0.44. They are, as for trip attraction, within the ranges of models presented in Gonzalez-Feliu et al. (2024). Concerning commodity models, industry/wholesaling leads to a good accuracy model, $R^2 = 0.63$ being considered high in this context (Gonzalez-Feliu and Sánchez-Díaz 2019). The other two models remain in the orders of magnitude of the freight attraction models defined here, but also of those observed in freight production in literature (Gonzalez-Feliu et al. 2024).

Craftsmen/service is in general the category with the lowest R^2 (except for freight production), mainly because of the heterogeneity of the category but also due the quantity and quality of data collected. Retailing/stores present solid models in terms of attraction but not in production, since this category is essentially receiving deliveries, most expeditions being returns and joint operations (i.e. a reception and an expedition for the same truck stop) do not present enough data for a detailed analysis. Industry/wholesaling presents the most accurate and robust models, mainly due to the high standardization of operations and practices in the sector.

The demand produced by industry and wholesaling has, in general, two main types of destinations: one is wholesaling or warehousing, and the other is retailing. A high part of the commodity expedited by this category is received by retailers and stores. Knowing the origins and the destinations of this demand is the first step in the deployment of new modes and organizations of transport that have a sustainable perspective. In this application, the demand transported by light commercial and tourism vehicles (which capacities are in general around 0.6 to 0.9 tons) are easily transferrable to cargo-bikes (recent developments see cargo-bike convoys able to transport up to 0.7 tons) or to electric vehicles up to 1.2–1.3 tons of capacity (a light vehicle has a capacity of 0.6 tons, a light commercial vehicle can raise this capacity to 1–1.2 tons). Knowing the frequencies and quantities transported allows then to define scenarios of development and anticipate planning and development for sustainable urban freight solutions.

6 Conclusion

This chapter proposed a comprehensive analysis of freight transport demand in Paris metropolitan region in a logic of segmenting demand to find the most suitable ways of transferring it to small, green vehicles such as cargo bikes and electric vehicles.

Results showed that demand needs to be characterized by considering sustainability patterns, which is not yet done in many works. The vehicle used to transport has an impact on the frequency, weight and volume of the deliveries, as well as on times taken to unload and deliver, and could also impact other characteristics, although data regarding those characteristics is not yet collected. The mode of management is also important, and the receiver's own account deliveries have characteristics that differentiate them from professional deliveries (either made by the shippers, who usually have transportation licenses, or by transport professionals).

The interests and utility of such models is that they allow to define and characterize the demand that can be switched to small vehicles, and then lead a discussion and planning issues concerning the needs of receivers and the possibilities of deploying greener vehicles, such as cargo-bikes (which can nowadays reach similar capacities than the vehicles considered for the segmented demand) or electric commercial vehicles.

This work, which is the first step in a more general demand segmentation, categorization and modelling logic, presents some limitations, since data do not allow us to examine some of the generation determinants. Moreover, only a part of the models is examined. Further developments see the deployment of a general, unified modelling framework, the proposal of complementary surveys to include explicitly sustainability patterns in freight demand modelling, and the use of such models in planning and analyzing transformative freight transport systems.

References

Ackoff R (1970) A concept of corporate planning. Long Range Plan 3:2–8
Amaral J, Holguín-Veras J, Roorda MJ, Reda AK, Gardrat, de Oliveira LK et al (2025) Freight trip generation in metropolitan areas: An international perspective. In: The 104th Annual Meeting of the Transportation Research Board, Washington D.C., January 5–9
Ambrosini C, Routhier J-l (2004) Objectives, methods and results of surveys carried out in the field of urban freight transport: an international comparison. Transp Rev 24:57–77
Ambrosini C, Routhier J-L, Sonntag H, Meimbresse B (2008) Urban freight modelling: a review. Innov City Logist Nova Sci Publ, New York, pp 197–211
Ambrosini C, Patier D, Routhier J-L (2010) Urban freight establishment and tour based surveys for policy oriented modelling. Procedia Soc Behav Sci 2:6013–6026
Ambrosini C, Gonzalez-Feliu J, Toilier F (2013) A design methodology for scenario-analysis in urban freight modelling. Eur Transp/Trasporti Europei 54:1–21
Bastida C, Holguín-Veras J (2009) Freight generation models: comparative analysis of regression models and multiple classification analysis. Transp Res Rec 2097:51–61
Benjelloun A, Crainic TG (2008) Trends, challenges, and perspectives in city logistics. In: Transportation and land use interaction, proceedings TRANSLU, vol 8, pp 269–284
Cattaruzza D, Absi N, Feillet D, & González-Feliu J (2017) Vehicle routing problems for city logistics. EURO Journal on Transportation and Logistics, 6(1), 51–79
Comi A, Delle Site P, Filippi P, Nuzzolo A, others (2012) Urban freight transport demand modelling: a state of the art. Eur Trans/Trasporti Europei 2012
Coulombel N, Dablanc L, Gardrat M, Koning M (2018) The environmental social cost of urban road freight: evidence from the Paris region. Transp Res Part D: Transp Environ 63:514–532
Crainic TG (2008) City logistics. In: Chen Z-L, Raghavan S (eds) State-of-the-art decision-making tools in the information-intensive age. INFORMS, pp 181–212
Gonzalez-Feliu J (2013) Multi-stage LTL transport systems in supply chain management. In: Cheung J, Song H (eds) Logistics: perspectives, approaches and challenges. Nova Science Publishing, pp 65–86
Gonzalez-Feliu J (2018) Sustainable urban logistics: planning and evaluation. Wiley
Gonzalez-Feliu J (2019) Logistics and transport modeling in urban goods movement. IGI Global
Gonzalez-Feliu J (2024a) Considerations on the use of approximations to estimate freight demand on the basis of a single explanatory variable with non-linear functional forms and generalization to statistical modelling. Theor Econ Lett 14:2158–2178
Gonzalez-Feliu J (2024b) Equivalences of freight trip generation functional forms: formulations, approximations and error formalization issues. Theor Econ Lett 14:617–631
Gonzalez-Feliu J, Peris-Pla C (2017) Impacts of retailing attractiveness on freight and shopping trip attraction rates. Res Transp Bus Manag 24:49–58
Gonzalez-Feliu J, Routhier J-L (2012) Modeling urban goods movement: how to be oriented with so many approaches? Procedia Soc Behav Sci 39:89–100

Gonzalez-Feliu J, Sánchez-Díaz I (2019) The influence of aggregation level and category construction on estimation quality for freight trip generation models. Transport Res E Logist Transport Rev 121:134–148

Gonzalez-Feliu J, Ambrosini C, Pluvinet P, Toilier F, Routhier J-L (2012) A simulation framework for evaluating the impacts of urban goods transport in terms of road occupancy. J Comput Sci 3:206–215

González-Feliu J, Cedillo-Campos MG, García-Alcaraz JL (2014) An emission model as an alternative to OD matrix in urban goods transport modelling. Dyna 81:249–256

Gonzalez-Feliu J, Semet F, Routhier J-L (2014) Sustainable urban logistics: concepts, methods and information systems. Springer

Gonzalez-Feliu J, Palacios-Argüello L, Suarez-Nuñez C (2020) Links between freight trip generation rates, accessibility and socio-demographic variables in urban zones. Arch Transp J Polish Acad Sci 53:7–20

Gonzalez-Feliu J, Ramirez-Rios DG, Pani A, Bouraoua S (2024) A categorization-modelling procedure to estimate urban goods transport demand with a minimum number of activity categories. Transport Res Proc 79:108–115

Holguín-Veras J, Jaller M (2013) Comprehensive freight demand data collection framework for large urban areas. In: Gonzalez-Feliu J, Semet F, Routhier J-L (eds) Sustainable urban logistics: concepts, methods and information systems. Springer, pp 91–112

Holguín-Veras J, Jaller M, Destro L, Ban X, Lawson C, Levinson HS (2011) Freight generation, freight trip generation, and perils of using constant trip rates. Transp Res Rec 2224:68–81

Holguín-Veras J, Campbell S, González-Calderón CA, Ramírez-ríos D, Kalahasthi L, Aros-Vera F et al (2018a) Importance and potential applications of freight and service activity models. In: Taniguchi E, Thompshon RG (eds) City logistics 1: new opportunities and challenges. Wiley, pp 45–63

Holguín-Veras J, Ramírez-ríos D, Encarnación T, Gonzalez-Feliu J, Caspersen E, Rivera-González C (2018b) The generation of freight and service activity. In: Browne M, Behrends J, Woxenius G, Holguín-Veras J (eds) Urban logistics: management, policy and innovation in a rapidly changing environment. Kogan Page Publishers, pp 19–51

Holguín-Veras J, Leal JA, Sánchez-Diaz I, Browne M, Wojtowicz J (2020) State of the art and practice of urban freight management: part I: infrastructure, vehicle-related, and traffic operations. Transp Res A Policy Pract 137:360–382

Holguín-Veras J, Ramirez-Rios D, Pérez-Guzmán S (2021) Time-dependent patterns in freight trip generation. Transp Res A Policy Pract 148:423–444

Lagorio A, Gonzalez-Feliu J, Pinto R (2018) Urban freight demand estimation: A probability distribution based method. In: Proceedings of the 7th international conference in information systems, logistics and supply chain, pp 415–422

Lagorio A, Gonzalez-Feliu J, Pinto R (2023) REMOVING ROADBLOCKS: converging quality and urban logistics to improve urban goods transport. Qual Prog 56:32–39

Marcucci E, Gatta V, Le Pira M (2023) Handbook on city logistics and urban freight. Edward Elgar Publishing

Nimtrakool K, Gonzalez-Feliu J, & Capo C (2018) Barriers to the adoption of an urban logistics collaboration process: a case study of the Saint-Etienne urban consolidation centre. City Logistics 2: Modeling and Planning Initiatives, 313–332

Pluvinet P, Gonzalez-Feliu J, Ambrosini C (2012) GPS data analysis for understanding urban goods movement. Procedia Soc Behav Sci 39:450–462

Sánchez-Díaz I, Holguín-Veras J, & Wang X (2016) An exploratory analysis of spatial effects on freight trip attraction. Transportation, 43(1), 177–196

Toilier F, Serouge M, Routhier J-L, Patier D, Gardrat M (2016) How can urban goods movements be surveyed in a megacity? The case of the Paris region. Transport Res Proc 12:570–583

Mixed Use of Urban Space for City Logistics

Marion Cottet

1 Problem Formulation

Urban space is a fundamental component of citizens' daily lives, structuring the ways in which people move, interact, and access essential services. It serves not only as a physical setting but also as a social and functional framework that supports everyday routines and community life. The way these spaces are designed and how the different activities they host are organized have a direct impact on the liveability of the city, influencing perceptions of safety, accessibility, and comfort.

Logistics plays a crucial role in the everyday functioning of cities. It ensures that shops are stocked, restaurants are supplied, construction sites receive materials, and healthcare facilities obtain equipment, in accordance with fluctuating demand. Efficient logistics is measured by its invisibility and seamlessness: when it functions well, it goes unnoticed. To preserve this performance in dense urban settings, city logistics need to be proactively integrated into urban planning and regulatory frameworks.

In the design and governance of cities, traditional urban planning tends to prioritize passenger transport, residential and commercial development, without integrating properly the consequential freight and service flows. This has led to a range of negative externalities: double parking, congestion—including additional greenhouse gas emissions and noise pollution, conflicts over curbside space, and inefficient use of valuable land. Over the past decade, the emergence of sustainable urban logistics plans (SULPs) has marked a positive evolution. These instruments have helped bring urban freight into the policy conversation, providing frameworks for

M. Cottet (✉)
Alliance for Logistics Innovation through Collaboration in Europe – ALICE, Brussels, Belgium
e-mail: marion.cottet@etp-alice.eu

A. Melkonyan-Gottschalk, Y. Li (eds.), *Urban Logistics Transformation*, Greening of Industry Networks Studies 15,
https://doi.org/10.1007/978-3-032-08387-6_13

diagnosing local challenges and articulating more coherent, context-specific strategies. However, while SULPs represent an important step forward, logistics is often still treated as a downstream operational issue, rather than as a structuring component of urban systems. Bridging this gap requires a paradigm shift: from logistics as an afterthought to logistics as a spatial and strategic dimension of urban life.

Recent trends in urban planning have increasingly emphasized the need to create more sustainable, less polluted, and less congested urban environments. This shift is driven by growing concerns about climate change, public health, and the quality of urban life. Local governments have responded by implementing a range of policy measures aimed at reducing the environmental impact of transportation and encouraging more sustainable mobility patterns. Notable examples include the establishment of low-emission zones (LEZs), which restrict access to certain areas for vehicles that do not meet specific environmental standards, thereby reducing air pollution and greenhouse gas emissions. In addition, cities are redesigning their street networks to prioritize non-motorized modes of transportation by introducing pedestrian zones, expanding networks of bicycle lanes, and enhancing public transport infrastructure. LEZs represent a concrete regulatory tool to enable more sustainable logistics practices, a lever to implement innovative solutions which require less space and energy, but which operational model might still be under construction.

Simultaneously, citizens' consumption behaviours are undergoing significant transformations, largely driven by the rapid expansion of e-commerce and the new practices it entails. Online shopping has become deeply embedded in everyday life, reshaping the way goods are purchased and delivered within urban areas. This shift has led to a substantial increase in home deliveries, frequent returns of packages, and a growing consumer expectation for fast, even same-day, delivery services. The logistical challenge of meeting these demands within constrained environments often conflicts with sustainability goals, as the proliferation of delivery vans needed to meet the demand can exacerbate congestion and pollutant emissions.

This reveals a fundamental challenge for urban logistics. On the one hand, demand for logistics services is surging, driven by e-commerce and other delivery-intensive sectors. On the other hand, cities are implementing stricter access regulations to promote sustainability, reduce emissions, and improve urban liveability. This creates rising pressure on land use, particularly in dense, high-cost areas, where logistics infrastructure competes with other urban functions. Urban planning lacks integration of such challenge.

In this context of scarcity of resources and space, multiple and mixed use of urban facilities resonates as a smart way to improve city logistics. This concept of mixed-use of space allows underutilized areas, such as parking lots, loading zones, or curbside, to be used for deliveries at specific times, while serving other functions the rest of the time—or divide the space among them. By making better use of existing space, cities can reduce delivery-related traffic, cut emissions, and improve access to goods, all without disrupting daily urban life. In this way, mixed-use space becomes a useful tool for building more flexible and sustainable urban logistics.

Mixed-use of space comes in several forms. The intent of this chapter is to focus on solutions with low need for infrastructure development, exploring how existing operating facilities, curbside, parking lots, and even open public space can be used as an enabler for more sustainable and efficient logistics, contributing to the liveability of cities.

2 Literature Discussion: Urban Logistics, Between New City Practices and Scarcity of Urban Space

Urban logistics plays a key role in sustaining city life by ensuring the flow of goods to businesses, services, and residents. As demand expands and urban populations grow, the importance of efficient and reliable delivery systems has only increased. As the volume and complexity of urban logistics continue to intensify, so do its externalities, raising interlinked challenges on environmental, social, and economic levels.

Regarding the environmental aspect, the impact of urban freight is quite important in cities. Estimates show that urban logistics is responsible for 20% of transport-related CO_2 emissions, and up to 50% of air pollutant particles. Beyond emissions, freight traffic can generate important noise pollution (Cerema 2015).

On a social level, urban logistics can potentially be the source of several negative externalities. Double parking, for example, can bother other users of the public space, and generate additional congestion. This disturbs citizens and peacefulness of public space, while also generating stressful working conditions for delivery drivers, who are already undergoing time-pressure restrictions. The design of urban space, including appropriate space to host logistics activities, plays an important role in generating such externalities.

From an economic perspective, last-mile logistics undergo significant challenges. A number often mentioned in literature, as originally cited by Goodman (2005), highlights that urban logistics can represent up to 28% of the overall cost of freight transportation of a merchandise. This is a consequent number which emphasizes the potential of improvement of the sector, combining both sustainability and profitability—in conjunction with the European Union's approach on competitiveness, resilience, and decarbonation (European Commission 2025).

Across Europe, cities are tackling the need to adapt the way they function due to climate change. Faced with air pollution, traffic congestion, and the urgent need to reduce carbon emissions, many are rethinking how space is used, how people move, and how goods are delivered. In this context, policies like low-emission zones (LEZs) and restricted access areas such as pedestrian zones are becoming more common. These efforts aim to make cities more sustainable and more pleasant to live in, but they also bring new challenges for urban logistics.

Low-emission zones limit access to certain areas of the city to the most polluting vehicles, from private cars to logistics trucks. As of 2022, more than 320 cities in

Europe had introduced some form of LEZ (Clean Cities Campaign 2022). Many are now planning to go further and create Zero-Emission Zones by 2030. Amsterdam, for example, has already started to implement its Zero Emission Zone for vans, trucks, and taxis in January 2025 (City of Amsterdam 2023). The consequence is that fewer motorized vehicles, and cleaner ones, circulate in the city—reducing air pollution, but also traffic and congestion. For delivery companies, it also means rethinking their operations: renewing their fleets, using new types of vehicles, and often relying on new processes, such as small distribution centres for last-mile delivery. Fewer cars in cities also signifies underuse of parking, both on and off-street, potentially liberating the space for other activities.

Pedestrian areas are also spreading in several European cities. Often situated in city centres, they favour accessibility to pedestrians and cyclists over motorized vehicles, which only have restricted access to these areas. They contribute to improving air quality, safety, and liveability of the space for citizens. An emblematic example is the superblocks in Barcelona, where several blocks are grouped into one area where car use is reduced, creating more room for walking, cycling, and community life (Rueda 2019). These types of areas require adaptation in the logistics processes, as motorized vehicles cannot access their delivery points (restaurants, shops, consumers) as easily. Operators need to adjust and use smaller, cleaner vehicles, changing delivery times (early morning or late night) or even delivery modes (parcel lockers, micro hubs at the outskirts of the restricted zone, etc.).

These new types of urban areas emphasize one already existing challenge for urban logistics, if not the biggest: the scarcity of space. Urban logistics are constrained by the lack of available space and high cost of land in dense city centres, where logistics functions compete with more profitable land uses such as residential development, shops, or office spaces. This competition has long pushed logistics facilities to peripheral areas, a phenomenon known as "logistics sprawl" (Dablanc and Rakotonarivo 2010). However, the rise of e-commerce and the demand for rapid delivery have renewed interest in reintegrating logistics infrastructure into urban cores. As Buldeo Rai et al. (2022) highlight, this shift has led to the development of "proximity logistics" (taking the shape of small facilities, e.g. micro-hubs), bringing back logistics in city centres, despite the persistent pressure from land cost. Despite the growing need, real estate markets still prioritize tenants who can pay higher rents, leaving few suitable options available in central locations for urban logistics (CBRE 2022).

Operators and urban planners need to be inventive to face this challenge. A noteworthy alternative lies in the activation of "idle real estate" (Schachenhofer et al. 2023), which are underutilized or time-restricted urban spaces such as (partly) vacant buildings, parking structures, or areas with limited daily activities. These residual spaces can be mobilized temporarily for logistics operations, when not used for their first purpose. In this chapter, this practice is referred to as "mixed use of space". The value of mixed use of urban space for logistics lies in its capacity to address the scarcity of space and high land cost by mobilizing existing, underutilized infrastructure. As car parking lots are becoming less utilized for private vehicles, the spare space becomes available for alternative uses, such as last-mile logistics. Other facilities, such as bus depots, are usually occupied only during specific time windows and can accommodate logistics operations at complementary times. In short, using urban space more flexibly can help accommodate logistics without necessarily requiring new construction or exclusive land use.

At the European level, this principle of concrete integration of logistics within urban space resonates with the ambitions of the New European Bauhaus (European Commission 2021), which promotes a more inclusive, sustainable, and aesthetically engaging vision for cities. While logistics has long been treated as a necessary but intrusive function, integrating it into mixed-use environments aligns with this vision of urban space as a shared living environment. Logistics should not be pushed out of sight, but reimagined so that it coexists with housing, green spaces, and public life. Doing so would not only improve the efficiency of logistics but also contribute to a more coherent and equitable use of urban space.

This ambition is further proven by the presence of this principle in European research and innovation programs. Projects recently funded under Horizon Europe Programme, such as DISCO,[1] URBANE[2], and UNCHAIN[3] projects, integrate the optimization of existing assets and urban space for city logistics. Other projects, such as MoLo Hubs,[4] funded by Interreg Noth Sea, also embrace an integrated approach, mixing services for mobility and logistics with other essential activities for citizens.

Many large-scale, mixed-use solutions also exist. Two emblematic projects are implemented in Paris' city centre, *Chapelle International* and *Immeublé Inversé* by Sogaris, a logistics real estate operator.

Located in Paris' 18th arrondissement, *Chapelle International* is 41,500 m^2 facility, referred to as a logistics hotel. The site integrates space urban logistics operations (centralized on the ground floor, with a possible integration with rail network) with many other services such as community amenities (urban farm on the rooftop, sports facilities), office spaces, or a data centre (which waste heat is used to heat up the logistics hotel and surrounding buildings). This logistic hotel is an example of harmonious integration in the city's environment (Sogaris 2018).

Still in Paris' city centre, Sogaris also repurposed a former six-level underground parking into a logistics hub, but not only. Operational since 2022, the 1600 m^2 hub, referred to as *Immeuble Inversé* ("Inverted Building"), is used for storage, order picking, and transhipment to cyclo-logistics. At street level, a small space welcomes the public, and parcels can be dropped off, picked up, or returned there. Combining both off-street logistics operations and services for the local community for better urban integration (Sogaris 2022a, b).

These are major infrastructures that concretely contribute to making urban logistics more integrated into the city. Their design promotes a functional coexistence

[1] https://discoprojecteu.com/

[2] https://www.urbane-horizoneurope.eu/

[3] https://unchainproject.eu/

[4] https://www.interregnorthsea.eu/molo-hubs

between logistics activities and everyday urban life needs. However, delivering such ambitious solutions requires years of planning, substantial public and private investment, and a high level of coordination across sectors. They are strategic urban interventions that reflect long-term commitments to reshaping how urban logistics integrate into cities. These solutions are complementary to smaller-scale solutions.

Fundamentally, upstream thinking and anticipation of urban logistics operations in urban planning documents is increasingly recognized as a critical lever.

In Paris, the city has taken a proactive approach to reintegrating logistics into its dense urban core through the implementation of "location perimeters". These are designated zones within which new developments are required to include a logistics component or preserve existing ones. More than sixty such perimeters have been defined across Paris (Dablanc 2023). This initiative marks a clear shift from viewing logistics as a nuisance to recognizing it as an essential urban function and integrating it well into the city.

In Villeurbanne, France, the ZAC des Gratte-Ciel project stands out as a logistics-oriented urban development initiative, where logistics functions are integrated from the earliest planning stages. It is led by the SERL group for Lyon Metropole and the City of Villeurbanne. In addition to its ambitions for urban densification, with 1800 new residents, 1600 students, and 40 shops, the project anticipates the logistical needs of a dense, mixed-use neighbourhood. From the early design phase, logistics considerations have been fully integrated, with dedicated delivery areas planned into the ground floors of buildings, and a space reserved for a future microhub. The goal is to support integration of logistics flows (retail shops deliveries, waste collection, access of craftsmen, etc.) to preserve the local quality of life. This proactive and long-term approach demonstrates a very innovative integration of urban logistics in new neighbourhoods (Banque des Territoires 2024).

While these types of solutions are of key importance to build the future of integrated, sustainable, and liveable cities, this chapter aims to focus on practical, short-term solutions that complement them. These include lower-investment, flexible uses of underutilized urban spaces, often temporary, that offer adaptable responses to effectively integrate urban logistics in dense cities.

3 Case Studies: Overview of Mixed Use of Space for Urban Logistics Activities

In Table 1, "ephemeral" refers to a solution that does not have any physical presence in urban space, outside of the few minutes where the logistics operations take place. "Installed" refers to solutions that require a fixed infrastructure to physically remain in place. This metric indicates the footprint of the solution on urban space.

"Transhipment" refers to the transfer of goods from one mode of transport to another during the logistics journey of goods. "Last-mile delivery", on the other hand, is the final stage in the delivery process where goods are delivered to the end

Table 1 Example of use cases of mixed use of space for urban logistics activities

Type	Name	Type of installation	Type of logistics operation	Maturity of the solution
Curbside	Loading and unloading zones	Ephemeral	Last-mile delivery	Operational
	Microhubs on curbside	Installed	Transhipment	Experimental
	Automated lockers as microhubs	Installed	Transhipment	Experimental
Open urban space	Street traffic lane as a temporary unloading zone	Ephemeral	Last-mile delivery	Operational
	Ephemeral beehive	Ephemeral	Transhipment	Experimental
	Autonomous driving parcel locker	Ephemeral	Last-mile delivery	Experimental
Parking lot	Mobile pick-up points in supermarket parking lots	Ephemeral	Last-mile delivery	Operational
	Urban logistics hub in underground parking	Installed	Transhipment	Operational
Operating facility	Bus depots as day-time transhipment hubs	Ephemeral	Transhipment	Operational
	Exhibition Centre parking lot as a transhipment hub	Ephemeral	Transhipment	Experimental & operational

consumer. This describes the step of the logistics chain where the mixed use of space solution is employed.

The expressions "experimental" and "operational" indicate the stage of maturity of the solution, whether it has only been used in an experimental context or is already fully operational. If a solution is marked as "experimental" in the table, it does not mean that it will not turn operational in the future.

3.1 Curbside

The term "curbside" defines the linear space between the road and the sidewalk. Historically dedicated to on-street parking, it is today shared between several activities: parking (motorized vehicles, e-scooters, bikes), public transport stops (buses, trams), parklets, art or vegetal installations, café terrasse… and of course, logistics (Roe and Toocheck 2017). The curbside can serve several needs of urban logistics operations, the most classical being loading and unloading zones, but more recently experimentations have been conducted including installations of micro-hubs or micro-consolidation centres.

3.1.1 Loading and Unloading Zones

The most simple and common example of mixed use of urban space for logistics are the loading and unloading zones. Often, their use is only dedicated to logistics operations for a few hours per day, usually in the morning and afternoon of working days. The time allocation between private car parking and loading and unloading activities allows a more efficient use of the space throughout the day. When shops and restaurants are open, unloading zones allow the logistics to park close to their final client, in a safe way. Outside of the delivery peak, residents or visitors can use the space to park their cars for the night or on non-business days.

This solution requires minimal infrastructure, usually only a vertical sign and pavement marking. However, it should be properly dimensioned for logistics operations (width, length, manoeuvre radius, etc.) to be relevant. Cerema (2009) has published an extensive guide on requirements to implement loading and unloading zones. They also need to be placed in accordance with the operational needs of the delivery companies (Fig. 1).

The conditions of parking (schedule of permitted parking for private cars, maximum time allowed for loading and unloading operations, stop time allowed) may vary from country to country, city to city, or even from street to street.

Studies show that in many cities, loading zones are frequently taken over by private cars, especially in busy areas, leaving delivery drivers with fewer options (Wangapisit et al. 2014). Only 10% of deliveries are performed on actual unloading zones; most of them are done through double parking or by parking on the sidewalk. To ensure compliance with loading zones regulation, both local police training and increased number of control are needed (Cerema 2015).

To facilitate the use and accessibility of these loading zones, many digital tools have been developed. They aim to help them identify the availability and occupancy of the loading zones in real-time, can also propose booking systems, and navigation guidance to the loading zone. They can be dedicated apps, or integrated APIs

Fig. 1 Loading and unloading zone occupied by a car in front of a shop in Lyon, France

(application programming interfaces). These digital tools are often connected to sensors or cameras placed on the loading zone to transmit information in real time. In some other cases, it is based on self-declaration of occupancy by the drivers.

3.1.2 Microhubs on Curbside

In 2022, the City of Paris, France, partnered with logistics real estate specialist Sogaris to experiment with a new model of light and temporary micro-hubs, installed directly in city curbside on former parking spots. The micro hubs were tested in two neighbourhoods (Arts et Métiers and Bastille) and acted as transhipment points for parcels between larger delivery vehicles and cargo bikes. Operated by Cargonautes and Ecolotrans, they also included a reverse logistics solution, as the micro hubs were also used as collection points for recyclable. With a light and aesthetic wooden design, the micro hubs fitted well in the streets and were accessible to the operators through a code lock.

The one-year pilot was not renewed, even though it participated in transporting sustainably 7 tons of recyclable waste and over 11,000 parcels. The experiment did not allow us to establish a viable economic model. Feedback highlighted lack of proper sealing; passers-by mistakenly used it as a trash depository (Sogaris 2022a, b; Ville de Paris 2022).

3.1.3 Automated Lockers as Microhubs

As part of an EU-funded project called URBANE, the city of Bologna, Italy, has implemented three automated, logistics operators-agnostic microhubs. Each one with a capacity of 40 boxes of different sizes. They are placed in three strategic locations (Porta San Mamolo, Via Calori, and Via Berlinguer) in the outskirts of the historical centre, a Limited Traffic Zone (LTZ). The microhubs serve as transshipment points from vans to light electric vehicles for the last-mile delivery of parcels. Transport operators TYP and Due Torri drop off their goods to be delivered in the LTZ at the microhub and can focus on their delivery rounds in the rest of the city. Then, the last-mile operator picks up the packages and delivers them in a consolidated and efficient manner within the LTZ (URBANE Project 2024).

The microhubs are managed by a third party, GEL Proximity, a digital platform that enables safe and secure data sharing, real-time monitoring, including availability and dispatching of the parcels in the different boxes, and communications between the different stakeholders (transport operators, last-mile service provider and final client). This system can also be used for business-to-consumer (B2C) activities, as it was also tested during the pilot.

This collaborative solution has improved logistical efficiency by reducing redundant vehicle trips and contributed to the reduction of $CO2$ emissions (−52% reported by TYP) (EIT Urban Mobility 2024). Beyond the great reduction in emissions and operational efficiency gained, one of the most distinctive features of this experiment is establishing successful cooperation among competing logistics operators.

Due to the success of the experiment, additional logistics operators expressed interest in expanding the solution to their activities beyond the scope of a co-funded project.

3.2 Open Urban Space

In city centres, streets tend to be narrow, with high density of shops and restaurants, resulting in a high need for logistics at the same time and place. In certain cities, it is also coupled with important car transit at specific times (e.g. peak hours). These needs require urban space but can be seen as complementary rather than conflicting.

Here, "open urban space" refers to any spare area of the urban space that can be optimized with urban logistics activities when it is underused. It can include road lanes and sidewalks.

3.2.1 Street Traffic Lane as a Temporary Unloading Zone

A great example is operating in Lyon, France. In the Presqu'île, Lyon's city centre bordered by two rivers, Rue Grenette is a strategic axis. It is a two-lane street, with one-way traffic flow. It makes a junction between three key parts of the city: Rhone River, Saone River, and the Presqu'île. At peak hours, it is a busy route, used by many cars to cross the city. Presqu'île is a dense commercial area, with numerous shops and restaurants. These activities generate a high demand for logistics operations, yet appropriate space is limited due to narrow streets, which result in double-parking of trucks, disturbance of overall traffic and dangerous working conditions for delivery drivers and other road users.

To better fit with the several needs of the areas (transit traffic and delivery of hotels, restaurants, and shops), a temporary use of the space has been implemented since 2015. From 9:30 to 16:30 on working days, the south lane is dedicated to delivery operations: trucks can stop for a maximum of 30 minutes to perform loading and unloading operations and safely deliver to the surrounding shops and restaurants (Cerema 2023) (Fig. 2).

This solution is indicated to users through a vertical sign placed along the street, as shown in Fig. 3. It is usually also marked on the road with yellow marks, prohibiting cars to park. A communication flyer was also distributed to familiarize the potential users and citizens with this new organization of the space.

The reallocation of the traffic lane exclusively for deliveries between 9:30 and 16:30 led to an 85% shift of loading activities onto the designated space, significantly reducing double-parking and delivery-related conflicts in this dense commercial street. Implemented with minimal infrastructure—simple horizontal and

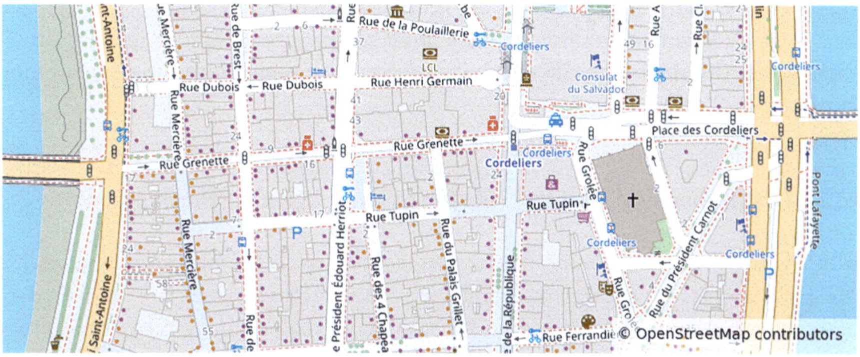

Fig. 2 Map of Rue Grenette in Lyon's Presqu'île (Open Street Map)

Fig. 3 Rue Grenette, Lyon, France (Marion Cottet)

vertical signage—the system has been well received by both delivery operators and shopkeepers. While no change was observed in the hourly distribution of deliveries, the intervention proved effective in organizing curb use and could be easily replicated in other streets with similar characteristics. Originally a one-year experimentation in 2015, the solution was turned into a permanent one. This initiative truly improved delivery conditions, as reported by delivery workers and local shop owners (LAET 2018).

A similar solution is implemented in Barcelona, but equipped with variable message signs (VMS) that can dynamically allocate space, enabling adaptable use of street space during critical periods, while preserving general accessibility (Huschebeck 2005).

3.2.2 The Ephemeral Beehive: Ad Hoc Urban Logistics Transhipment

In 2023–2024, in Lyon, France, several stakeholders collaboratively launched *La Ruche Éphémère*, an experimental model of mobile urban logistics hubs designed to facilitate last-mile delivery via cargo bikes, without requiring any permanent infrastructure or storage facilities. Coordinated by Centsept, the initiative involved a wide range of partners including Fends la Bise, Cargonautes (developer of the Cyke software), Résonance, Laboratoire Aménagement Économie et Transport (LAET), the logistics operators XPO Last-Mile France and Brasserie Dulion, all with operational support from the City of Lyon, Metropole of Lyon. The core idea of the Ruche is to enable a seamless transhipment between a utility vehicle and one or more cargo bikes in a temporary, shared space. These could be loading and unloading zones or larger urban space where a truck and cargo bike could easily fit without disturbing the other users of the space. Two configurations were tested: fixed Ruches situated at pre-defined locations, and mobile ones that shifted location based on delivery needs and the driver's route, requiring high synchronization. For example, the fixed location was a wide area between a bicycle lane and a large, paved road (Fig. 4). A mobile location was often coupled with a pre-existing delivery point of the truck, which allowed mutualization of the stop and a gain in efficiency (Paquet et al. 2024).

Operations are precisely coordinated: deliveries are pre-sorted and scheduled via a shared IT system Cyke, and both the truck driver and the bike couriers arrive within a narrow time window, avoiding the need for any intermediate storage. Once the parcels are transferred (typically within 20 minutes), the *Ruche* disappears, freeing the space for other uses. Across the operation of 27 *ruches*, the system transferred over 3 tonnes of goods, avoided approximately 19 kg of CO_2 emissions, and delivered 317 parcels to 45 recipients.

Coordination between players was key, especially since most spaces used were not formally reserved: the city of Lyon simply gave informal approval, and each operation had to ensure safety and avoid conflicts of use with pedestrians or traffic (Paquet et al. 2024).

The experiment demonstrated the environmental potential of this solution, reducing CO_2 emissions and road congestion without relying on dedicated logistics real estate, and optimizing idle urban space. However, it also exposed operational feasibility but with challenges, including timing mismatches and a still-fragile economic model. Low parcel volumes, high coordination and preparation costs, limited interoperability between digital systems, and the absence of strong regulatory incentives (e.g. delayed LEZ enforcement) further limited shippers' engagement.

Fig. 4 Fixed location of La Ruche Ephémère

3.2.3 Autonomous Driving Parcel Locker

In Helsinki, Finland, an experiment was launched in 2024, in the context of two Horizon Europe projects, URBANE and DISCO, the Ruoholahti microhub pilot. Coordinated by Forum Virium Helsinki, the city's innovation unit, the pilot brought together a diverse group of logistics actors, including DB Schenker, DHL Express, A2B, LMAD (last-mile autonomous delivery), and TwinswHeel. The microhub worked as a shared urban consolidation point, where parcels arriving by conventional vans were transferred to low-emission delivery modes: cargo bikes and autonomous delivery robots (ADRs) for distribution in the surrounding area (Forum Virium Helsinki 2025).

A noteworthy part of this pilot is the use of ADRs for B2C operations. Local citizens could order online and receive a notification to come pick up their parcels at a pre-defined location, where they would find the ADR, stationed for 1-hour time slots. Consumers could access the parcels thanks to a PIN-code locker system. Creating an ephemeral solution for out-of-home delivery, directly in urban space.

Evaluation showed significant operational and environmental benefits. With an average delivery time of approximately 22 minutes, ADVs enabled the delivery of around 100 parcels while limiting travel to 0.6 km per trip. This shift resulted in a 3% reduction in CO_2 emissions compared to the baseline scenario. The solution also achieved 100% delivery safety and a high customer satisfaction rate of almost 90%.

Digitalization played a strong part in the pilot; blockchain technology was used to support secure data exchange between stakeholders in a safe and secure manner. Difficult winter conditions, specific to climates such as Helsinki's, however, generated some operational troubles such as motor dysfunctions due to the cold temperatures (URBANE Project 2025; Forum Virium Helsinki 2025).

3.3 Parking Lots

The increasing adoption of active mobility and public transport, combined with regulatory measures such as low-emission zones, has led to a decline in the use of traditional parking infrastructure in city centres. This opens the possibility of repurposing underutilized parking spaces for alternative functions, notably urban logistics, through shared and flexible use models.

3.3.1 Mobile Pick-Up Points on Supermarket Parking Lots

The collaboration between IKEA UK and the supermarket chain Tesco is an interesting example of mixed use of parking lots for urban logistics.

The pilot was launched in 2022, and 12 locations were selected to experiment with the use of mobile pick-up points, for IKEA consumers to collect their online purchase from designated zones within Tesco car parks. The experiment was so successful that in July 2024, the online retailer announced that it had reached 100 pick-up sites (IKEA 2024). This solution allows us to consolidate deliveries to centralized and strategic places, reducing the number of at-home deliveries, also through financial incentives (Insight DIY 2023). By utilizing Tesco's existing space, these mobile pick-up points integrate seamlessly into consumers' existing mobility routines.

This partnership highlights the potential of leveraging underutilized parking lots for mixed urban services, shifting consumers' delivery choices, and grouping their trips to reduce overall emissions.

3.3.2 Urban Logistics Hub in Underground Parking

Another interesting case of sustainable and collaborative urban logistics is the Cordeliers Urban Logistics Hub located in the heart of Lyon's Presqu'île district. Launched in 2012, this 320 m^2 facility is integrated into the ground floor of the Cordeliers car park, in a very strategic location of the city centre, and is managed by the semi-public entity Lyon Parc Auto (LPA). The hub includes dry storage areas, a refrigerated unit, a charging installation for electric vehicles, and space for loading and unloading vehicles.

Initially operated solely by Deret, through a two-year agreement, the hub handled early-morning deliveries from 20 km outside of Lyon using hybrid trucks, and last-mile distribution was conducted by smaller electric vehicles. Still observing underuse of the facility during daytime hours, LPA initiated a successful mutualization with Ooshop in 2015. Ooshop, Carrefour's express delivery service, is non-competitive with Deret. Within its upstream flows, Deret started to consolidate dry goods from Ooshop's platform in a neighbouring city, while Ooshop's cold products are delivered separately. Final-mile deliveries are performed by two bi-temperature electric vans operating within a 2 km radius (LAET 2016). In 2024, after a request for proposals issued by LPA, two other operators joined the hub: Toutenvélo, a cargo bike delivery cooperative, and OuiCompost, a service provider for collecting and composting food waste (France Mobilités 2024).

The Cordeliers urban logistics unit in Lyon illustrates the tangible benefits of mutualized last-mile delivery in dense urban areas. The system has led to a reduction of 14 tonnes of CO_2 annually and cut local pollutants and negative externalities by more than 50%. It ensures regular operations with reduced traffic constraints and no noise nuisance, thanks to the use of electric vehicles. For Ooshop, delivery productivity improved by 20%, with time per client dropping from 19 to 13 minutes, and fuel consumption reduced by 9%. The model also supports better working conditions for staff and demonstrates economic viability, pointing to the potential for replication in other city locations (LAET 2016).

Their inclusion illustrates the adaptability of urban logistics hubs in supporting diversified, low-impact urban services beyond conventional freight distribution. This Urban Logistics Hub is a great example of successful collaboration among public and private partners, effectively working towards integrated and sustainable urban logistics. In 2024, building on the success of this first hub, LPA has replicated this solution of Urban Logistics Hub in two other underground parking (LPA 2024).

3.4 *Operating, Underused Facilities*

In addition to the use of public spaces, logistics operations can also be integrated into existing real estate infrastructures. Originally designed for other purposes, it optimises their use during periods of underutilization. Here are two examples of cases where logistics operations are embedded into buildings, which main activities are not logistics.

3.4.1 Bus Depots as Day-Time Transhipment Hubs

In Paris, France, the public transport operator, RATP, noting the emptiness of their bus depots during the daytime when all their vehicles are roaming around the city, decided to capitalize on the space, a rare commodity in city centres—especially in a dense capital such as Paris. In 2021, RATP launched a request for proposal on several of its bus depots to make their space available for decarbonized logistics operations. Situated in both the city centre and suburbs, the bus depots can serve as transhipment stations during daytime and serve their first function during nighttime. Amazon, along with other transport operators such as Ecolotrans and Chronopost, was selected to experiment with this innovative model in 2021, and tested it out in three depots. All three operated in different bus depots. The experiment took the shape of a four-year leasing agreement recently renewed to continue exploiting the space (RATP 2021).

In practice, in the morning, once the RATP buses are out of the depot, the pre-sorted parcels from e-commerce purchases arrive by truck at the facility. There, parcels are transhipped to cargo bikes. The parcels are delivered by cargo bikes in a 3–4 km radius around the depot. The drivers come back once to reload and perform a second round per day. At night, the facility is completely emptied to accommodate the buses.

The logistics drivers are allowed to use the existing facilities and equipment (toilets, etc.), and pre-existing slopes and heights were already compliant with the needs of the logistics trucks and cargo bikes. Only a few infrastructure adjustments were necessary to secure the space for both usages. Safety markings and signs were added to clearly identify the spaces dedicated to buses or logistics operations. At night, no trace of Amazon activities remains at the bus depot. The cargo bikes are parked and recharged in a nearby parking lot (within a five-minute bike ride) (Pasquet 2025).

RATP has replicated this solution in several other bus depots, in and around Paris City centre, with several companies. This collaboration between public and private stakeholders has proven to be successful, but did come with discussions, adjustments, and iterations about respective expectations and processes in order to find an adequate operational model. Administrative procedures also showed to be paced differently between private and public entities, but nothing that would jeopardize the collaboration.

3.4.2 Exhibition Centre Parking Lot as Transhipment Hub

A similar example takes place in Thessaloniki, Greece. In the context of the Horizon Europe-funded project DISCO, an experiment took place in TIF HELEXPO, a facility hosting events, congresses, etc., strategically situated near the historical centre of the Greek city. To optimize its available space and complement its other activities, the facility converted an area for logistics operations. ACS, a Greek logistics service, participated in the experiment to operate in the exhibition centre.

The TIF-HELEXPO premises are situated in a very strategic position in the city, right next to the historical centre of Thessaloniki. The infrastructure is also only temporarily used due to its original purpose: hosting events, congresses, etc., which, by nature, are occasional. The opportunity to use the existing space for complementary activities was therefore logical.

In this case, the first step consists of experimenting with the use of this spare space. A few changes were needed to set up the experiment, including the relocation of the ACS facility to TIF Helexpo, where specific infrastructures were added, such as cameras and movable barriers to allow logistics operations at all times (24/7 access needed). These modifications allowed both TIF Helexpo activities and logistics operations to perform business-as-usual, while ensuring the safety and efficiency of all activities.

In the context of this experiment, an existing tool, developed by research institute CERTH, called WareM&O, was used. It allows to assess the availability of space in warehouses and book them through smart contracts.

This experiment showed positive outcomes, including more safety for the logistics services provider, gains in efficiency, and improvement of logistics operations. The exhibition centre also capitalized on its underused space, which required only a small investment to make the infrastructure compliant with logistics activities. However, it also faced several operational challenges. The spaces selected initially were either located on the first floor, creating difficulties for logistics operations that required ground-level access for loading and unloading, or were too small to accommodate ACS's operational needs. Moreover, many of the available areas lacked essential infrastructure, such as restrooms, office space, and designated loading and unloading zones. After discussions between TIF Helexpo and ACS, a slightly larger space was identified, though it was situated in a less accessible area. The location required some infrastructure upgrades: traffic control barriers, automated camera-based access systems, and a reprogrammed ticketing system enabling ACS vehicles equipped with RFID tags to enter the site outside of standard hours (DISCO project 2025). These adjustments helped create the conditions necessary for the pilot to function effectively, and shortly after the end of the experimentation, ACS and TIF Helexepo reached an agreement to continue this collaboration beyond the experimental phase.

These examples show interesting opportunities for public–private partnerships for specific, complementary uses of urban infrastructure. It contributes to the maximization of the use of the existing space, while necessitating low alterations of the building.

4 Analysis on Current Practices and Enablers for Effective Mixed Use of Space

The use cases presented in this chapter illustrate how mixed-use approaches of urban space, when carefully designed and implemented, can lead to multiple positive outcomes. These include measurable reductions in greenhouse gas emissions, noise pollution, and traffic congestion; as well as improved delivery efficiency, better working conditions for logistics operators, and enhanced coexistence between freight activity and other urban functions. Such solutions offer a pragmatic response to the historical exclusion of logistics from traditional urban planning, providing flexible and adaptive frameworks that reclaim and reallocate public space more intelligently. When supported by clear governance, stakeholder engagement, and adequate regulation, they not only mitigate the externalities of urban freight but also contribute to more liveable, resilient, and functionally integrated cities. Their potential impact could be significantly amplified if systematically embedded in urban planning documents and long-term mobility strategies. The following analysis examines the key levers that enable the successful integration of mixed-use urban logistics solutions, with the aim of informing more coherent and sustainable urban policy frameworks.

The review of these use cases reveals that the successful implementation of mixed-use urban logistics solutions depends on an interplay of spatial, regulatory, operational, and collaborative factors. These solutions operate within the constraints of limited urban space while aligning with broader urban goals such as reducing emissions and improving liveability.

One important lever lies in the hands of local authorities, through regulations. Regulatory frameworks, such as low-emission zones or vehicle access restrictions, tend to generate the necessity and legitimacy of solutions such as microhubs, micro consolidation centres, or other ad hoc solutions, as they encourage transshipment to lower-emission vehicles. When regulatory ambition is high, it creates both the demand and justification for the use of such a solution, while also increasing the chances of its economic viability. Transparent, flexible, and co-developed regulations are more likely to achieve broad compliance and innovation uptake.

Public institutions, as well as public–private entities, have another powerful lever in their hand: public land. As shown in several use cases, they have the possibility to issue tenders, request for proposals, and provide the space to stakeholders aligned with their ambition to facilitate space optimization and mixed-use urban environment. This role extends beyond permitting, as they can set conditions for use, encourage shared models, sustainability, and ensure alignment with broader spatial and mobility strategies.

However, regulatory alignment and land availability alone do not guarantee effectiveness. Physical and technical conditions must also be met for urban logistics infrastructure to be functional. Appropriate location is paramount: microhubs and loading zones must be sited close to relevant areas (e.g. outskirts of LEZ), and they must be physically adapted to the vehicles that will use them (e.g. adequate dimensions for parking and delivery manoeuvres on unloading zones, ground floor access, and appropriate height and slopes for microhubs). Proper signalization of such solutions has revealed to be very important for explicit space repartition and guidance of delivery drives and a clear statement of use conditions. Small, low-investment adjustments are sometimes required to properly adapt existing infrastructure and enable the mix of uses, such as the installation of cameras or access barriers. Workers also need to have access to basic commodities. About enforcement, lack of police control has proven detrimental to the effective use of loading and unloading zones, as private cars tend to occupy the space during delivery time windows.

Across several of these cases, digital tools emerge as real enablers of coordination, efficiency, and performance monitoring. These range from app-based interfaces displaying location and availability of loading zones with booking options, advanced platforms to monitor a microhub, digital twins for scenario testing, to an online platform to book available idle space for logistics operations through blockchain smart contracts. The integration of real-time, standardized data increases the reliability and efficiency of operations but requires true interoperability, ensuring secure data sharing and easy integration in the existing system to facilitate uptake by logistics operators. However, these technological integrations require standardization and true interoperability to be efficiently used.

Collaboration among stakeholders (municipalities, logistics service providers, property owners, citizens, etc.) emerges as a foundational but sometimes challenging component. The most effective models embrace an iterative, co-creative approach, allowing for multiple rounds of testing, feedback, and adjustment. They also invest in relationship-building: regular communication, trust-building measures (e.g. secure data sharing), and even symbolic gestures such as shared branding or co-hosted events. These ways allow to overcome barriers such as different administrative paces between public and private entities, alignment of goals and expectations, etc. Another notable point is that in some cases, the collaboration between only non-competitive stakeholders facilitated the process, while in other, a trusted-third party and secure data sharing system allowed to reassure companies to perform consolidation among competitors.

Most of the use cases presented in this chapter started with an experimental phase, at a smaller scale. This allowed us to identify barriers and enablers and find the right operational frameworks for all involved entities. Once the concept is proven, it is easier to scale it up, as done in many of the above-mentioned cases. Also, it encourages new stakeholders to replicate or join the developed solution.

The solutions studied are not isolated innovations, but rather part of a wider logistics ecosystem. Their successful integration into local delivery networks is crucial. They need to be strategically placed and answer an actual need, such as transhipment to low-emission vehicles to access a LEZ, as stated previously. Otherwise, their economic model tends to be too weak to be viable. At the planning stage, a thorough understanding of logistics needs is essential to ensure that the location and available space are suitable, well integrated into the urban context, and conducive to efficient operations.

Mixed use of urban space for logistics does not go unnoticed by citizens. Public awareness campaigns have been shown to increase the social acceptability of such solutions, and adequate integration in urban space, both visual and functional, also plays a role. When infrastructure is aesthetically adapted to its surroundings and clearly communicates its function, it is more likely to be accepted by the public. Even simple improvements, such as better signage or visible information panels, can help increase usage and reduce resistance.

Some of the presented solutions have proven effective on an operational level, while some still have difficulties finding their economic models. Across all cases, a wide range of public and private stakeholders is involved at various levels. Their engagement is shaped by regulatory frameworks, access to land, and, crucially, a shared willingness to test new ways of working in support of more sustainable logistics practices.

5 Recommendations and Guidelines for Implementation

Plan with logistics in mind. Many innovative solutions for urban logistics are emerging, but there is a need to effectively plan the ground for these innovations. Upstream thinking and planning are necessary to create a fertile ground for these innovations to be successful. Urban projects, such as the redesign of curbside, new pedestrians' zones, establishment of restricted areas, or even (re)development of entire neighbourhoods, require a proper consideration of urban logistics' needs associated with the local activities—whether they are shops, restaurants, offices, habitation buildings, etc. If urban logistics needs are not considered at the planning stage, they will most likely become issues and threaten the liveability of the space.

Acknowledge the necessary conditions. The potential of urban space is considerable for urban logistics, which only require small amount of time for effective operations, but the settings need to be appropriate locally, with dedicated spaces, clear signage, adequate conditions (e.g. height, width, slopes, sufficient turning radius, etc.), enforcement (e.g. police control to prevent unauthorized use of loading zones), to therefore ensure safe conditions both for the delivery workers and the other users of public space.

Ensure compatibility. Due to the diverse nature of urban logistics activities and their economic models, the usage of underutilized facilities seems like a promising and appropriate solution. It requires ensuring that the different activities are compatible time-wise (e.g. sharing space at different times of the day) and nature-wise (e.g. goods transported can be temporarily stored in the same space). Clear instructions and appropriate signs for the different users are required to avoid conflicts and prevent security issues.

Set rules that enable. Such solutions perform better when in the right regulatory framework. Low-emission zones, or timed access windows serve as enablers for urban logistics solutions, as they create both the incentive and the operational stability needed for adoption. Well-designed regulations not only encourage more sustainable practices but also improve the viability of shared logistics infrastructure.

Start small. Experimentation is a great way to start implementing such solutions. Starting with a smaller-scale pilot project allows us to validate assumptions, strengthen stakeholders' confidence, and co-build viable economic and operational models. Iteration should be expected and planned for, both in infrastructure and in collaborative arrangements. Monitoring is important to assess environmental, economic, and social impacts, to identify improvements needed, but also to communicate on successes.

Collaborate to innovate. Each mixed use of space opportunity requires a tailored solution to effectively fit in the existing space, while still including standardized processes to gain efficiency. It should be based on a collaborative relationship, shared vision, and clear communication between diverse stakeholders, including logistics providers, real estate owners, technology developers, and local authorities. Public authorities play a key role by enabling collaboration through calls for proposals, partnerships, and support for experimentation. Digital integration is foundational for efficient and reliable collaboration, with interoperable platforms that enable real-time planning, curbside management, and coordination between operators. The use of real-world data, combined with technologies like blockchain, digital twins, and predictive analytics, enhances trust and performance when well-integrated into operators' existing systems.

Make it citizen-friendly. Citizen acceptance is important and should not be neglected; therefore, clear communication campaigns and thoughtful, aesthetic integration of logistics solutions into the urban environment are key. They ensure these infrastructures fit seamlessly into citizens' daily life.

In conclusion, the mixed use of urban space for logistics is not only a technical or spatial challenge, but a deeply collaborative and political endeavour. When done well, it can transform logistics operations from a burden into a recognized asset in public space, supporting more liveable, sustainable, and resilient cities.

References

Banque des Territoires (2024) La ZAC Gratte-Ciel: La logistique en quartier dense à Villeurbanne. Available at: https://www.banquedesterritoires.fr/realisation/transformation-ecologique-urbaine-lyon (Accessed: 12 May 2025)

Buldeo Rai H, Kang S, Sakai T, Tejada C, Yuan Q, Conway A, Dablanc L (2022) Proximity logistics: characterizing the development of logistics facilities in dense, mixed-use urban areas around the world. Transp Res A Policy Pract 166:41–61. https://doi.org/10.1016/j.tra.2022.10.007

CBRE (2022) Urban logistics and rental growth: evidence from building-level data. Available at: https://www.cbre.com/insights/viewpoints/urban-logistics-and-rental-growth-evidence-from-building-level-data (Accessed: 12 May 2025)

Cerema (2009) Aménagement des aires de livraison : guide pour leur quantification, leur localisation et leur dimensionnement. CERTU. Available at: https://www.cerema.fr/fr/centre-ressources/boutique/amenagement-aires-livraison (Accessed: 12 May 2025)

Cerema (2015) La *logistique urbaine : connaître et agir*. Centre d'études et d'expertise sur les risques, l'environnement, la mobilité et l'aménagement (Cerema). Available at: https://www.cerema.fr/fr/centre-ressources/boutique/logistique-urbaine-connaitre-agir

Cerema (2023) Partage temporel des voiries urbaines pour optimiser l'espace public. Cas d'usage. Available at: https://doc.cerema.fr/Default/doc/SYRACUSE/594498/partage-temporel-un-outil-d-optimisation-des-voiries-et-espaces-publics

City of Amsterdam (2023) Zero-emission mobility in Amsterdam: Implementation agenda 2023–2026. Available at: https://assets.amsterdam.nl/publish/pages/867636/zero-emission_implementation_agenda_2023-2026.pdf

Clean Cities Campaign (2022) The development trends of low and zero-emission zones in Europe. Transport & Environment. Available at: https://cleancitiescampaign.org/wp-content/uploads/2022/07/The-development-trends-of-low-emission-and-zero-emission-zones-in-Europe-1.pdf (Accessed: 14 May 2025)

Dablanc L (2023) Land-use planning for a more sustainable urban freight. In: Marcucci E, Gatta V, Le Pira M (eds) Handbook on City logistics and urban freight. Edward Elgar Publishing, pp 246–266. https://ideas.repec.org/h/elg/eechap/19924_12.html

Dablanc L, Rakotonarivo D (2010) The impacts of logistics sprawl: how does the location of parcel transport terminals affect the energy efficiency of goods' movements in Paris and what can we do about it? Proced Soc Behav Sci 2(3):6087–6096. https://doi.org/10.1016/j.sbspro.2010.04.021

DISCO project (2025) Thessaloniki use case presentation by Vasilis Giannoudis (CERTH) Presented at POLIS-ALICE webinar series: mixed use of Urban Space for City Logistics, 9 April 2025. Online. Available at: https://www.etp-logistics.eu/polis-alice-webinar-series-mixed-use-of-urban-space-for-urban-logistics/ (Accessed: 15 May 2025)

EIT Urban Mobility (2024) Cooperation amongst competitors in Bologna's parcel distribution. Available at: https://marketplace.eiturbanmobility.eu/best-practices/cooperation-amongst-competitors-in-bolognas-parcel-distribution (Accessed: 15 May 2025)

European Commission (2021) New European Bauhaus: Beautiful, sustainable, together (COM(2021) 573 final). Available at: https://eur-lex.europa.eu/legal-content/EN/TXT/?uri=CELEX:52021DC0573 (Accessed: 28 May 2025)

European Commission (2025, January) Competitiveness compass. Available at: https://commission.europa.eu/topics/eu-competitiveness/competitiveness-compass_en (Accessed: 24 May 2025)

Forum Virium Helsinki (2025) Lessons learned from the Microhub pilot collected in a Concept Report. Available at: https://forumvirium.fi/en/publication/lessons-learned-from-the-microhub-pilot-collected-in-a-concept-report/ (Accessed: 15 May 2025)

France Mobilités (2024) Espace Logistique Urbain Cordeliers. Available at: https://www.francemobilites.fr/projets/espace-logistique-urbain-elu-cordeliers (Accessed: 12 May 2025)

Goodman RW (2005) Whatever you call it, just don't think of last-mile logistics, last. SupplyChainBrain. Available at: https://www.supplychainbrain.com/articles/601-whatever-you-call-it-just-dont-think-of-last-mile-logistics-last Accessed: 25 May 2025

Huschebeck M (2005) Multi-use lane in Barcelona (Spain). POLIS Network. Available at: https://www.polisnetwork.eu/wp-content/uploads/2019/06/barcelona_casestudymultiuselanees2.pdf

IKEA (2024) IKEA announces its 100th Tesco click and collect point with 100 more on the way. Accessible at: https://www.ikea.com/gb/en/newsroom/corporate-news/ikea-announces-its-100th-tesco-click-and-collect-point-with-100-more-on-the-way-pub96af59a0/ (Accessed: 10 May 2025)

Insight DIY (2023) IKEA announces national rollout of Tesco collection points. Accessible at: https://www.insightdiy.co.uk/news/ikea-announces-national-rollout-of-tesco-collection-points/13111.htm (Accessed: 10 May 2025)

LAET (2016). Espace Logistique Urbain des Cordeliers. Laboratoire Aménagement Économie et Transport (LAET). Available at: http://tmv.laet.science/formation/ELUcordeliers.html (Accessed: 26 May 2025)

LAET (2018) Lyon—Rue Grenette. Transport Marchandises en Ville Laboratoire Aménagement Économie et Transport (LAET). Available at: https://tmv.laet.science/formation/Grenette.html (Accessed: 30 May 2025)

LPA (2024) LPA annonce cinq nouveaux projets pour son ELU Cordeliers. Lyon Parc Auto (LPA). Accessible at: https://www.lpa.fr/wp-content/uploads/2024/10/LPA-CP-ELU-20241017.pdf (Accessed: 10 May 2025)

Paquet V, Michaux C, Abdelhai L., Toillier F, Delahaye F (2024) Le Centsept. Report: Expérimentation La Ruche Éphémère—Hub mobile de cyclologistique

Pasquet J (2025) Use of a bus depot for logistics operations—Amazon logistics. Presented at POLIS-ALICE webinar series: Mixed use of urban space for city logistics, 9 April 2025. Online. Available at: https://www.etp-logistics.eu/polis-alice-webinar-series-mixed-use-of-urban-space-for-urban-logistics/ (Accessed: 15 May 2025)

RATP (2021) Transformation des dépôts de bus pour favoriser la mixité urbaine et la logistique du dernier kilomètre. Available at: https://ratpgroup.com/fr/engagements-et-innovation/responsabilite-societale/favoriser-mixite-urbaine-et-biodiversite/; https://www.trm24.fr/la-ratp-transforme-ses-espaces-bus-en-centres-logistiques/; https://www.leparisien.fr/info-paris-ile-de-france-oise/transports/grand-paris-des-depots-bus-ratp-prets-a-accueillir-des-livraisons-damazon-et-chronopost-18-04-2021-IUONT7QWPRBQJIX7YK37BDVGDM.php

Roe M, Toocheck C (2017) Curb appeal: curbside management strategies for improving transit reliability. National Association of City Transportation Officials (NACTO). Available at: https://nacto.org/wp-content/uploads/NACTO-Curb-Appeal-Curbside-Management.pdf

Rueda S (2019) Superblocks for the design of new cities and renovation of existing ones: Barcelona's case: A framework. https://doi.org/10.1007/978-3-319-74983-9_8

Schachenhofer L, Kummer Y, Hirsch P (2023) An analysis of underused urban infrastructures: usage opportunities and implementation barriers for sustainable logistics. Appl Sci 13:7557. https://doi.org/10.3390/app13137557

Sogaris (2018) Chapelle International. Available at: https://www.sogaris.fr/fiche/chapelle-international/ (Accessed: 21 May 2025)

Sogaris (2022a) L'Immeuble Inversé. Available at: https://www.sogaris.fr/fiche/limmeuble-inverse/ (Accessed: 21 May 2025)

Sogaris (2022b) Les micro-hubs—Logistique urbaine innovante. Available at: https://www.sogaris.fr/publication/micro-hubs-logistique/ (Accessed: 21 May 2025)

URBANE Project (2024) The URBANE living lab in Bologna. Presented at POLIS—ALICE greening urban freight webinar series 2024 #1: Physical internet led urban logistics, 27 March 2024. Online. Available at: https://www.etp-logistics.eu/greening-urban-freight-through-physical-internet-27th-march-2024/ (Accessed: 15 May 2025)

URBANE Project (2025) Helsinki, Finland—URBANE lighthouse living labs. Available at: https://www.urbane-horizoneurope.eu/demonstration-cities/lighthouse-living-labs/helsinki-finland/ (Accessed: 15 May 2025)

Ville de Paris (2022) *Des expérimentations sur la logistique urbaine*. Available at: https://www.paris.fr/pages/des-experimentations-sur-la-logistique-urbaine-27710 (Accessed: 21 May 2025)

Wangapisit O, Taniguchi E, Teo J, Qureshi A (2014) Multi-agent systems modelling for evaluating joint delivery systems. Procedia Soc Behav Sci 125. https://doi.org/10.1016/j.sbspro.2014.01.1489

Finding the Right Space for Urban Logistics: An Accessibility Analysis for Walking and Cycling to Parcel Lockers

Víctor Ferran and Lorena Axinte

1 Introduction

Growing urban logistics activities are having a significant impact on urban spaces and congestion, driven, in part, by the rising demand for home deliveries and the rapid expansion of e-commerce, particularly over the past decade. Parcel lockers, if strategically deployed, can help alleviate some last-mile logistics challenges. In theory, parcel lockers could save emissions as logistics service providers can consolidate deliveries and reduce the number of kilometres driven, while also avoiding failed deliveries (Buijs and Cipriano 2023). In practice, the displacement factor—the shift in transportation patterns from courier deliveries to recipients collecting their parcels—plays a key role. When recipients use their cars—particularly in suburban and rural areas, emission savings are lost. This is why the impact of parcel lockers is neither inherently positive nor negative, as much depends on the use of parcel lockers. Still, current systems of parcel lockers usually lack a holistic strategy to cover the needs of the demand and the different areas where population concentrates. To overcome this issue and achieve the potential benefits of parcel lockers, this chapter presents a structured framework for designing and integrating an open parcel locker system in an urban logistics context.

V. Ferran (✉)
Bax Innovation, Barcelona, Spain
e-mail: v.ferran@baxcompany.com

L. Axinte
Bax Innovation, Barcelona, Spain

Cardiff University, Cardiff, Wales, UK
e-mail: l.axinte@baxcompany.com

A. Melkonyan-Gottschalk, Y. Li (eds.), *Urban Logistics Transformation*, Greening of Industry Networks Studies 15,
https://doi.org/10.1007/978-3-032-08387-6_14

2 Background and Literature Review

Parcel lockers are quickly becoming a ubiquitous service, mirroring the prevalence of ATMs in our daily lives. Much like ATMs revolutionised banking accessibility, parcel lockers can be a game-changer in the realm of package deliveries. While this transformation should be benefiting inhabitants equitably, regardless of their location or circumstances, accessible parcel locker networks have certain environmental implications. Their location, as well as their usage, is a main determinant of the impact on urban spaces and the public realm. Even when placed on private land, lockers need to be replenished/emptied, requiring careful planning to avoid congestion and safety issues. As various providers might want to install their single label, close networks, too high density can potentially result in public backlash, especially if noise and aesthetics are not considered.

For these reasons, public authorities across the world have started thinking about the best ways to balance between offering better and more accessible logistics services to their inhabitants, while working on wider aims such as increasing the quality of life. Approaches range from proactively setting rules and regulations to reactively addressing market failures as they arise (Buijs and Cipriano 2023). Both have their merits and drawbacks, highlighting the complex regulatory landscape that surrounds parcel locker implementation.

In Groningen, the city chose an ex-ante approach, aligning with ongoing and future changes such as the ones inscribed in its sustainable urban logistics plan (e.g. reaching zero emission logistics in the city centre by 2025), or its mobility vision (e.g. active travel as the main transport mode, facilitated by mobility hubs).

To get a grasp of the state of the art, a benchmarking exercise helped the City of Groningen understand best practices from across the world. The benchmarking exercise was mostly based on desk research, complemented with punctual interviews (e.g. with City of Baerum) and expert consultations (e.g. with E-CONSULT) for further clarification. Although the benchmarking was initially going to focus solely on frameworks, policies, and regulations, the lack of relevant information was overcome by including also unofficial guidelines, studies, and pilots deemed potentially useful. Results are based on case studies from five different countries (Austria, Norway, UK, Singapore, USA), as summarised in Table 1.

Acknowledging the quickly advancing topic, the following pages try to summarise the main findings of this comparative analysis. The summary does not intend to be all-encompassing, but rather to streamline the most important aspects that planners, policy makers, practitioners, and service providers could consider while implementing parcel lockers.

Groningen is amongst the first (few) public authorities that adopted a proactive approach and developed ex-ante regulations intended to set up clear rules for logistics service providers. The Norwegian VIV network is amongst the few other examples we could find. VIV concluded that a joint procurement exercise for self-service collection points would allow them to use resources more efficiently, across administrative boundaries. The approach could also simplify the process for service providers, offering equal treatment of offers.

Table 1 List of references of best practices for parcel lockers

Organisation/country	Document (year)	Type	More info
Federal Ministry Republic of Austria. Climate action, environment, energy, mobility, innovation and technology	White label parcel boxes guidelines (2022) (Rektor (FH) Prof. (FH) Mag. Dr. Andreas Breinbauer 2021)	Guidelines	Link
	Criteria for the use and choice of location for parcel boxes in communities (Mag. Gerda Hartmann, Ratgeber: Einsatzkriterien und Standortwahl von Paketboxen in Gemeinden 2023)	Guidelines	Link
The Norwegian VIV network (municipalities of Oslo, Bærum, Asker, and Drammen, Viken County, and the Norwegian public roads administration)	Common policy for self-service collection points—placement principles, criteria, and case management (2021–2022)	Common multilevel regulatory framework	Upon request
Knapskog, M, Caspersen, E., Briseid Kraft, I, Herheim, H.	Policies on parcel lockers: Strengthening the sustainability potential for last mile delivery solutions through multi-level governance networks (2022)	Academic paper	Upon request
UK government	Planning application (n.d.)	Planning application	Link
Centre for London	The Active Last Mile: How can we boost out-of-home deliveries? (Millie Mitchell et al. 2022)	Report/ guidelines	Link
Infocomm media development authority, Singapore	Parcel Locker Alliance (n.d.)	Pilot programme	Link
New York City Department of Transportation (NYC DOT)	LockerNYC (2023)	Pilot programme	Link

The London Plan also mentions that new developments should consider in their Delivery and Servicing Plans measures to reduce failed deliveries, such as parcel lockers. No further guidance is offered for other types of areas.

More often than not, most public authorities adopt a reactive approach through ex-post regulation, acting only when and if the market fails (e.g. in the case of a parcel locker placed in a protected area). In between the two ends of the spectrum, Baerum aims to request a planning application not only for new facilities, but also for parcel lockers that have already been installed.

Business and operational models are a widely discussed (and contentious) topic in the deployment of parcel lockers. Most public authorities would prefer an open network which is carrier-agnostic and can be used by any courier firm to avoid duplication of single-label lockers. In an ideal case, the network is also open to other businesses and even individuals for peer-to-peer shipments. Centre for London recommends such open models, especially in strategic locations, such as areas with poor air quality of low traffic neighbourhoods. In VIV's case, a parcel locker provider should be able to show a written agreement with at least four other service providers and ensure sufficient capacity for all. VIV would sign an agreement with 1–3 suppliers for 4 years, with the possibility of extension for another 2 years. Other open networks include LockerNYC operated by GoLocker in New York; WienBox, operated by A1 in Vienna; as well as Singapore's locker network deployed, owned, and operated by Pick. Concerns regarding such models refer to certain providers being favoured and turning the system into a monopoly.

Despite the public authorities' preference, closed networks are much more common. These tend to be single/private label, being used by a specific parcel locker provider (and, potentially, to its chosen logistics partners).

Potential functions and types of users (Table 2) are closely connected to the type of business and operation models. Generally, public authorities prefer both categories to be as diverse and wide as possible.

Generally, public authorities can more easily impose requirements when parcel lockers are placed on public grounds. In VIV, locations are co-determined by municipalities, the landowner, and the selected supplier. If the municipalities deem necessary, the supplier must deploy parcel lockers also in locations with less financial basis to ensure a good degree of coverage for residents. In Austria, building on public land requires the approval of the municipality, and rental income may be generated. Similarly, in the UK, lockers placed on streets need to be granted planning permission from the local authority. A fee is determined on a case-by-case basis.

To avoid planning permissions, while also facilitating trip chaining, many providers place parcel lockers in private areas (e.g. shopping centres, gas stations, transport hubs). In Austria, examples also include sites which are only accessible to a certain group of people or only with appropriate keys or digital means of identification (e.g. entrance areas of residential complexes, courtyards, locations of self-storage providers, company buildings, etc.).

Two of the use cases consulted (VIV with additional aspects from Oslo, and the Austrian guidelines) offered detailed infrastructure requirements for the parcel locker providers, as summarised in Table 3.

Given that aesthetics are very important, especially in old city centres, the integration of parcel lockers with existing elements could provide a more pleasant urban environment, increasing their acceptance. Centre for London offers the examples shown in Fig. 1 for place-based integration.

To ensure a universally accessible service, public authorities also include certain accessibility requirements (Table 4).

Table 2 Potential functions and users of parcel lockers

Potential functions	Potential users
Pick-up/drop-off	Courier express parcel services
Web shop	Local businesses
Returns (products)	Other service providers
Returns (packaging)	Online shops/marketplaces
Locker/short-term storage	Private individuals

Table 3 Infrastructure requirements for parcel lockers

VIV	Parcel lockers must operate without a power supply They should be neutral and adapt to surrounding installations in terms of colour and size Their ICT solutions should have a universal design Parcel lockers should not be used as an advertising surface The supplier must ensure that posters and tagging are removed within 3 days
Oslo	Parcel lockers should be given the same colour as the other elements they are to be combined with They should be placed next to the wall and not as loose, independent elements. When deploying two or more modules, they can be placed with the back against each other Ideally, municipalities should develop an overall strategy for placement, colour, and design
Austria	Parcel locker sites should provide: Stable internet connection 230 V power connection Levelled floor, being able to bear a load of up to 1000 kg/m^2 A roof, even though the installation location does not have to be frost-proof or protected from the weather Adequate light Sufficient space in front of the parcel box for operation Possibility for extension Installation site size recommendations: A basic unit with 2 modules, one of them with a terminal, for at least 10 compartments would measure approx. 1000x650x2000 mm (WxDxH) An additional inspection space of at least 200 mm over the total width of the locker should exist The ratio of boxes to flats is around 1:3, e.g. 1 box for 3 flats
New York	Security features: All locker locations will be equipped with two security cameras LED lighting Anti-theft mechanisms on locker compartments Installation sites chosen to: Provide adequate circulation space Avoid conflicts with street furniture Ensure smooth pedestrian flow Avoid obstructing windows, fire escapes, or public art

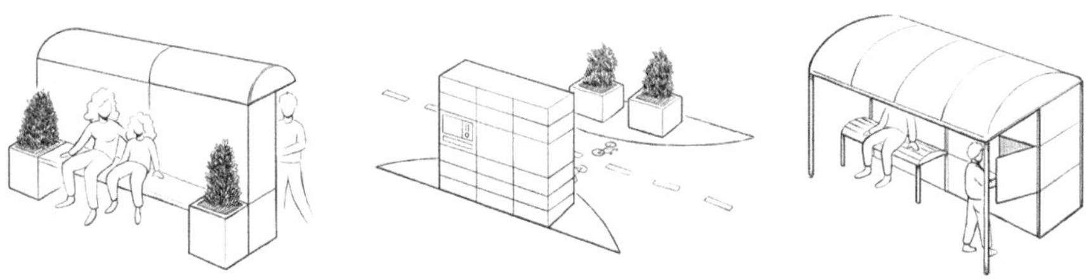

Fig. 1 Parcel locker integration in public space (visual inspired by Millie Mitchell/Centre for London)

Table 4 Accessibility requirements for parcel lockers

VIV	The planning application should assess parking, exits, and manoeuvring areas The supplier is responsible for cleaning, snow removal, and site maintenance Parcel lockers must have year-round operation and work in all weather conditions
Austria	CEN/TS 16819 technical specification: postal services—parcel boxes for end use EN 17210:2021 functional requirements: accessibility and usability of the built environment

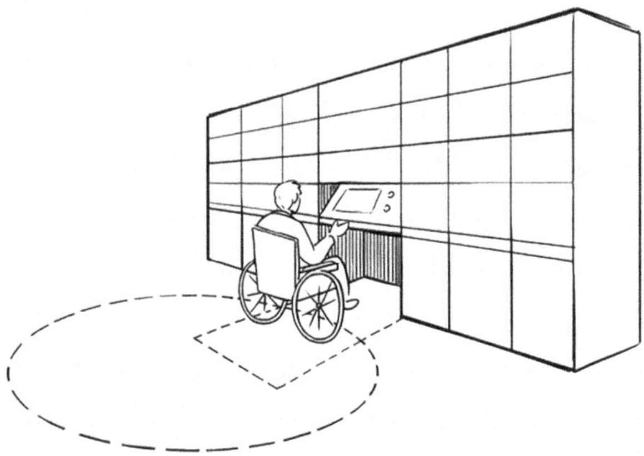

Fig. 2 Barrier-free design of the installation site (visual inspired by Bernhard Hrushka architecture B4)

The Austrian Guidelines provide some useful visual examples in conformity with the European standards and regulations (Fig. 2).

3 Methodology

The first part of the literature review focuses on identifying and benchmarking best practices in the planning, deployment, and management of parcel lockers, drawing on experiences from various cities and initiatives across Europe and beyond. The second part takes a proactive approach; the City of Groningen wanted to ensure that parcel lockers are available and accessible for most, if not all, of their inhabitants. Given that the location and density of lockers can impact the receivers' choice of transport, we conducted a series of spatial analyses to determine the most suitable locations for an open network of parcel lockers. The process has been developed together with the City and the University of Groningen, and has been shaped by the data available, as well as the continuously evolving needs.

The analyses followed three main steps:

1. Understanding the city context:
   - Demand: Where do most people live?
   - Infrastructure: Where are people most likely to walk and cycle?
   - Priorities: Where would the city prefer to locate parcel lockers?
2. Finding the most accessible 10 public spots.
3. Filling up the gaps of the current private networks.

Analysis Applied to Groningen's Case

Figure 3 shows the city's neighbourhoods and the population distribution. The more intense colours reflect the denser areas, pointing towards potential higher demand for logistics services.

To determine the spaces where people are most likely to use active transport, we used the space syntax betweenness centrality analysis. Figure 4 shows the segments of the network that are most used by bicycles. Central streets and streets that connect different neighbourhoods tend to have high centrality values. This means that the spatial configuration of the network would carry a higher flux of users, connecting different origins and destinations within a given radius (10 km in this case) faster.

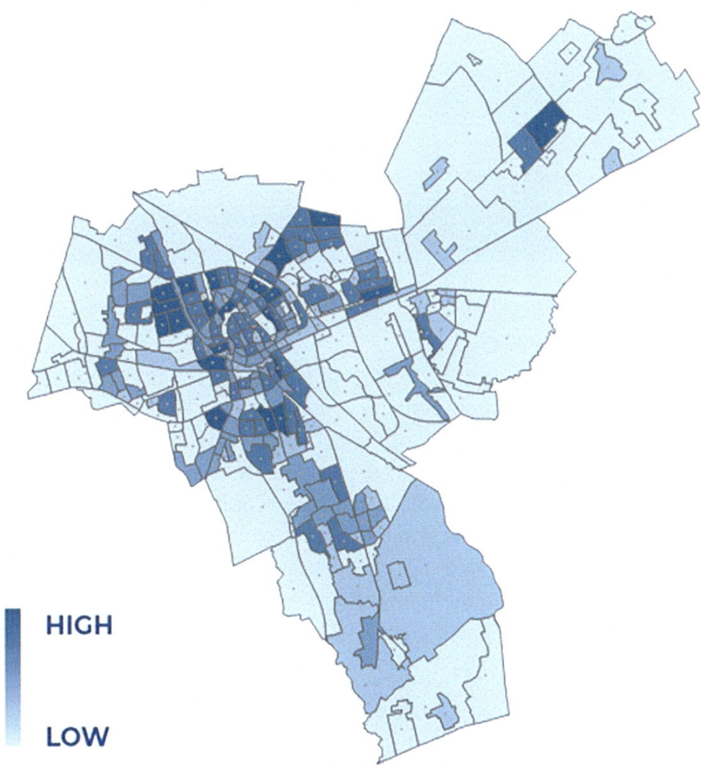

Fig. 3 Neighbourhoods and population density Groningen

Fig. 4 Streets most likely used by cyclists

The kernel density of the betweenness centrality (Fig. 5) shows where the higher densities of the most used segments of the network are located in the city. The highest values cover the central area of the city, as well as the area near the canals, where traffic is concentrated on fewer roads, especially near bridges.

Completing the contextual information, we also considered the city's preferred locations. Groningen is aiming to develop a strategy for transport hubs, integrating both mobility and logistics solutions whenever possible and needed. Figure 6 shows the types of hubs that could host parcel lockers:

- Buurthub or neighbourhood hub (green): 26 units.
- Wijkhub or bigger hub (red): 5 units.
- Park & Ride (yellow): 5 units.
- Stations (orange): 5 units.

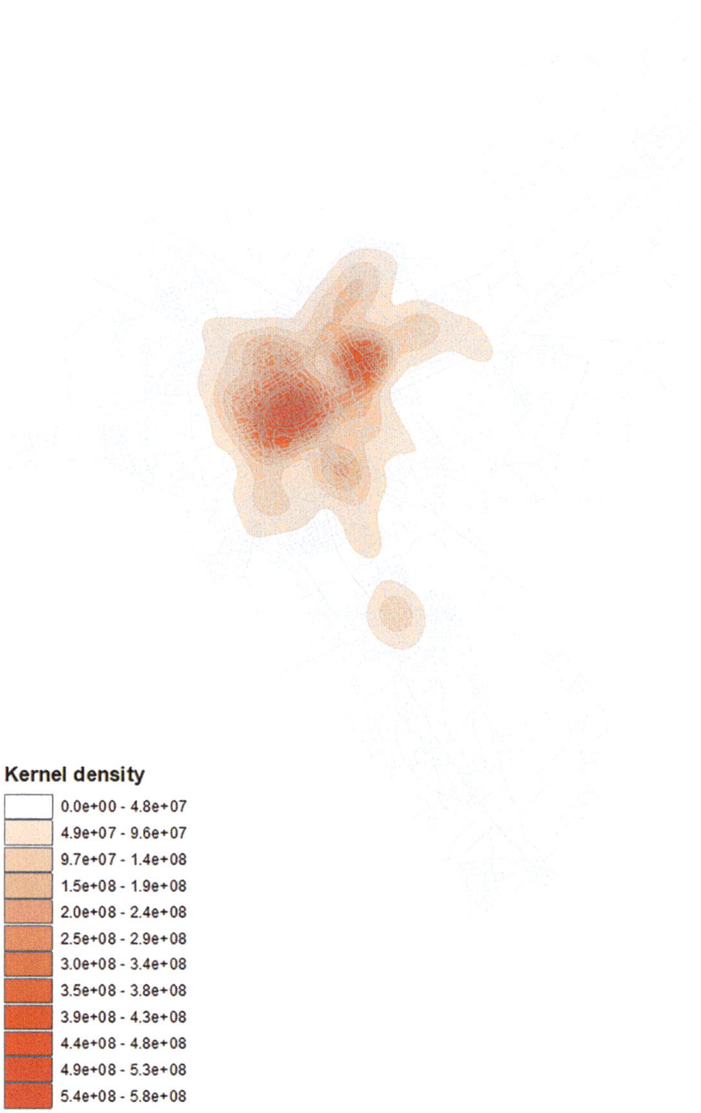

Fig. 5 Density of streets most likely used by cyclists

As a second step, we used location-allocation models to assess the most appropriate 10 locations that could serve most of the city's population. This analysis was carried out twice: the first time selecting all possible locations, including neighbourhood hubs (Fig. 7).

Figure 8 shows the coverage of these 10 selected lockers at five, then and 15-min cycling intervals. Figure 9 is a visualisation of this coverage overlaid with the population density distribution, showing how the most densely populated

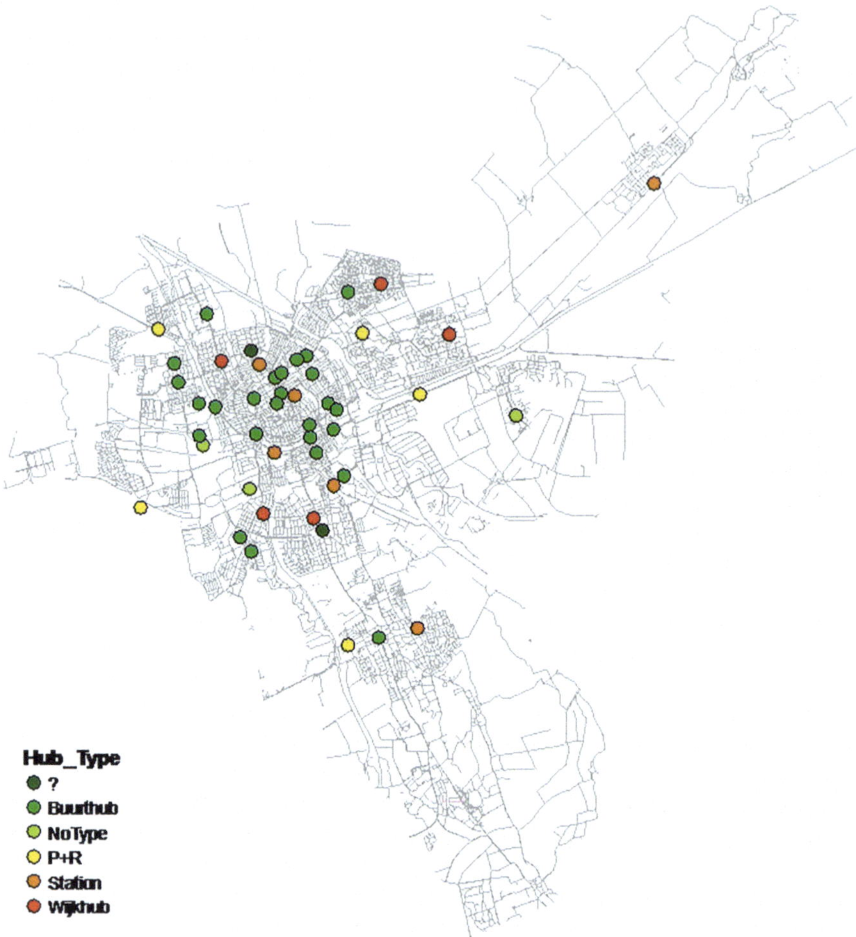

Fig. 6 Types of mobility hubs developed by the city of Groningen

neighbourhoods would be covered by these optimal locations. Since Groningen is a flat city, we do not consider its geography, but it can easily be integrated whenever relevant.

It's important to note that in both cases, the points resulting from our analysis overlap with the two largest current providers (Figs. 10 and 11), showing that their locations are carefully chosen. These results guided us in analysing the potential gaps in the current private offer for policy lockers and other pick-up drop-off points (PUDOs).[1]

[1] PUDOs is a more generic term, encompassing parcel lockers and other collection and delivery points, such as partnering shops and individuals collaborating with LSPs (typical mostly in the Dutch context).

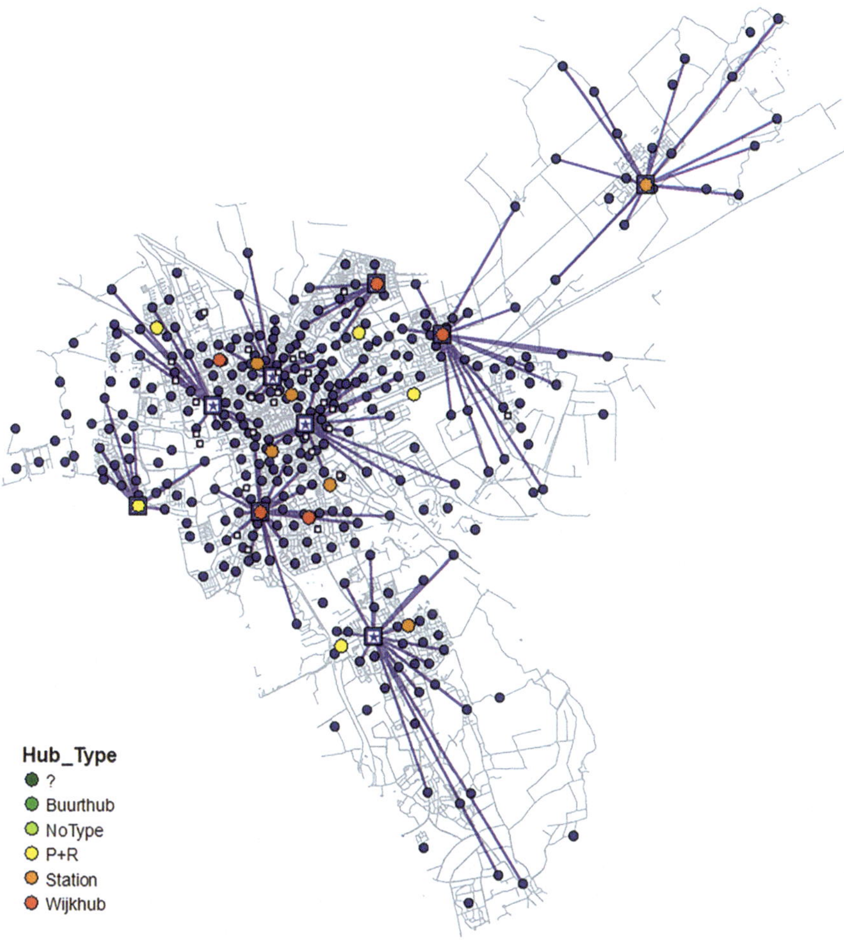

Fig. 7 Optimal location of ten lockers in the city

The PUDO networks of some of the biggest logistics service providers (PostNL & DHL) are already accessible to most of the population within a 10-minute bike ride (see annex for PostNL and DHL current PUDOs bike coverage). For this reason, we tried to identify any potential gaps. Our results show that the situation is different for parcel lockers, with areas such as Ten Boer and Haren not well covered by parcel lockers.

Furthermore, pedestrian access for both PostNL (Fig. 10) and DHL (Fig. 11) PUDOs is much more limited, highlighting the potential spots that should become a priority for the city to improve pedestrian access. Overlapping the coverage of the current and all the possible new locations (in green) provided by the city of Groningen, the following new locations could help to fill the gaps:

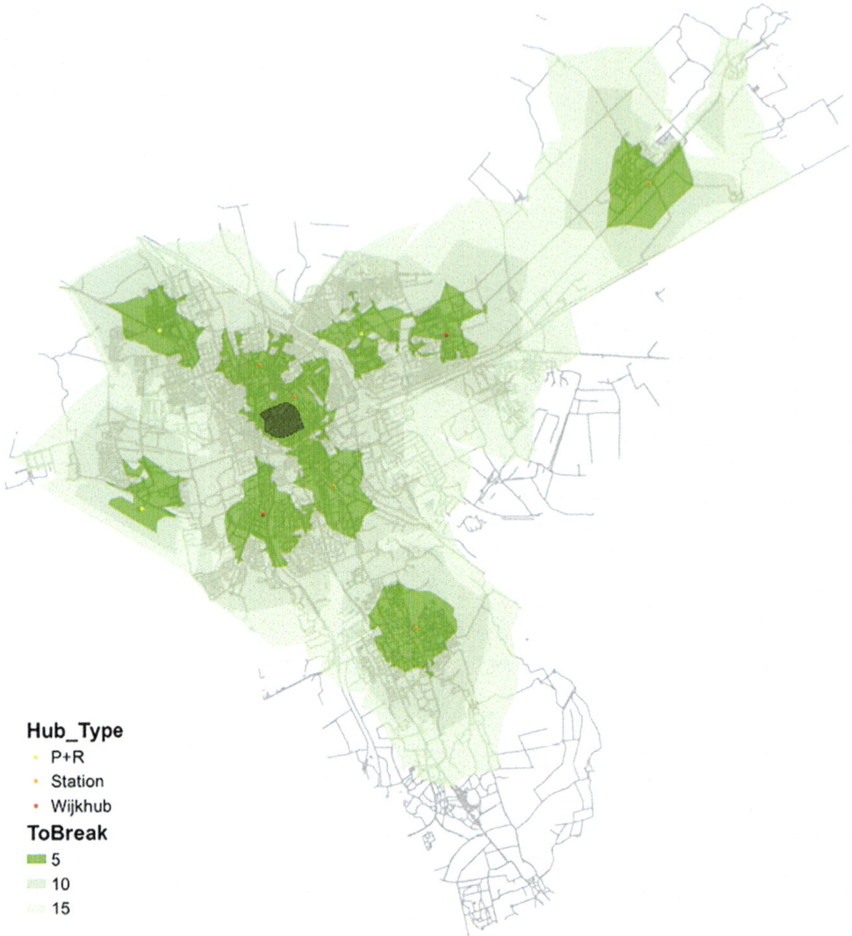

Fig. 8 Area coverage for ten optimally located lockers at a five-minute interval for cyclists

1. Meerstad.
2. S.O.J. Palmelaan.
3. Martiniplaza. This point would not overlap with any existing PUDO. Although it would not cover an area of high population density, it would be close enough to serve areas that are not currently well served (Fig. 12).

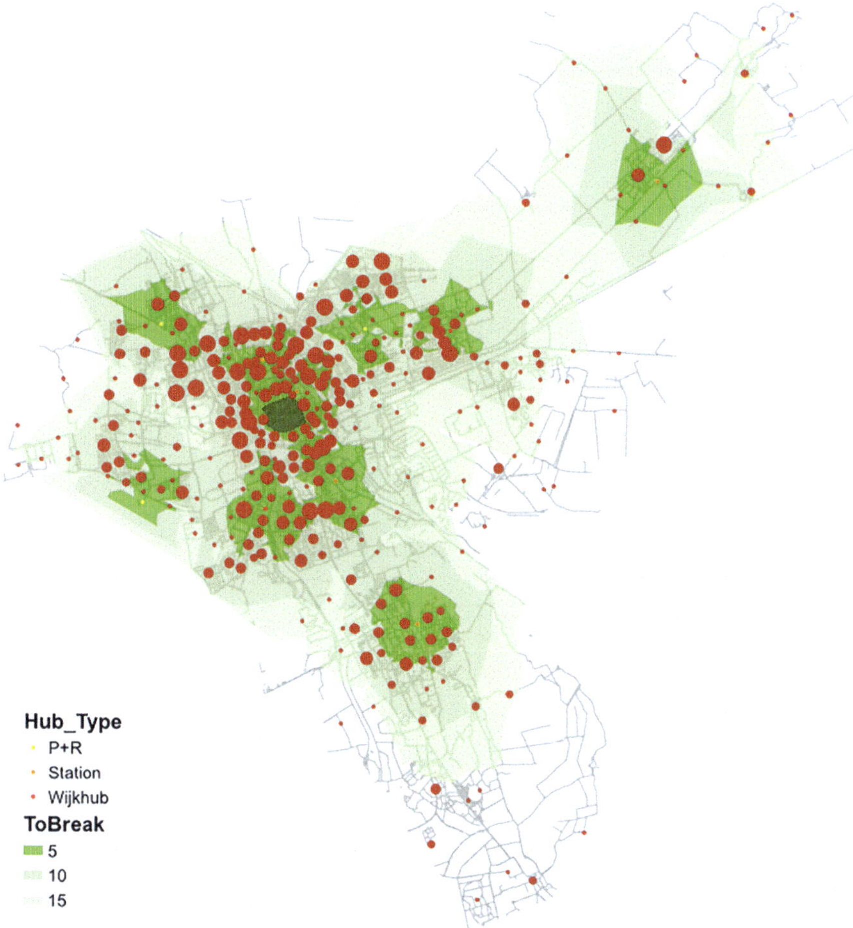

Fig. 9 Area coverage for lockers at a 5-min interval for cyclists and population density

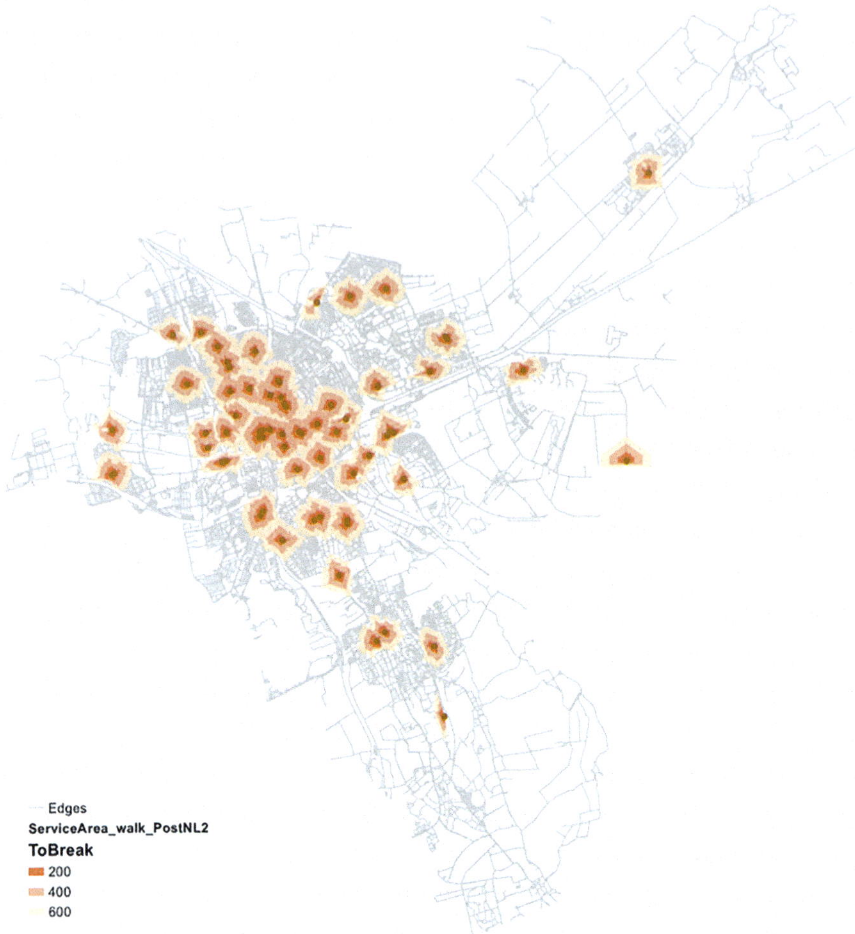

Fig. 10 Area covered by PostNL lockers at 5 min intervals for pedestrians

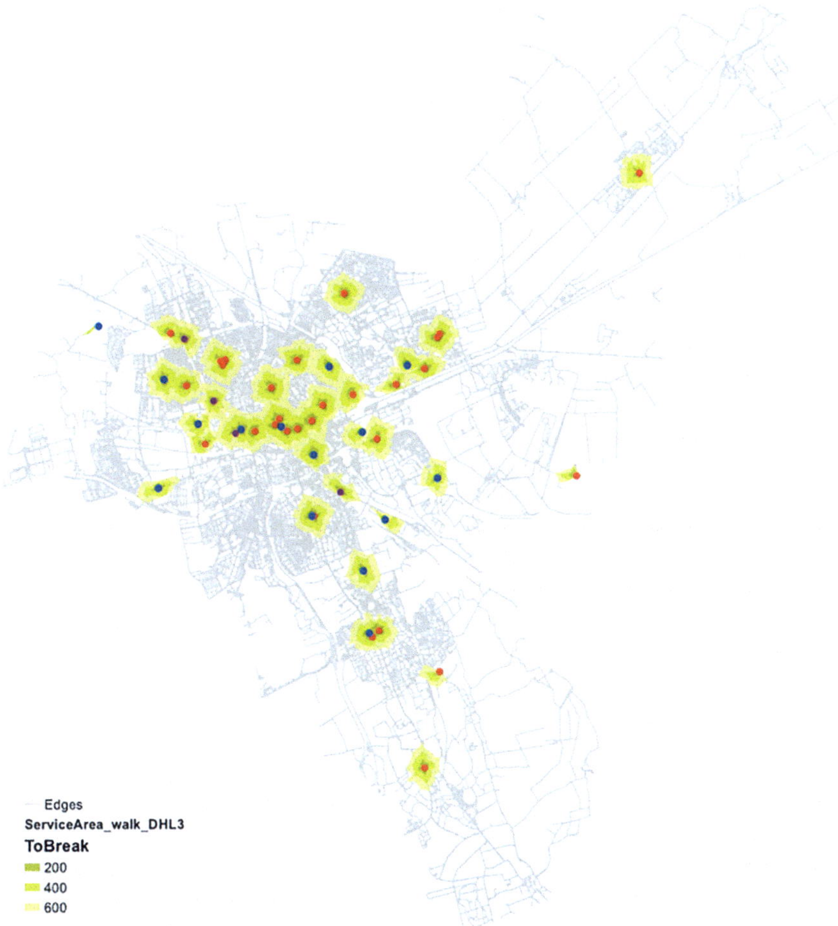

Fig. 11 Area covered by DHL lockers at 5 min intervals for pedestrians

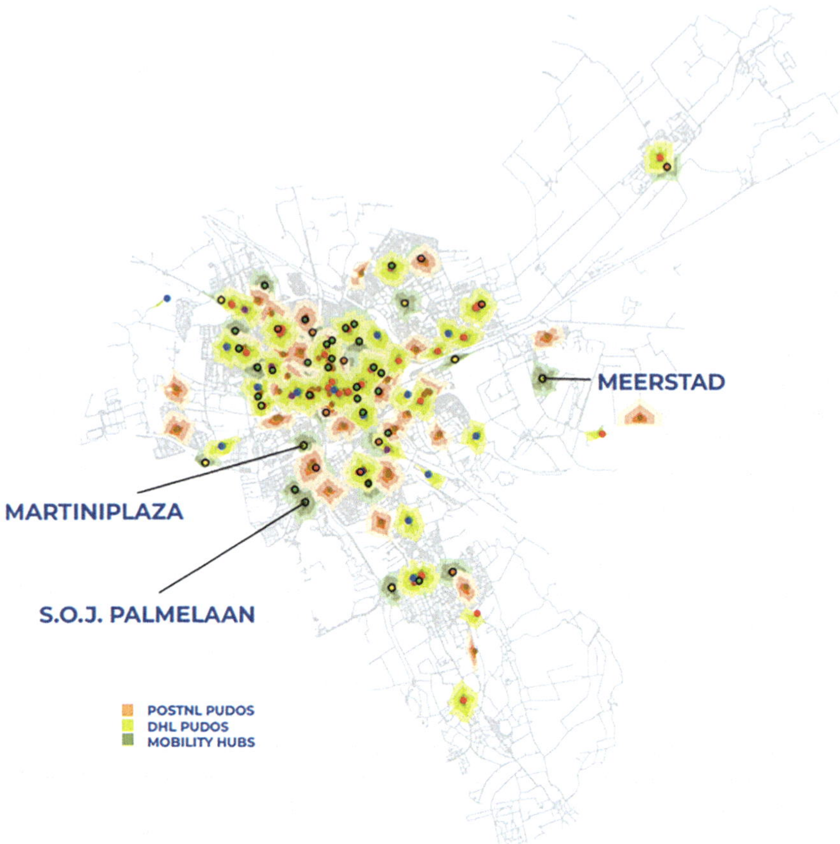

Fig. 12 Priority locations for parcel lockers to complement the existing private offer

4 Conclusions

The benchmarking exercise demonstrates that despite the large-scale implementation of parcel lockers across the world, there is no universal approach for public authorities to follow. Nonetheless, the structured approach presented in this study demonstrates how spatial analysis and international benchmarking can inform the sustainable integration of parcel lockers into urban logistics planning. By applying this methodology in the City of Groningen, we identified ten optimal public locations for parcel lockers that would maximise accessibility for most residents within a short walking or cycling distance. This targeted placement supports the city's goal of promoting active travel and reducing reliance on car-based parcel collection, which, in turn, can help decrease the distance travelled and associated emissions.

Aiming to reduce complexity for public authorities in general, and Groningen in particular, the following recommendations were structured into *key* and *nice to have*. They should allow municipalities to determine priorities and reach a common ground during consultations and negotiations with logistics service providers.

Key Recommendations

1. Public authorities should leverage their power to regulate public spaces, either through a dedicated framework for parcel lockers or through other connected policies (e.g. accessibility requirements, aesthetics of historical city centres, etc.).
2. Negotiating for an open parcel locker system would allow for a more diverse range of providers, users, and uses. This could have negative implications for the number of vehicles used to deliver and pick up parcels. Although consolidation in a specific hub and delivery by one single company could reduce the issue, it could also lead to a de facto monopoly.
3. Installation requirements might vary, but cities should always ensure that parcel lockers are safe to use (e.g. well-lit locations), accessible (e.g. allowing enough manoeuvre space for wheelchair users), and avoid conflicts with other space users (e.g. affecting pedestrian flows).
4. Given the scarcity of data on urban logistics, data reporting agreements could help public authorities get a better grasp of urban freight conditions, while also allowing them to track usage of parcel lockers and make any changes necessary.

Other Nice to Have Aspects to Consider

1. Ideally, parcel lockers should be able to operate without a power supply, using a sustainable energy source such as solar.
2. Paying attention to aesthetics is important, especially to preserve the built heritage in historic city centres.
3. Universal design could increase user accessibility, regardless of age and abilities, ensuring more equitable access to this service.
4. Parcel lockers could be modular, offering the possibility to expand during peak periods such as Christmas, and downsize to make space for other uses.

The different layers of spatial analyses provide useful insights for the City of Groningen on where to locate parcel lockers to increase user accessibility. First, mapping the city context helps to understand where most people live and where the higher demand for deliveries might be coming from, which streets most people are likely to walk or cycle on, and how it all connects with the mobility hubs already planned by the city. In this way, Groningen can make sure that mobility hubs are multi-use and can serve both mobility and logistics in an integrated way.

Second, the models developed help optimise the distribution of parcel lockers across the city. We highlight ten locations with high potential to increase accessibility for the city's most populated areas within short walking or cycling distances. This can be particularly useful if a new brand is interested in creating an alternative network.

Third, we analyse the current distribution of parcel lockers and other pick-up drop-off points to understand the existing gaps and underserved demand. While some of the largest LSPs have well-developed networks accessible for cyclists, they are harder to reach by pedestrians. Our analysis reveals specific gaps in the current private parcel locker networks, particularly in the three locations of Meerstad, S.O.J. Palmelaan, and Martiniplaza, where coverage is insufficient. By prioritising underserved areas such as these, public authorities can ensure more equitable access to parcel lockers, encouraging sustainable travel behaviour and potentially increasing user satisfaction.

While comprehensive usage data (e.g. reductions in trips or increased user satisfaction) are not yet available due to the recent implementation of these recommendations, the spatial models and the benchmarking provide a strong evidence base for regulatory action. The approach enables municipalities to proactively regulate and plan parcel locker deployment, rather than reacting to market-driven installations that may not align with broader mobility or sustainability goals.

A similar exercise can be replicated for any other city interested in developing a network of parcel lockers in public spaces, or in influencing the locations chosen by logistics service providers. Similarly, such studies can support providers in understanding where their services might be most needed. The analysis is not limited to parcel lockers, though, and can be redesigned for any other service of interest.

Acknowledgement The contents presented in this chapter have been developed within the framework of the ULaaDS project, which received funding from the European Union's Horizon 2020 research and innovation programme under grant agreement No 861833. Bax, the City of Groningen, and the University of Groningen were all partners in the project. The initial results were jointly developed by the current authors, Paul Buijs and Klaas Schurer from University of Groningen.

References

Buijs P, Cipriano R (2023) Policy approaches for placing parcel lockers in public space. Res Transp Bus Manag 49:101064. https://doi.org/10.1016/j.rtbm.2023.101064

Bundesministerium für Klimaschutz, Umwelt, Energie, Mobilität, Innovation und Technologie (BMK). (n.d.-a) Einsatzkriterien und Standortwahl von Paketboxen in Gemeinden: Ratgeber. Retrieved June 6, 2025, from https://www.bmk.gv.at/themen/mobilitaet/transport/gueterverkehrslogistik/publikationen/ratgeber-paketboxen.html

Bundesministerium für Klimaschutz, Umwelt, Energie, Mobilität, Innovation und Technologie (BMK) (n.d.-b) White Label Paketboxen Leitfaden. Retrieved June 6, 2025, from https://www.bmk.gv.at/themen/mobilitaet/transport/gueterverkehrslogistik/publikationen/white-label-paketboxen.html

Locker Alliance (n.d.) About Locker Alliance. Retrieved June 6, 2025, from https://www.lockeralliance.net/about-locker-alliance

Mag. Gerda Hartmann MJ (2023) Ratgeber: Einsatzkriterien und Standortwahl von Paketboxen in Gemeinden. Wien

Mitchell M, Bosetti N, Harding C (2022) The active last mile: how can we boost out-of-home deliveries? Centre for London. Retrieved from https://centreforlondon.org/wp-content/uploads/2022/12/Centre-for-London-Active-Last-Mile.pdf

New York City Department of Transportation (2023, August) NYC DOT launches initiative to reduce truck deliveries and environmental impact of packages. Retrieved from https://www.nyc.gov/html/dot/html/pr2023/dot-launch-initiative-packages-enviromental-trucks.shtml

Rektor (FH) Prof. (FH) Mag. Dr. Andreas Breinbauer, Fd-I (2021) Leitfaden "White Label Paketboxen".Wien

UK Government (2015) The Town and Country Planning (General Permitted Development) (England) Order 2015, Schedule 2, Part 7, Class C—Click and Collect Facilities. Retrieved from https://www.legislation.gov.uk/uksi/2015/596/schedule/2/part/7/crossheading/class-c-click-and-collect-facilitie

Developing and Enhancing Out-Of-Home Delivery Systems to Address the Challenges of the Rapid Growth of E-commerce: Outlook and Strategies from French Local Authorities: The Case of Valence Romans Mobilités

Edouard Ballois, Pauline Buosi, Alexandre Galzy, Cécile Gautronneau, Martin Munier-Jolain, Kevin Pereira, and Clarissa Reali

1 Addressing the Impacts of the Rising of E-Commerce: A Growing Preoccupation

1.1 The Dynamics of Evolution of E-Commerce in France

After a slight decrease in 2022, the parcel market in France has registered a clear recovery in 2023. Close to 1.6 billion parcels were carried, representing a progress of 4.2% as compared with the previous year (ARCEP[1]). This growth happens after a year of decrease and echoes to a positive dynamic that has settled throughout the past few years.

This recovery is largely carried by the rebound of the importation flows, which had dropped in 2022 due to prolonged sanitary restrictions in China, affecting the whole global logistics chain. In 2023, the volumes of imported parcels have risen by 3.7%, reaching 374 million items. This level was similar to the one observed in 2020, at the peak of COVID-19 crisis (ARCEP).

This dynamism is largely linked to the expansion of e-commerce, where transactions have increased by 72% between 2017 and 2023 (FEVAD)[2]. This high increase

[1]ARCEP is the France's Electronic Communications, Postal, and Print media distribution Regulatory Authority.

[2]FEVAD is the French Federation of e-commerce and distance selling.

E. Ballois · P. Buosi (✉) · A. Galzy · C. Gautronneau · M. Munier-Jolain · K. Pereira · C. Reali
Interface Transport, Lyon, France
e-mail: pbuosi@interface-transport.com

A. Melkonyan-Gottschalk, Y. Li (eds.), *Urban Logistics Transformation*, Greening of Industry Networks Studies 15,
https://doi.org/10.1007/978-3-032-08387-6_15

showcases the rising power of e-commerce, which keeps disrupting the distribution and logistics sectors, fostered by the evolution of technologies (Fig. 1).

In this context, economic stakeholders are pushed to diversify their offer for delivery and pick-up options. This perspective aims to both enhance the quality of service offered to customers and find alternatives related to the costs and constraints of home delivery, notably affected by the failed deliveries due to the recipient being absent.

Addressing this double objective, out-of-home deliveries have started to develop in France since the early 2000s. The pick-up points at local shops, initiated by La Redoute[3] and Les 3 Suisses,[4] followed by the first automatic parcel lockers such as E-Box or Citissimo, have coincided with the rapid growth of e-commerce.

The main operators are:

- Specialized stakeholders, notably PickUp, Mondial Relay, and Relais Colis, dominate the market in terms of network coverage, registering, respectively, 17,000, 13,000, and 9000 pick-up points at local shops on the continental French territory.
- Express retailers or online retailers, who have developed their own network of pick-up points (Colis Privé, UPS, FedEx, DHL, etc.) and/or of parcel lockers (Amazon, etc.) (Table 1).

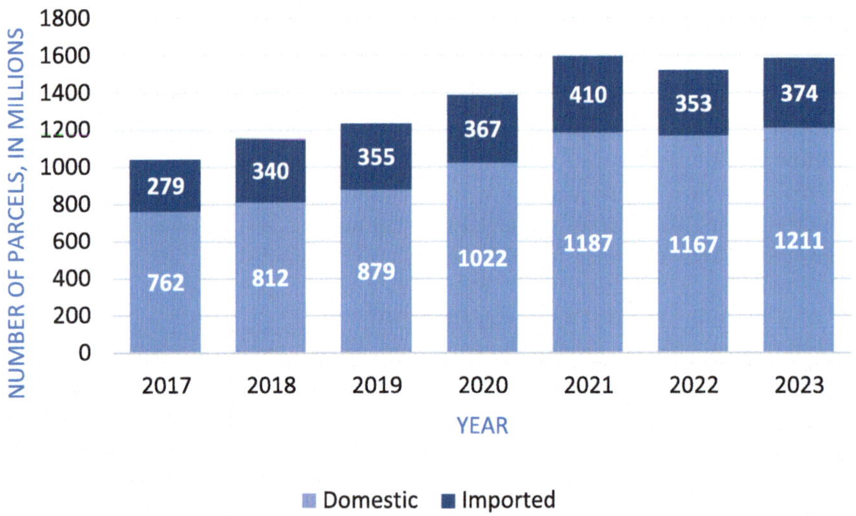

Fig. 1 Number of parcels in circulation in France between 2017 and 2023 (from ARCEP 2023)

[3] La Redoute: currently a French e-commerce website, La Redoute was one of the initiators of the French out-of-home delivery much before the generalization of the internet by offering a service based on commands through a catalogue of products (clothing, home, etc.).

[4] Les 3 Suisses: used to offer a similar service as La Redoute. It has stopped editing its catalogue in 2014 after 82 years of service in France, due to the internet making the business model obsolete.

Table 1 Out-of-home delivery market in France (Interface Transport 2025)

Network name(brand)	Number of pick-up shops	Number of parcel lockers	Associated networks
PickUp	17,000	4300	–
Mondial Relay	11,000	7000	GLS, FedEx
Relais Colis	9000	–	DHL
Colis Privé	5500	1500	DHL
Vinted Go	7000		Welco
UPS	5000		–
Geodis—GLS	Unknown		Pick me
FedEx—Amazon locker	Unknown		–

For partner retailers, this service allows them to generate additional income alongside their core business and to increase their visibility among consumers by attracting a new customer base. However, if not properly managed, this activity can lead to challenges such as storage space saturation and overcrowding in their shops. In addition, there is a high turnover rate among partner retailers, often driven by compensation that is seen as largely symbolic compared to the workload involved: managing parcel flows, ensuring availability, dealing with time constraints, and handling logistical complexity. These imbalances can quickly make the model unsustainable for some retailers, resulting in high turnover and instability in the territorial network of pick-up points.

At the consumers' level, in 2024, home delivery remains the favorite delivery option in France, selected by 54% of online shoppers. Pick-up delivery confirms its attractivity, especially for its convenience, limited cost, and broader time range to pick-up parcels. Among out-of-home delivery options, pick-up points at local shops outpace by far the Post Office (preferred by 11% of buyers) and parcel lockers, clearly progressing, chosen by close to 28% of e-consumers (FEVAD 2024).

Home delivery flows generate a lot of additional distance to cover, due to the difficulties of operators to rationalize their delivery rounds and to a higher proportion of failed deliveries. It generates congestion issues within city centers related to the increase of delivery vehicle flows, more greenhouse gases, and transport related pollution (atmospheric, noise pollution, etc.), and a reduced rentability for operators.

On the contrary, out-of-home delivery allows for consolidating parcel deliveries destined to several consumers at a unique pick-up point, limiting the risk of failed deliveries thanks to the extended time range that is offered to pick-up parcels. It hence contributes to the limitation of nuisances related to BtoC delivery. However, its environmental benefits strongly depend on the mode of transportation used by individuals between their home and their pick-up point, especially in rural areas (Fig. 2).

Fourty-five percent of e-consumers chose to go pick-up their parcel by walk if their pick-up point (at local shop or lockers) is located near their home. This means that the density of the pick-up points' network (their amount and geographical

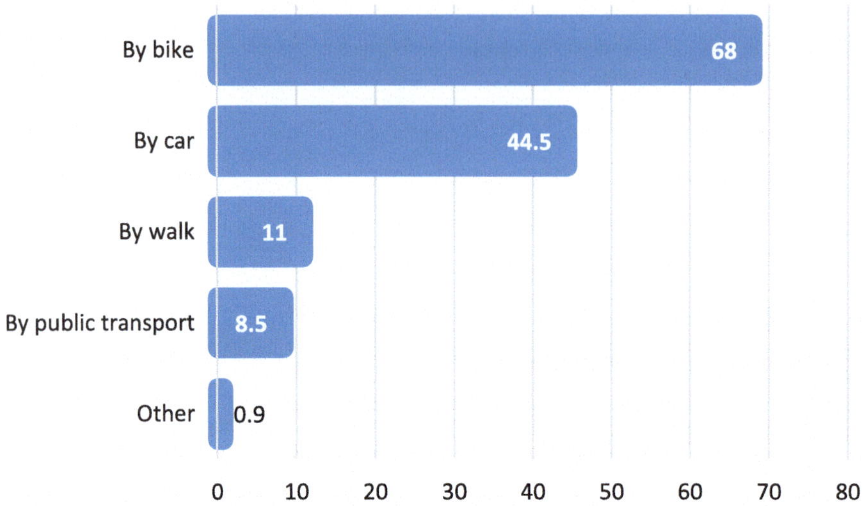

Fig. 2 Use of modes of transport to pick-up parcels, expressed in percentage (SendCloud 2024)

distribution) has a direct influence on the ecological footprint of the delivery. The more these points are and the better they are distributed, the more consumers can join them by walk, reducing the CO_2 emissions related to motorized commutes.

1.2 For Local Authorities, A Topic at the Crossroads of Numerous Issues

Out-of-home delivery, although still being little documented, constitutes an increasingly visible reality in public space. The atomization of delivery flows—in time as well as in space—accentuates its impacts, as much on the occupation of the public roads as on the commercial and logistical dynamics. In this context, the expectations of local authorities lie at the crossroads of several issues, sometimes hardly compatible:

- Stabilize a vision of urban development in a context of rapid mutation: the first challenge faced by local authorities is the stabilization of a coherent vision regarding the development of out-of-home delivery. Nevertheless, this exigence of clarity encounters a reality marked by significant changes. The intensification of flows related to e-commerce, the increased competition between logistics operators and the continuously emerging new practices make the logistics landscape particularly unstable. These mutations fall within short temporalities, complexifying the perspective of any sustainable projection or planning. To deal with these dynamics, the local authorities must demonstrate their agility while affirming long-term guiding principles.

- Conciliate local economic development and e-commerce: another core issue, the articulation between the development of the local commerce and the continuous growth of e-commerce. In city centers especially, the local shops suffer some fragilization, exacerbated by the competition with online marketplaces. Yet, e-commerce can also be a source of opportunities for these local actors. The opening of new selling channels, the welcoming of parcel lockers, or the role of pick-up points can all benefit the merchants, allowing them to integrate—instead of suffering from—the out-of-home delivery logistics. This suggests a personalized guidance and coordination between economic actors, operators, and local public authorities.
- Optimize the logistics chains and limit nuisances: the question of optimization of the logistics chains and of reduction of nuisances related to deliveries remains at the core of the preoccupations. Should we favor home delivery or out-of-home delivery? The answer largely depends on the local context—urban, peri-urban or rural—as well as on the profile of the recipients. In dense urban areas, home delivery often encounters high failure rates, causing additional flows, while in out-of-home delivery, the performance depends on the modalities of pick-up chosen by the recipients: dedicated commutes, motorized or not, accessibility level of parcel lockers or local shops pick-up points. There is then a need to adapt the solutions to local realities while ensuring the limitation of environmental impact and social disturbance of delivery flows.

What levers promote sustainable and shared urban logistics?

In this complex situation, several levers can be activated to address the expectations of the local authorities while taking into account the needs of the other stakeholders: logistics operators, local shops pick-up points, and final customers:

- Defining clear principles for urban development, integrating logistical issues in urban planning documents.
- Foster the implementation of mutualized solutions (shared lockers, urban consolidation hubs …) and of soft mobility delivery modes.
- Support the local initiatives gathering local merchants and the ecosystem of delivery services (pick-up points, decarbonized urban logistics).
- Adapt the policies to the local specifications and to the behavior of the customers.
- Implement a collaborative governance, open to experimentation, evaluation and adjustment of measures.

This article focuses specifically on the BtoC deliveries, setting aside CtoC[5] models that, even if they sometimes rely on similar networks (pick-up points, parcel lockers), have not been the subject of specific attention within the framework of this analysis.

[5] CtoC: Customer to Customer.

2 Which Good Practices Activated by Local Authorities? Overview of France and Europe

2.1 Key Insights from French Studies: The Transport Mode to Pick-Up Parcels more Decisive than the Pick-Up Mode Itself

The notion of out-of-home delivery in the present study, as opposed to home delivery, is defined as the whole set of solutions deployable "in third place" and proposing a pick-up/sending parcel service: pick-up points in local shops, parcel lockers or post offices.

According to FEVAD, home delivery represents more than half of the order collection's modalities in France (54%), with 34% of "standard" delivery and 20% of "express" delivery. Home delivery thus remains predominant on the national scale, as well as in Europe. The position of France is, however, more singular regarding the solutions activated in terms of out-of-home delivery, as pick-up in local shops keeps being the preferred option as compared with parcel lockers (even though the recent dynamics of development favor parcel lockers), whereas this ratio remains balanced in the other European countries (E-shopper barometer 2022).

Beyond the figures inherent to the sector and e-commerce consumption on a national scale, only few data or studies are available to characterize the offer and the demand of out-of-home delivery on a targeted territory. Furthermore, the growth of e-commerce raises questions about the environmental impact of deliveries. If out-of-home delivery is often regarded as more sustainable than home delivery, it remains challenging to determine, since there are numerous parameters to consider.

The CEREMA[6] and the LAET[7] in France are the entities assessing these questions. We can quote, for instance, the "ADM" survey on "household decoupled purchases" realized in the metropolitan area of Lyon, which has enabled to gathering the first local results on the impact of deliveries related to pick-up modalities in the overall assessment. It has allowed for stress, in particular, that 2/3 of the pick-up of purchases are related to a dedicated commute (pick-up being the sole reason of it), with half of this share being realized with an individual motorized vehicle.

Following the same logic, a recent study from the French agency for ecological transition (ADEME 2025) has shown that the environmental impact of a parcel depends on a plurality of factors: overpacking, vehicles' loading rates, but also the mode of transportation used by the customer in case of out-of-home pick-up. Indeed, out-of-home delivery leads theoretically, with the mutualization of deliveries, to a

[6] CEREMA: French public institution dedicated to supporting public policies, under the supervision of the ministry for ecological transition and regional cohesion.

[7] LAET: the Transport, Urban Planning, and Economics Laboratory is under the supervision of the Lumière Lyon 2 University, the National Center for Scientific Research (CNRS) and the National School of Public Works of the State (ENTPE, school for sustainable regional planning).

reduction of CO_2 emissions. This, however, only comes out as an obvious fact if the consumer reaches its pick-up point by walk, bicycle or public transport.

Yet, almost one third of customers use their car only to pick-up a parcel according to these national estimations, which can cancel the expected environmental benefits. On the contrary, a home delivery integrated as part of a well optimized delivery route can turn out to generate a lower environmental impact.

The consumers' preferences vary: while 78% prefer home delivery if there is no additional fee, 71% would choose out-of-home delivery if economically more interesting. However, only 45% perceive it as a more ecological option, often without taking into account their own mode of transportation (SendCloud 2024).

To limit the impact of deliveries, the ADEME recommends activating several levers: to limit overpacking, reduce distances, optimize the loading rate of vehicles and to favor pick-up options that are compatible with soft mobility.

As a conclusion, no delivery option is intrinsically virtuous regarding recent data and considering the difficulties of generalization of results. Environmental performance depends on a broad range of factors, as much technical as logistical or behavioral. It represents a collective challenge: involving at the same time e-retailers, logisticians, territories and consumers.

2.2 *Perspectives from European Literature: Inspiring Local Innovative Initiatives Mostly Kept at Local Scale*

Affordable, flexible, and sustainable in view of the increasing number of customers opting for it, OOH delivery, launched in a dynamic of continuous growth in all parts of Europe over the past few years, incentivizes the local authorities to consider its related issues, aligning their political vision to the urban, economic, and environmental challenges that it subtends. Where a majority of decision makers would opt for a reactive approach, whether by rejecting the perspective of occupying public space with a well-managed private activity or taking the role of a simple "auditor" examining potential requests initiated by the distributors, some of them stood out through their proactivity, mobilizing action levers allowing them to control, at least partially, the nuisances related to OOH (Dobber and Buijs 2023).

As part of the EU-funded project ULaaDS, Groningen (Netherlands) has, for instance, chosen to act as a service provider[8]filling the gaps of the existing private network of pick-up points. By crossing data and cartographies, it has targeted strategic accessible public spots, where no option was proposed yet by private stakeholders for OOH at a walking or cycling distance, to minimize the risk for customers to generate CO_2 emissions and traffic by joining their pick-up point by car.

[8] In the sense that it develops an offer of service that has not been made available by other stakeholders since then. It aims at serving both the distributors and final customers, the latter being more likely to pick-up their parcel using soft mobility mode as the public pick-up point is now located near their home.

Another interesting approach is the promotion of an open lockers' network. Denmark was one of the first European countries establishing a comprehensive framework of parcel lockers. Its "Nærboks" or "nearby locker" has developed as a semi-open network, welcoming deliveries from multiple distributors (NordPost, DHL, Bring). To optimize the lockers filling, no pre-allocation of compartments was made between the different operators, enhancing the capacity utilization. In its analysis, the Norwegian Centre for Transport Research suggests going further, by promoting an alternative approach that would consist in using consolidation terminals where a collective distribution of parcels would be made to the lockers—rather than each distributor making its own——hence potentially reducing the last mile logistics' traffic (Hovi et al. (2023); Norwegian Centre for Transport Research 2023).

We can also mention the alternative OOH solution experimented in Valladolid (Spain) as part of the URBANE project: an in-truck delivery service, consisting in delivering customers directly in their car instead of using pick-up points, in park and ride car parks. Avoiding dedicated travels, it has resulted in a reduction of 15–50% of CO_2 emissions according to the vehicle (with a distance travelled by the delivery vehicle reduced by 30–50%).

Through plural European funded projects targeting greener deliveries—URBANE, GREEN-LOG, TRACE, GREENTURN, CODE ZERO being part of the most recent—and national or local initiatives, the thought has been open to differentiated and a locally adapted logic related to OOH delivery. From data management through technical and technological optimization (use of AI, IoT, blockchain technologies, robots, automated or modular vehicles…) to regulatory actions, it has offered a wide range of possibilities and guidance for policymakers, whose role and decisions often appear as decisive in the implantation of successful solutions to contain the potential spread of CO_2 emissions due to the growth of OOH delivery; yet only sporadically implemented.

3 A Master Plan to Produce Knowledge and Mobilize the Stakeholders' Ecosystems

3.1 An E-Commerce Expansion Facing Limited Response from Local Authorities

At the local scale and in the frame of territorialized initiatives aiming for more sustainable urban logistics, e-commerce is often mentioned as a key element of urban logistics, but also as a potential generator of tensions in urban spaces.

Different typologies of answers to this issue have been identified, and are part of dedicated actions, such as:

- Communicating and raising awareness among e-commerce users on its impact: for example, in the Paris metropolitan area, a pilot was launched in 2018 to follow the consumption behaviors of a panel of customers.

- Elaborating a Master Plan for out-of-home deliveries pick-up points. On the side of local authorities, benefits from OOH deliveries pick-up points can be found in the service offered to its citizens. At the same time, it is a means to contain the negative externalities generated by e-commerce parcel transportation, such as congestion and air pollution. Thus, the Master Plan for OOH deliveries pick-up points is a meaningful tool to be used by local authorities as it helps to understand and quantify the local needs in terms of OOH delivery pick-up points, but also serves as an operational tool proposing territorialized and shared solutions. Those solutions mostly consist in recommendations to expand the existing OOH pick-up points network but also in the identification of structural operational issues that could improve the system in the coming years, in urban, suburban, and even rural areas. Several local authorities have engaged in such initiatives, like Valence Romans Mobilités[9] but also the Le Havre metropolitan area.

Several local authorities have already taken up this issue and initiated innovative approaches to better understand the logistics challenges associated with e-commerce. The Lyon Metropolitan Area, for example, has been a pioneer by funding a household survey specifically focused on e-commerce practices within its territory, providing valuable data to inform public policy.

In this context, the InterLUD program emerges as a particularly relevant tool. It not only facilitates the pooling of knowledge and experience at the national level but also provides a structured action framework to support local governments in implementing concrete solutions tailored to the specificities of their territories.

3.2 The Rise of a National Program in Favor of Urban Logistics

Created at the end of 2022, in relation with the dispositive of Energy saving certificates (EEC), the InterLUD program aims at supporting local authorities with the appropriation of logistical issues on their territory. It constitutes, in that sense, a local adaptation to the sustainable urban logistics strategy initiated at the national level by the intergovernmental committee of logistics (CILOG).

Although the InterLUD program addresses in priority the groups of municipalities, it allows the development of local action plans that produce indirect effects on different types of stakeholders: carriers, wholesalers, merchants, real estate investors, vehicles' constructors ...

In France, the national program InterLUD has involved about 50 territories since 2022. The signature of a convention with the program allows the systematization of territorial diagnostics to shed light on local logistics and freight transportation issues, with the identification of several types of action: support the coherence with

[9] French name for the local authority in charge of mobility. Can be translated by "Valence Romans Mobility"

freight transport regulations, strengthen the network and ergonomic of delivery areas, preserve urban land in city centers to develop more virtuous logistics organizations, etc.

Focus on Energy Saving Certificates
The mechanism of Energy saving certificates (EEC), created in 2005, consists, notably, in fixing energy consumption quotas to energy suppliers. The latter can then acquire additional certificates by financing actions allowing for to reduce energy consumption. The EEC programs, hence, permit to financing some actions targeting energy savings. These actions consist mainly of establishing best practices by raising awareness, educating, or providing equipment, allowing a shift in habits. The law n° 2010–788 from the 12 July 2010 makes possible the delivery of EEC, in the form of a financial contribution to the programs related to Energy demand management.

3.3 Producing Knowledge to Apprehend OOH Deliveries Offer and Demand

Different methods and analyses exist to elaborate a diagnosis of the offer and demand for OOH deliveries pick-up points in a territory.

On the offers' diagnosis, the objective is to identify, locate, and quantify the different OOH pick-up points and their respective network operator (UPS, DHL, etc.). This requires reviewing the data shared by the network operators online. Then, a comprehensive approach can be built, thanks to the diverse information collected: accessibility of the pick-up points (address, opening hours), number of multi-operators' pick-up points, etc. Additionally, interviews with dedicated stakeholders can be engaged to complete the overview: network operators, logistics service providers, and local shop associations.

Looking at the demand, it can be defined thanks to fieldwork: thanks to interviews with local shops that are pick-up points and end users, an analysis of the needs and behaviors of OOH deliveries' users can be done. This methodology was the one followed to consolidate knowledge in the study made for the Valence Romans Mobilités (VRM) territory, which is described below.

4 The Feedback from a French Local Authority Confirming the Needs to Enhance Knowledge and Monitoring on OOH Deliveries

4.1 The Territory of Valence Romans Mobilités

Valence Romans Mobilités is the local authority in charge of mobility in the living area of Valence and Romans-sur-Isère (population 258,000). That area, which stretches across two counties in south-eastern France (Drôme and Ardèche), has a varied typology of territories (mixing urban, peri-urban and rural areas). Most of the population and activities are concentrated in the two urban centres of Valence (population 122,000, which extends beyond the Rhône River, into the Ardèche county) and Romans-sur-Isère (population 43,000), as well as in the suburban communities that have registered strong recent demographic growth. The rest of the area is rural, mostly composed of villages. From the point of view of mobility and economic attractiveness, the area operates independently from the metropolitan areas of Lyon and Grenoble, both of which can be reached in around 1 h.

Valence Romans Mobilités has been developing a proactive urban logistics policy during the last 10 years, as confirmed by the Master Plan for Goods Mobility, elaborated and approved in 2016. To take better account of this issue, the plan sets out 15 actions supported by all the stakeholders involved, both public and private (Fig. 3).

In 2022, Valence Romans Mobilités wanted to go further by signing a tripartite agreement with the partners involved in the InTerLUD program, in which the local authority undertakes to continue implementing the action defined by the Master Plan for Goods Mobility, in particular by launching a study to draw up a Master Plan for Out-of-Home Deliveries.

4.2 Out-of-Home Delivery Offer on the Territory

On the Valence Romans Mobilités territory, we have had the opportunity to carry out two different studies linked to the out-of-home delivery points. First, a diagnosis of the offer and demand in 2023, and then an update of the offer in 2025, which highlighted many interesting aspects, looking at the rapid evolution of OOH delivery pick-up points that can be observed.

In 2022, the territory recorded 200 different OOH delivery pick-up points, including:

- 129 pick-up points in local shops (65%)
- 47 post offices (23%)
- 24 lockers (12%)

Figure 4 summarizes the issues of offer and demand in Valence Romans.

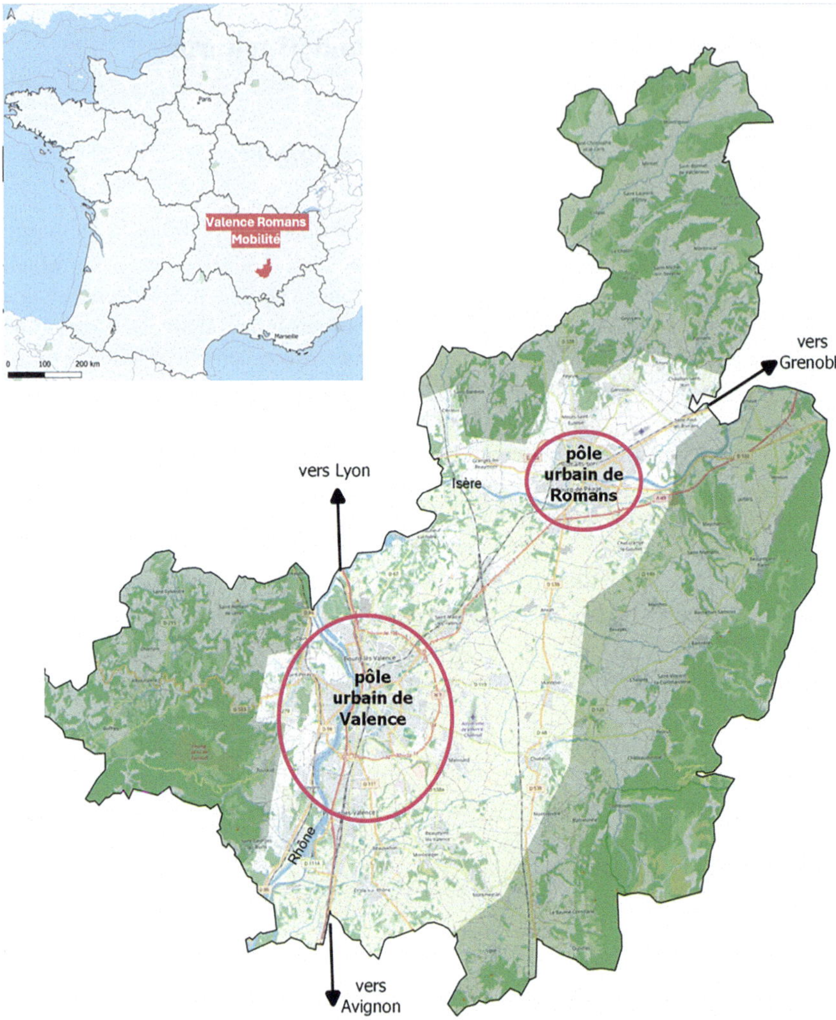

Fig. 3 Location of the urban centres in Valence Romans Mobilités

4.3 Leveraging Commuter Routes to Improve Delivery Efficiency

On the territory, our analyses of home-work journeys (commonly known as commuting) show that 50% of commutes made by people living in suburban and rural areas are destined for the central towns of Valence and Romans-sur-Isère. Given that the modal share of car use remains high, particularly in peri-urban and rural areas, there is a significant need to optimize and enhance alternative modes of transportation in the area. In terms of e-commerce, and in order to limit the need for specific

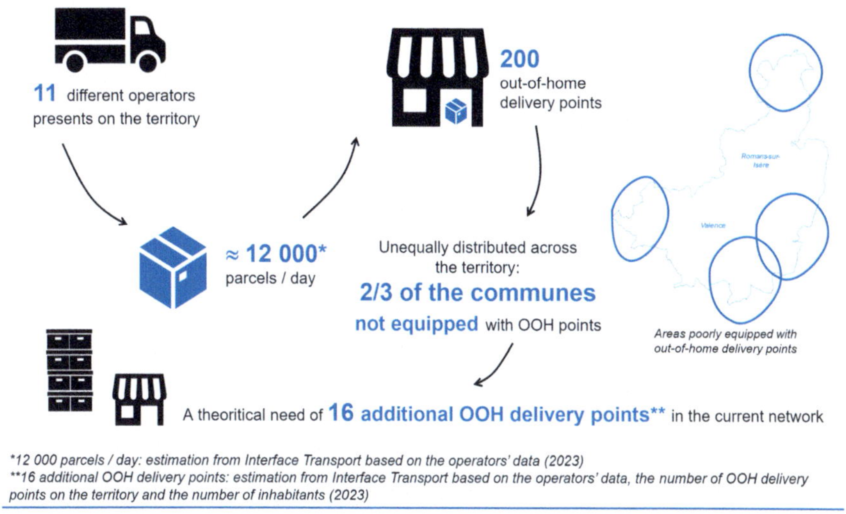

Fig. 4 Synthesis of the OOH delivery stakes, in the VRM territory (Interface Transport 2023)

journeys, the strategy could therefore be to locate new pick-up points in local shops and lockers close to the main routes linking the two main towns to the suburban and rural areas. These include the N532 (linking Valence to Romans, via the high-speed train station) and the D538 (linking the north and south of the area via Romans). The mobility dynamics are therefore something very interesting to investigate to understand and plan the development of the out-of-home delivery pick-up points network (Fig. 5).

4.4 *A Fast-Changing Network in the Short Term*

In 2025, the local authority in charge of mobility (VRM) decided to update the diagnosis of the out-of-home delivery offer on its territory by focusing solely on the registration of delivery points. The resulting outcomes are particularly remarkable in apprehending the e-commerce evolution dynamics. This way, we can highlight the following information:

- 306 out-of-home delivery points have been identified on the territory of Valence Romans Mobilités in 2025, against 200 in 2023, representing an increase of 50% in 2 years.
- 59 out-of-home delivery points (almost 30%) are no longer in service (including 20 that are due to the shop closure itself) while 165 have opened during this same period (2023–2025), representing 54% of "new points" within this new out-of-home delivery offer.

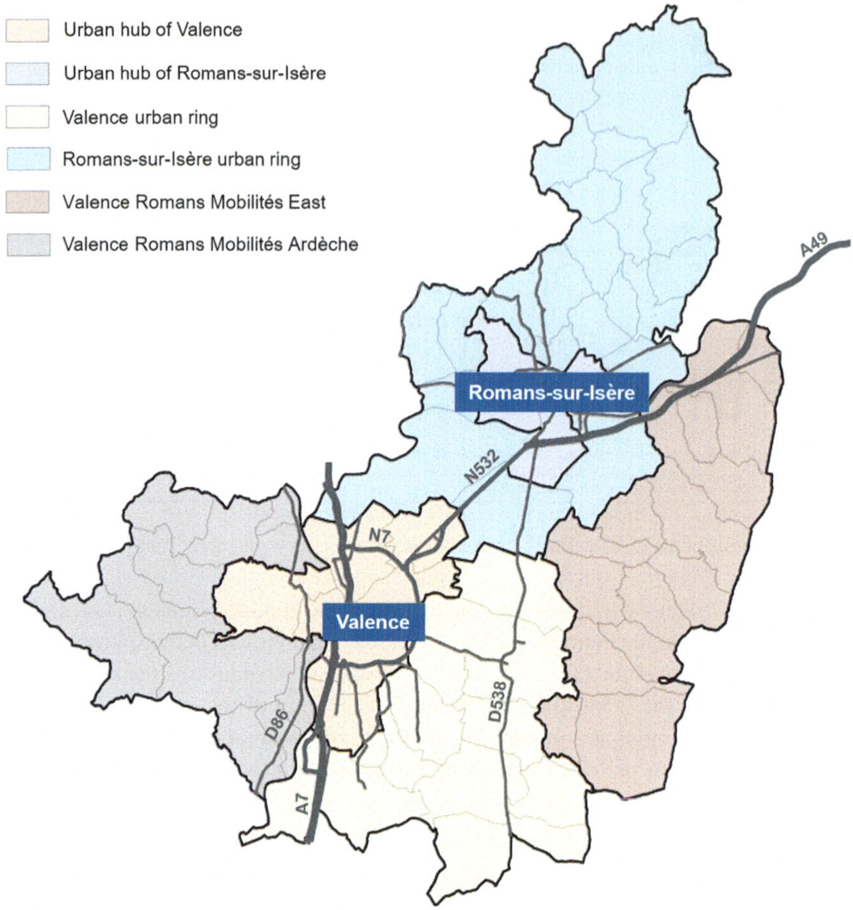

Fig. 5 Map illustrating the different urban areas of the Valence Romans Mobilités territory (Interface Transport 2023)

- A strong development of parcel lockers, which amount has been multiplied by 4: 97 in 2025, against 24 in 2023, accounting now for 32% of all out-of-home delivery points on the territory (against 12% in 2022).
- A progress of multi-network pick-up points: 60% of the out-of-home pick-up points, apart from lockers, are affiliated to several networks, against only 27% in 2023. This practice is rarely officially accepted by the operators because of the strong competition between the different networks. Nevertheless, the identification of several operators on the same site occurs when the merchants do not strictly respect the nature of their contracts, working with several networks, or when the same site welcomes side by side several parcel lockers from different operators (Fig. 6).

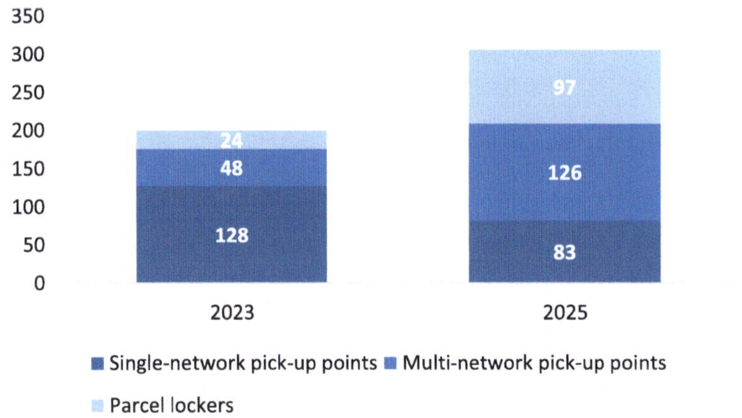

Fig. 6 Evolution of the OOH delivery points on the territory of Valence Romans Mobilités

From a local point of view, the evolution of the out-of-home delivery network between 2023 and 2025 relies on two approaches:

- a polarization of the offer that persists in the centralities of Valence and Romans,
- an expansion of the territorial coverage of out-of-home delivery pick-up points in peri-urban and rural areas, which had been identified as isolated in 2023 and are no longer so today (Fig. 7).

4.5 *The Tensions in the Centralities Confirmed in 2025*

This work on the update of the offer, and the realization of additional interviews with the main operators, has allowed us to point out, in the short term (2023–2025), the new dynamics driven by the operators' strategies.

One of the main highlights is the emergence of parcel lockers, which are still in the minority in terms of offer but are now highly popular among both network operators (opportunity to strengthen the network in the short term) and customers (24/7 availability and customer experience). In order to strengthen their network, network operators are taking proactive steps with private stakeholders (private parking owners, shopping centers, etc.) to develop lockers in a highly competitive environment. Compared with pick-up points in local shops, these parcel lockers are not affected by turnover, as the above data show. Thus, network operators are inputting a lot of effort (e.g., recruiting sales experts) to find new locations, not relying on public spaces.

Engaged in a complete logistical reorganization, Mondial Relay initiated in 2024 a deep transformation of its pick-up points' network. The company has started to considerably reduce the number of pick-up points in local shops, to the benefit of a redesigned shipping model, increasingly relying on automatic parcel lockers.

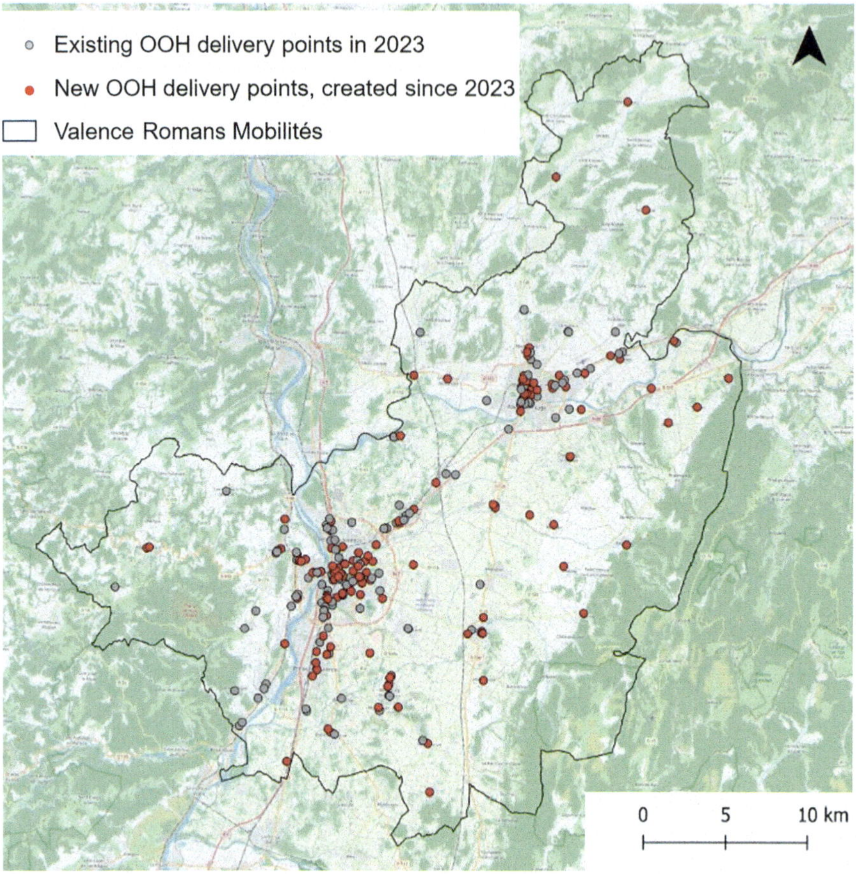

Fig. 7 Evolution of the out-of-home delivery network of Valence Romans Mobilités between 2023 and 2025 (Interface Transport 2025)

According to several press sources, close to 3500 local shops pick-up points should be removed by the end of 2025, representing nearly 25% of the historical network of Mondial Relay. Among them: 2500 would not be replaced, especially in the areas where the volume of parcels is judged as insufficient or hardly profitable, and 1000 would be redeployed elsewhere or transformed to enhance the efficiency of the network. This reduction would aim at rationalizing the network and concentrating the flows on more active pick-up points or on 100% automated solutions. Mondial Relay would hence pursue the massive deployment of automatic parcel lockers, to target notably: 7500 active lockers, against 4000 at the beginning of 2023, and an objective of 10,000 lockers at horizon 2026 (France 3 Régions 2025). This refocusing of Mondial Relay on automatic parcel lockers showcases the turning point mentioned earlier in the shipping model for out-of-home deliveries. If it addresses the expectations of the urban consumers on both topics of rapidity and autonomy, it also leads to a reconfiguration of the relations between logistics and local commerce, with contrasted effects according to the territories.

In addition, as the network evolves, there is still a need to consolidate the pick-up point offer in order to meet demand. These needs are still particularly acute in the central cities (Romans, Valence in our example) and main transit areas; the densification of the network seems to be a decisive factor in meeting demand, particularly during peaks in activity and seasonal periods.

5 Conclusion and Recommendations

This chapter presents the characteristics and particularities of the evolution of e-commerce in France over the past few years, in parallel with the will of local authorities to better understand and manage the externalities related to these consumption practices, and more widely to make urban logistics more sustainable. The presented analysis focuses in particular on the establishment of a Master Plan of out-of-home delivery points on a territory, as a lever activated by Valence Romans Mobilités. These are the early stages of a topic that is still rather undocumented. Nonetheless, several lessons learned can be highlighted on the basis of this work.

5.1 How to Determine the Best Location of Out-of-Home Delivery Points?

In terms of best practice, several trends seem to stand out to meet the demand. They are listed in the strategic document from Valence Romans Mobilités.

Several places are targeted in priority:

- The territorial centralities, embodied by the cities of Valence and Romans-sur-Isère: even if these latter are currently the better equipped in terms of density of out-of-home delivery points, all of the operators approached bear witness to a need of consolidation and renewal of the offer under these centralities to address the current tensions and saturation challenges observed within partner shops. These needs were significant in 2022 as well as in 2025.
- The structuring road axis: the roads targeted here are the ones welcoming daily commutes (home-to-work journeys). In this regard, they constitute, for their local environment, sites of interest in the perspective of developing the offer of out-of-home delivery solutions. The parcel lockers appear to be the better adapted pick-up solution for these types of places.
- The multimodal places: train stations, park and ride facilities, or carpooling stands as the transport network nodes on which parcel pick-up solutions can allow to integrate existing travel chains, while promoting, in addition, these

spaces to integrate new services for the users. The parcel lockers seem to be the solution to favour in these places.

- The connecting towns: several sectors have been identified in the frame of collaborative workshops involving the establishment of the Master Plan, as intermediate polarities on the territory (most populated towns after Valence and Romans). These municipalities have sometimes seemed underequipped in terms of offer and present, in parallel, a relative commercial density, making them attractive for the implementation of out-of-home delivery points from the operators' point of view. Furthermore, they represent potential transit points from some areas identified as sparsely equipped with OOH delivery points. The solutions to endorse here can be pick-up points in local shops, but also parcel lockers, according to the possibilities. Between 2022 and 2025, some of these places have seen their offer evolve and do not necessarily need any more implementation points, which does not allow us to generalize clear principles regarding them.

5.2 What Levers to Activate?

The operational adaptation of the Master Plan of out-of-home delivery points of Valence Romans Mobilités is structured around four distinct modes of action, co-constructed during the workshops.

The connection between merchants and network operators, in order to consolidate the amount of out-of-home delivery points on the territory, especially within the most sparsely equipped municipalities, as well as on the structuring road axis. It consists, for the municipalities (identified as places for strategic development), of sensitizing, especially the local merchants, to the opportunity of becoming an out-of-home delivery point and to give them the contacts of the network operators.

The animation of a workgroup about out-of-home delivery allows us to pursue the dynamic engagement with the local stakeholders (municipalities, network operators, merchants, recipients) in the long term.

For Valence Romans Mobilités, it consists of:

- Diffusing the results from the surveys on home-to-work journeys. This allows us to identify the most frequented axis of travel, which are the privileged places to establish out-of-home delivery points,
- Registering the new establishment of out-of-home delivery points and following the consolidation of the network over time,
- Identifying new offers and new services that would be established in the territory.

This workgroup must allow for to maintenance of the mobilization of the different stakeholders and to realign the efforts of opening out-of-home delivery points according to the evolution of delivery practices.

In the medium and long term, Valence Romans Mobilités also considers acting as a facilitator to promote the introduction of parcel lockers in the new development projects, aligned with the Master Plan. To do so, there is a need to raise awareness among the relevant services within the local authorities and the municipalities, as well as with the private project developers, to deploy this kind of service.

Finally, in the medium and long term, the production of a new set of knowledge by Valence Romans Mobilités on the practices related to households' remote shopping will allow for better anticipate the needs of a territory in terms of out-of-home delivery, to sensitize the recipients to the responsible purchase practices and to evaluate the results of the actions engaged in the long term.

5.3 For a Pragmatic and Reasoned Approach

Over and above the principles set out above, this case study demonstrates the extent to which the subject of out-of-home delivery points is particularly dynamic, with some very significant developments recorded over a short period of time. For example, between 2023 and 2025, half the number of out-of-home delivery points in the Valence Romans Mobilités area will have been renewed, and the proportion of lockers will have increased significantly over the same period. The task of defining appropriate, sustainable development principles is therefore a particularly complex one for local authorities, which have to reconcile consumers' expectations with private development strategies driven by network operators.

In this context, and with the aim of taking effective action, several levers have been identified in France and in other European countries concerned by these issues, and we believe that there is a consensus on these from the point of view of local and regional authorities:

- Work to raise collective awareness of the challenges of e-commerce and related local issues. This sharing makes sense for identifying the expectations of the ecosystem and targeting potential areas for improvement.
- Consolidate local knowledge to develop a systemic approach to mobility: the question of supply and demand for OOH delivery pick-up points is key, as is understanding the dynamics of mobility (commuting flows, use of active modes, etc.) and consumer practices.
- Anticipate the integration of out-of-home delivery solutions into future urban development projects. Depending on the nature of the projects and the associated targets, it may be decisive to determine the size of future parcel collection requirements and, where appropriate, to include services to optimize or encourage virtuous practices. These actions can help to ensure that development projects are effectively considering these new issues linked with e-commerce and OOH delivery.

References

ADEME (2025) Online business: environmental impacts of logistics, transport and travel [in French]. https://librairie.ademe.fr/societe-et-politiques-publiques/6261-7873-commerce-en-ligne-impacts-environnementaux-de-la-logistique-des-transports-et-des-deplacements.html

APUR (2022) Drive piéton, dark kitchens, dark stores. New forms of food distribution in Paris [in French]. https://www.apur.org/fr/nos-travaux/drive-pietons-dark-kitchens-dark-stores-nouvelles-formes-distribution-alimentaire-paris

ARCEP (2023) Mail, parcel and related activities markets in France—2023 [in French]. https://www.arcep.fr/cartes-et-donnees/nos-publications-chiffrees/observatoire-courrier-colis/marches-courrier-colis-activites-connexes-france-2023.html

Augereau V, Dablanc L, Curien R (2023) Delivery relays in e-commerce logistics. HAL Open Science [in French] https://hal.science/hal-04151561v1/document

Dobber R, Buijs P (2023) Policy approaches for placing parcel lockers in public space. In: Proceedings of the 9th international physical internet conference (IPIC 2023). https://pure.rug.nl/ws/portalfiles/portal/697940772/IPIC2023_-_Full_paper_Dobber_Buijs_final.pdf

FEVAD (2024) How do French people get their online orders delivered? [in French]. https://www.fevad.com/comment-les-francais-se-font-ils-livrer-leurs-commandes-sur-internet/

France 3 Régions (2025) Mondial Relay closes its relay points: rural traders in difficulty with the arrival of lockers [in French]. https://france3-regions.franceinfo.fr/grand-est/haut-rhin/mulhouse/mondial-relay-supprime-ses-points-relais-des-commercants-ruraux-en-difficulte-face-a-l-arrivee-des-lockers-3149777.html

Hovi IB, Pinchasik DR, Dong B, Strømstad H, Brunstad ØL (2023) Packaged lockers as a delivery solution: usage and impact patterns during network expansion, TØI Report 1959/2023. Institute of Transport Economics, Oslo. [full text in Norwegian] https://www.toi.no/getfile.php/1375937-1688040469/Publikasjoner/T%C3%98I%20rapporter/2023/1959-2023/1959-2023_Summary.pdf

SendCloud (2024) Out-of-Home Delivery Survey: Customers Expect Convenience with Delivery Options [in French]. https://www.sendcloud.com/fr/enquete-2024-les-livraisons-hors-domicile-de-plus-en-plus-populaires/

Urban Collection Points for People-Friendly, Environmentally Responsible, and Circular Logistics

Sébastien Horemans

1 Introduction

At the end of the COVID crisis, many analysts expected a return to physical retail at the expense of e-commerce. However, this has not been the case, as confirmed by the European E-commerce Report 2024, published by Ecommerce Europe and EuroCommerce in collaboration with the Centre for Market Insights of the Amsterdam University of Applied Sciences. The report highlights that in 2023, over 75% of Europeans were active online shoppers, with the share of e-commerce purchases reaching 15.1% of total retail sales in Europe. These figures underscore that e-commerce represents a long-term structural shift in retail, one that will continue to shape consumption patterns and logistics needs in European city centers (Ecommerce Europe and EuroCommerce 2024). Fashion items such as clothing and footwear, along with beauty products and books, remain the most popular product categories in European e-commerce. These categories share key logistical advantages: they are lightweight, non-perishable, easy to package, and generally not fragile. This synergy between product characteristics and logistics efficiency explains their dominant position in online sales.

This shift toward more sustainable consumption patterns is reshaping both the retail landscape and the logistics sector. The rise of second-hand and refurbished goods is particularly evident in the fashion and electronics markets, where platforms like Vinted, Leboncoin, and specialized refurbishment programs from major retailers—such as IKEA's take-back initiatives—are becoming mainstream. Consumers are increasingly seeking not just convenience, but also alignment with their environmental and social values when shopping online.

S. Horemans (✉)
Pick & Smile, Boves, France
e-mail: shoremans@pickandsmile.com

A. Melkonyan-Gottschalk, Y. Li (eds.), *Urban Logistics Transformation*, Greening of Industry Networks Studies 15,
https://doi.org/10.1007/978-3-032-08387-6_16

These developments place new demands on logistics systems, especially in the last mile. Traditional parcel delivery models, optimized for new, uniformly packaged goods, are often ill-suited to the specific challenges posed by second-hand and refurbished items. For example, returns and reverse logistics are more complex when handling unique, used, or repackaged goods. Moreover, the logistics of circular economy products often require additional services, such as quality control, repair, or preparation for resale.

Out-of-Home (OOH) delivery solutions—including pickup points and automated lockers—are emerging as particularly well-suited to these new demands. By enabling consolidated deliveries and facilitating reverse logistics processes, these systems not only reduce the environmental footprint of deliveries but also help address the logistical complexities associated with product returns and reuse.

As the circular economy becomes more embedded in mainstream consumption habits, logistics operators, retailers, and urban planners will need to collaborate more closely to develop infrastructure and services that support this transition. This includes optimizing pickup points to handle returns, providing storage for bulkier items, and developing digital systems that can seamlessly integrate forward and reverse logistics.

Ultimately, the convergence of e-commerce growth, innovative delivery solutions, and circular economy principles presents a unique opportunity to build urban logistics systems that are not only more efficient but also more environmentally responsible and socially inclusive. According to Kantar's Fast E-commerce Insights (Kantar 2024), while economic benefits remain the primary motivation for consumers engaging with circular economy purchases (cited by 72% of respondents), environmental concerns follow closely, influencing 38% of consumer choices. This highlights that sustainability is no longer a marginal factor, but a significant driver in shaping modern purchasing behavior.

2 Home Delivery and Out-of-Home Delivery

The delivery landscape in Europe has undergone a profound transformation in recent years. Home delivery, where parcels are delivered directly to the recipient's residence, remains the dominant method, preferred by 76% of online shoppers according to the Geopost E-shopper Barometer 2023 (Geopost 2023). However, alternative methods are growing rapidly in popularity. Out-of-Home (OOH) delivery refers to collection options where customers retrieve parcels from designated locations, such as parcel shops or automated lockers. The same survey indicates that 24% of buyers now choose to collect their parcels at a pickup point, including 23% who specifically prefer automated lockers.

The increasing use of OOH delivery reflects consumers' need for flexible logistics solutions that fit their personal and professional routines. Pickup points, often hosted within local shops or service providers, have become well-established, while automated lockers—introduced later—are expanding quickly across Europe. The central principle of OOH delivery is to consolidate multiple deliveries at a single location, reducing the number of individual trips by delivery vehicles, and thereby cutting emissions and alleviating urban congestion.

However, this shift has also introduced tensions within the logistics ecosystem. Many logistics providers are now prioritizing automated lockers over staffed pickup points to increase operational efficiency and profitability. Yet this strategy can come at a social cost. In France, for instance, one leading logistics provider abruptly closed two-thirds of its staffed pickup points, leading to public backlash and negative media attention. The move was intended to address the underutilization of lockers, but many French consumers still prefer collection points staffed by human attendants for added convenience and personal service.

These dynamics are prompting a rethinking of urban logistics strategies. Beyond environmental objectives, OOH delivery offers opportunities to strengthen local economies by supporting small businesses and creating new jobs. If designed thoughtfully, the expansion of OOH delivery networks could contribute to more sustainable, socially inclusive, and economically vibrant urban logistics systems. Out-of-Home (OOH) delivery encompasses two primary forms: parcel lockers—automated machines providing self-service access to parcels—and pickup points located within existing local businesses such as convenience stores, florists, or newsagents. While both models serve the goal of consolidating deliveries and reducing urban congestion, they offer distinct benefits and face unique challenges. Parcel flows through OOH delivery encompass both Business-to-Consumer (B2C) and Consumer-to-Consumer (C2C) shipments. According to DHL (2024), OOH deliveries represented approximately 15% of total parcel deliveries in 2020 and are projected to grow to 29% by 2025, reflecting the accelerating adoption of parcel lockers and pickup points across Europe as preferred delivery methods. It should be noted that automated lockers are suitable for parcels weighing less than 30 kg, but not for parcels with weight over 30 kg.

The C2C flow is set to play an increasingly important role in OOH deliveries, as this mode of delivery is particularly suited to the constraints of this type of transaction:

- Low delivery costs.
- Flexible pickup and delivery times.
- Possibility of developing ancillary services (product returns).

Parcel lockers provide unmatched convenience through 24/7 availability and minimal staffing requirements, making them attractive for consumers with unpredictable schedules. Pickup points, on the other hand, generate direct economic benefits for local shops through commissions on parcel handling, increased foot traffic, and opportunities for additional sales. Moreover, staffed pickup points often provide valuable social interactions, which can be especially important in smaller communities or for elderly customers. Rather than choosing between one or the other, cities and logistics providers can benefit by developing OOH delivery networks that leverage the strengths of both solutions.

3 Pick-Up Point

Although there is lack of official statistics, the number of Pick-Up Points in Europe is currently estimated at ranging between 250,000 and 400,000, based on key industry players. Approximately 80% of these points are located in stores or businesses with different core activities (e.g., florists, newsagents, supermarkets, etc.). These drop-off points are operated by major players. They are mainly carriers such as Geopost, DHL, GLS, Inpost, Geodis, and Fedex, as well as retailers such as Amazon and Vinted Go. More specific operators are also working on the distribution of fresh/food products (Table 1).

It should be noted that pure players in the field of pickup points are beginning to appear on the market, whose sole specialty is this service, such as SmilePickup in Belgium, Spain, and France, and SmilePinguin in Canada. Smile Pickup is a network of pickup points for parcels weighing over 30 kg, and currently has 280 pickup centers in Belgium, Spain, and France. These parcels, commonly known as "large parcels," particularly concern furniture, but also bulky parcels destined for the industrial sector.

Urban and peri-urban Pick-Up Points offer an opportunity to integrate with a wide range of services that cities and citizens need. In addition to parcel pickup and drop-off, management of product returns for new purchases, back-up storage for local businesses, e.g., maintenance companies and craftsmen, service points, and charging facilities for cargo bikes or electric light commercial vehicles.

Table 1 Advantages and limitations of Pick-Up Points for different stakeholders

Type of stakeholder	Advantages	Limitation
Property owners & Puck-up point operators	Generate traffic and therefore potential sales from customers who come to pick up their parcel, and are likely to buy a product or service. Create a social link through the exchange of goods when the parcel is delivered.	For some outlets, limitations on storage space when this is not the main activity.
E-commerce customers	Avoid the risk of having their parcel stolen from their mailbox. Create a social link through the exchange of goods when the parcel is delivered, which can be very valuable, especially for senior citizens Avoid receiving the usual delivery notices, which are much more related to the burden of delivery than to the absence of the recipient.	Limitations on opening hours
Logistics operators	Not having to invest in logistics premises Not having to recruit employees to manage parcel delivery	Lower profitability

4 Automated Parcel Machine (APM)

Automated Parcel Machines (APM), commonly known as parcel lockers, have experienced significant growth across Europe since the early 2000s. These self-service lockers allow consumers to retrieve or drop off parcels at their convenience, typically using QR codes or PIN codes. The key appeal of APMs lies in their simplicity: they require only a basic electrical connection and access to a mobile or wired network. For property owners, such as retailers or residential buildings, they are relatively easy to host, as they occupy limited space and generate modest rental income.

Major operators of parcel locker networks in Europe include Geopost, GLS, InPost, and DHL. Unlike staffed pickup points (PUPs), where the hosting location earns a small commission per parcel handled, APMs usually operate on fixed rental agreements, irrespective of the number of parcels processed. This model offers predictability for operators but disconnects host earnings from usage levels (Table 2).

Most parcel lockers are designed to accommodate small parcels, typically weighing just a few kilograms. While larger compartments or "XL crates" are being piloted to handle bulkier items, profitability challenges have limited their widespread adoption. However, for specific retail sectors—such as furniture and sporting goods—automated lockers provide valuable out-of-hours pickup options. Retailers like Leroy Merlin, IKEA, and Decathlon have integrated lockers into their customer service models for precisely this reason. The following figures show two types of parcel lockers.

Table 2 Advantages and limitation of APMs for different stakeholders

Type of stakeholder	Advantages	Limitation
Property owners	Generates modest rental income and occasional foot traffic	Require a vehicle and driver to stop for the entire load
E-commerce customers	Flexible hours Selecting a location offering the possibility to pick up a parcel without additional trips	Limited space for specific sizes when returning goods (no solution for a large package) No customer queries service Sometimes leads to handling errors: scanning the barcode and placing an unpackaged product in the locker.
Logistics operators	Easy to deploy with minimal regulatory barriers Low operational costs	No solution for large packages

APMs for small parcels—photo taken in a residential area in France

Parcel lockers for extremely large items—photo taken in Munich Transport Logistics exhibition 2025

Despite these benefits, the transition to automated lockers has not been without controversy. Since 2023, several logistics companies have actively pursued the replacement of traditional pickup points with APMs to reduce costs. In a notable case, InPost, through its subsidiary Mondial Relay, abruptly closed two-thirds of its staffed pickup points in France in early 2025. The decision triggered public backlash and media criticism, particularly from small retailers and local artisans who relied on the commission from parcel handling as an important source of supplemental income.

While APMs clearly offer efficiency advantages for logistics operators and increased convenience for some consumers, the transition raises broader questions about social impact, particularly regarding employment and support for local businesses. Balancing operational efficiency with the social and economic roles of pickup points remains a key challenge for sustainable urban logistics development. Quadient is notably developing a network of APMs in Europe, again as a dedicated activity, without being a carrier.

5 Furniture Recycling and Logistics Innovation in the Circular Economy

The return and processing of end-of-life furniture and associated waste have become critical issues in Europe, particularly in light of the rapid expansion of e-commerce. The European Union has enacted a series of measures to enhance recycling and promote the circular economy in various sectors, with the French framework for furniture waste management serving as a noteworthy example.

Since 2017, the French regulation on Furniture Waste (Déchets d'Éléments d'Ameublement, D.E.A.) has reshaped the operational landscape of the furniture industry. Anchored in Decree No. 2017-1607 and the associated Order of 27 November 2017, this regulation introduced specific technical and financial requirements aimed at increasing recovery rates and reuse potential. The regulation has had a significant influence on market dynamics, compelling stakeholders to adapt their logistics, collection, and processing operations in a highly competitive environment (DHL 2024).

According to the French Environment and Energy Management Agency (ADEME), 2.69 million tonnes of furniture were placed on the market in 2018, yet only 990,000 tonnes were collected for treatment (République Française 2017). In 2020, Ecomaison (formerly Eco-Mobilier), the national body responsible for managing the D.E.A. stream, collected 842,000 tonnes—representing a 4% decline from 2019 due to COVID-19 disruptions. Of the treated waste in 2020, 4.9% was reused, 46% recycled, and 42% used for energy recovery. This marked a decline in circular performance from 2019, when recycling rates reached 51% and reuse 6% (ADEME 2019).

Targets introduced by the decree required, from 2021 onward, that 1.5% of household furniture waste and 5% of professional D.E.A. be made available to social economy actors, with at least 60% of that amount eligible for reuse. In 2022, the regulation mandated that 50% of separately collected D.E.A. be recycled and 90% recovered through reuse, recycling, or energy valorization. By 2023, 40% of all D.E.A. had to be separated from general waste streams, significantly raising the bar for sorting and logistics systems (Ecomaison 2020).

Key stakeholders affected by these regulations include domestic producers, EU-based distributors, extra-EU importers, white-label retailers, and online platforms. In response, the supply chain has had to evolve. Returned furniture items—typically bulkier, more fragile, and less hygienically packaged than new products—require specialized handling. This challenge has led to the emergence of XL pickup and drop-off points, with companies such as SmilePickup and C Chez Vous offering dedicated D.E.A. collection services when customers receive new items (Legifrance 2017).

In October 2021, Ecomaison launched a nationwide call for tenders to strengthen its regional collection infrastructure. This initiative underscored the growing importance of waste logistics and provided new opportunities for service providers and logistics firms (SmilePickup 2025).

Online furniture sales in France are also surging. While large furniture had a mere 10% online market share in early 2020—compared to 15% in the UK or US—forecasts suggest that up to 50% of all furniture and home decor purchases will be made online within the next 20 years (Ecomaison 2021). This growth implies a fivefold increase in the volume of furniture waste entering the D.E.A. stream, stressing an already under-capacitated collection and processing infrastructure. Efficient logistics for reuse, particularly for moisture-sensitive and damage-prone items, will be essential.

To meet evolving environmental requirements, the sector must embrace ecodesign strategies. This includes manufacturing furniture that is easy to disassemble, repair, and reintroduce into circulation—a design ethos already prevalent in the electronics industry. Ecomaison has initiated a "collaborative ecodesign" approach to promote such practices (Xerfi 2021).

Technological innovation will also be key. Enhancements in sorting capacity, hygiene protocols, and reverse logistics infrastructure—such as automated platforms and specialized transport vehicles—are expected to contribute significantly. Crucially, these trends present not only regulatory challenges but also business opportunities. Circular economy practices, particularly reuse and second-hand markets, have generated strong revenue streams. Companies like Advinta, eBay, and Vinted have capitalized on this momentum. Pickup points (PUPs) now cater to their clients, handling large volumes of parcels and offering scalable alternatives to lockers, especially for bulky items (Ecomaison 2025; Vinted 2025).

PUPs can also serve as temporary storage for repairable appliances, further reinforcing their role in circular systems. Unlike lockers, they can easily handle XL shipments and items requiring space, ventilation, or specific handling. In sum, the circular transformation of the furniture sector is only beginning. With the right infrastructure and logistics support, pickup points can play a pivotal role—not only in supporting environmental goals but also in reshaping the logistics of e-commerce.

6 Multifunctional Pickup Points as Urban Service Hubs

As cities seek more sustainable and efficient models of urban service delivery, pickup and drop-off points (PUPs) are emerging as critical infrastructure nodes. Beyond their original function in e-commerce logistics, these spaces are increasingly being conceptualized as multi-functional urban hubs. Strategically located, human-operated PUPs offer the opportunity to integrate a wide array of services, enabling synergies between circular economy practices, sustainable mobility, and community needs.

6.1 Ad Hoc Shared Warehouse for SMEs and Local Business

In dense urban environments and commercial centers, auxiliary storage plays a pivotal role in supporting local businesses, especially small and medium-sized enterprises (SMEs). During peak sales periods or logistical disruptions, businesses often require temporary off-site storage to handle overflow stock or manage reverse logistics flows. For example, fragrance brands offering return incentives—such as vouchers in exchange for used bottles—frequently face daily return volumes that exceed on-site capacity. Utilizing local pickup and drop-off points (PUPs) as micro-hubs enables efficient consolidation and retrieval of returned goods, often by the same courier delivering new stock. This not only minimizes transportation overheads but also supports circular business models by promoting reuse and recycling practices (Mangiaracina et al. 2015).

Such proximity logistics are increasingly vital. Many urban businesses suffer from spatial constraints, limited staff availability during delivery hours, or erratic workloads. Pickup points serve as cost-effective and flexible delivery alternatives, especially for craftspeople and maintenance technicians who are rarely available to receive items directly (Allen et al. 2018). In the building sector, for instance, materials and equipment must often be delivered outside standard working hours. Traditional parcel lockers are unsuitable for large or fragile items. Pickup points equipped with secure, human-managed drop-off spaces (SAS) can accommodate these deliveries. This presents an opportunity to expand service offerings and employment while addressing unmet needs in urban logistics infrastructure (Crainic et al. 2009).

6.2 Urban Hubs for Sustainable Last-Mile Delivery

The operational efficiency of soft-mobility solutions—such as eCargoBikes, electric microvans, or pedestrian carts—is constrained by their limited range, typically not exceeding 3 kilometers before costs become prohibitive. To extend their viability, PUPs can act as urban departure hubs, enabling last-mile deliveries using clean, lightweight vehicles. Strategically locating such hubs across city centers and suburbs allows operators to pool deliveries and reduce redundant trips (Gonzalez-Feliu et al. 2014).

Starting in October 2025, France and Poland will pilot eCargoBike-based deliveries from designated pickup points, with vehicles capable of carrying loads of up to 250 kg. Similar initiatives have been trialed in Spain, where FM Logistic's CityLogin project deployed electric scooters with modular containers to serve urban delivery loops in Madrid (FM Logistic 2025).

These initiatives not only reduce greenhouse gas emissions but also optimize route density. PUPs, with their pre-existing physical footprint and customer traffic, are ideally suited to host such delivery systems—serving both retail logistics and circular economy applications.

6.3 A Platform for Essential Urban Services

The gradual disappearance of local amenities—such as banking branches, healthcare services, or postal counters—has contributed to the social decline of many urban and peri-urban neighborhoods. As a response, multifunctional PUPs can provide decentralized access to essential services, acting as community anchors and revitalizing dormant urban zones (Dablanc and Montenon 2015).

In addition to parcel handling, these facilities can integrate services such as key duplication, small repairs, administrative assistance, and basic health services. Importantly, their operation typically requires two full-time employees, offering meaningful employment opportunities, particularly for individuals with limited formal qualifications or those re-entering the workforce (Muñuzuri et al. 2015).

7 Recommendation

Pickup points (PUPs) present clear advantages in terms of service efficiency, environmental sustainability, and local employment. However, fully realizing their potential requires advanced, data-driven logistics systems. The integration of artificial intelligence (AI)—particularly in last-mile delivery—enables a more adaptive and precise approach to out-of-home (OOH) delivery management.

AI tools can process real-time data on demand patterns, traffic conditions, and operational constraints to support dynamic routing, resource optimization, and scenario planning. These capabilities translate into tangible outcomes such as:

- Reduced greenhouse gas and air pollutant emissions.
- Decreased urban congestion.
- Lower logistics and delivery costs.

A critical application of AI lies in identifying optimal locations and configurations for PUPs. This begins with the collection of comprehensive datasets, including demographic profiles, commercial activity, accessibility, and transport links. Public data sources often serve as the foundation (McCormack and Goodchild 2020).

To ensure the reliability of predictive models in urban logistics, data preprocessing is a critical step. Raw datasets must be thoroughly standardized and cleaned to resolve inconsistencies, correct anomalies, and address missing values. Only after establishing a coherent and high-quality data foundation can machine learning algorithms—such as Random Forest or XGBoost—be effectively applied (Chen and Guestrin 2016). These models are well-suited for handling complex, multivariate datasets and are particularly valuable in identifying high-demand zones and optimizing the spatial configuration of pickup points (PUPs).

Beyond site selection, maintaining high levels of customer service also depends on accurate forecasting of operational needs. This includes anticipating both material and human resource requirements at each PUP location, which in turn relies on precise estimations of parcel flow. By analyzing projected delivery volumes, companies can allocate staffing, schedule deliveries, and ensure smooth handling of incoming orders.

Traditionally, forecasting models have relied predominantly on historical data to predict order volumes. While useful, this approach has clear limitations, particularly in its inability to account for disruptive external factors—such as public holidays, labor strikes, or health crises—that can significantly influence consumer behaviour and delivery patterns. To improve responsiveness, more sophisticated models that incorporate real-time data and contextual variables are increasingly necessary. It should also take into account the proximity of commercial zones, such as shopping centers, or key infrastructure, such as highway exits. The objective is to allow the algorithm to pinpoint the most efficient and accessible locations, enabling couriers to deliver parcels as close as possible to large numbers of consumers. This facilitates minimal travel distances for each customer retrieving a parcel—whether during their commute, while shopping, or on their way home.

References

ADEME (2019) Déchets d'éléments d'ameublement: Données 2018. *Agence de la transition écologique*

Allen J, Piecyk M, Piotrowska M, McLeod F (2018) Understanding the impact of E-commerce on last-mile light goods vehicle activity in urban areas. Transport Studies Unit, University of Westminster, London

Chen T, Guestrin C (2016) XGBoost: a scalable tree boosting system. In: Proceedings of the 22nd ACM SIGKDD international conference on knowledge discovery and data mining. ACM, New York, pp 785–794

Crainic TG, Ricciardi N, Storchi G (2009) Models for evaluating and planning city logistics systems. Transp Sci 43(4):432–454

Dablanc L, Montenon A (2015) Impacts of E-commerce in urban logistics: analysis and policy recommendations. Eur Transp Res Rev 7(3):25–38

DHL (2024) Out-of-home delivery & returns trends 2025. Available at: https://www.dhl.com/global-en/home/insights/topics/ooh-delivery-report.html (Accessed: 23 June 2025)
Ecomaison. Rapport d'activités 2020. https://ecomaison.com/ (accessed June 2025)
Ecomaison (2021) Appel à candidatures: Renforcement du réseau de collecte
Ecomaison. Ecoconception collaborative. https://ecomaison.com/innovation/ (accessed June 2025)
Ecommerce Europe & EuroCommerce (2024) European E-commerce report 2024. Centre for Market Insights, Amsterdam University of Applied Sciences. https://www.ecommerce-europe.eu/research/european-ecommerce-report-2024/
FM Logistic. CityLogin: Urban logistics with zero emissions. https://www.fmlogistic.com (accessed June 2025)
Geopost (2023) E-shopper Barometer 2023. Available at: https://geopost.com/en/insights/e--shopper-barometer-2023/ (Accessed: 23 June 2025)
Gonzalez-Feliu, J., Semet, F., Routhier, J.L.: Sustainable urban logistics: concepts, methods and information systems. Springer-Verlag Berlin Heidelberg (2014)
Kantar (2024) Fast E-commerce insights. Available at: https://www.kantar.com/campaigns/fast-ecommerce-insights (Accessed: June 2025)
Legifrance (2017) Arrêté du 27 novembre 2017 relatif aux exigences techniques et financières de la filière DEA. J Officiel de la République Française
Mangiaracina R, Perego A, Seghezzi A, Tumino A (2015) Innovative solutions to reverse logistics challenges: a literature review. Int J Phys Distrib Logist Manag 45(5):568–590
McCormack E, Goodchild A (2020) Delivering goods in urban areas: are autonomous delivery robots part of the solution? Transport Lett 12(8):540–547
Muñuzuri J, Cortés P, Guadix J, Onieva L (2015) Solving the last mile issue: receiving practices in e-commerce retailing. Transp Res E Logist Transport Rev 75:1–18
République Française (2017) Décret n° 2017-1607 du 27 novembre 2017 relatif aux déchets d'éléments d'ameublement. J Officiel de la République Française
SmilePickup. Services de collecte DEA. https://smilepickup.fr (accessed June 2025)
Vinted. Rapport sur l'économie circulaire en Europe. https://vinted.com (accessed June 2025)
Xerfi (2021) Meubles et décoration : perspectives de marché à horizon 2040. Xerfi Precepta

Extended Producer Responsibility (EPR) in Urban Logistics: Enabling Circular Economy Through Efficient Waste Take-Back Schemes

Tom Vöge and Julia Hobohm

1 Introduction

1.1 Background

Extended producer responsibility (EPR) schemes have emerged as a key policy tool of the circular economy to shift waste management responsibilities upstream to producers. However, integrating EPR into urban logistics systems presents a range of challenges related to coordination, infrastructure, data tracking, and consumer participation.

At the same time, urban logistics is undergoing a significant transformation driven by the rapid growth of e-commerce and the pressing need for sustainable development. One major challenge is managing the increasing volume of waste generated by urban consumption, particularly packaging, electronics, textiles, and batteries. As cities strive to become more circular and climate-neutral, the logistical infrastructure required to enable product return, reuse, and recycling must evolve along with it.

1.2 Problem Formulation

EPR schemes are designed to internalise the environmental costs of products by making producers financially and operationally responsible for their post-use phase. However, the implementation of EPR in urban logistics systems remains fragmented

T. Vöge (✉) · J. Hobohm
GRS Batterien Service GmbH, Hamburg, Germany
e-mail: voege@grs-batterien.de

A. Melkonyan-Gottschalk, Y. Li (eds.), *Urban Logistics Transformation*, Greening of Industry Networks Studies 15,
https://doi.org/10.1007/978-3-032-08387-6_17

and under-optimised. The role of logistics providers, e-commerce platforms, and municipalities in enabling effective waste take-back schemes under EPR remains poorly defined and insufficiently integrated into urban planning and policy frameworks.

The traditional linear model of urban logistics, based on the principle of take, make, use, and dispose, has proven inadequate for addressing the growing volumes of waste generated by e-commerce and urban consumption. In particular, the absence of efficient take-back systems for end-of-life products such as batteries, electronics, textiles, and packaging contributes to unsustainable levels of urban waste, greenhouse gas emissions, and resource depletion. As cities become more densely populated and e-commerce continues to expand, the pressure on urban logistics systems increases, exacerbating these challenges.

1.3 Objectives

This section aims to investigate the potential of EPR schemes as a strategic tool for advancing circularity in urban logistics systems. Specifically, it seeks to explore how EPR-driven waste take-back systems can support the integration of reverse logistics into urban freight operations and how this contributes to more resource-efficient, low-carbon cities.

A core objective is to assess the intersection of EPR and urban logistics, identifying synergies and gaps in policy, infrastructure, and stakeholder coordination. The section also seeks to contribute new insights to the evolving field of circular urban logistics by analysing existing practices, challenges, and innovations at the local and European levels.

Additionally, the section aims to develop conceptual and practical guidance for embedding EPR more effectively into urban planning and transport strategies. This includes examining the operational design of take-back systems, the role of producer responsibility organisations (PROs), and the impact on last-mile delivery networks. Through this focus, the section ultimately aims to enhance the academic and practical understanding of how EPR can drive systemic transformation in urban logistics to support a circular economy.

2 Literature, Current Practices, and Knowledge

2.1 EPR in Theory and Policy

The emergence of EPR as a key policy mechanism has been widely studied within the context of environmental governance and waste management. Initially conceived by Thomas Lindhqvist (2000) in the 1990s, EPR aims to shift the responsibility of post-consumer waste from municipalities to producers, thereby incentivising product design for durability, repairability, and recyclability (Bocken et al. 2016). Over the past two decades, the EPR concept has evolved into a central pillar of circular economy policy (Rizos et al. 2017), especially in Europe, where it is enshrined in regulations for packaging, electrical and electronic equipment, batteries, and, more recently, textiles.

At the European policy level, the Waste Framework Directive (2008/98/EC, revised in 2018) provides the legal foundation for EPR implementation across Member States. The directive requires the establishment of EPR schemes for key waste streams, notably batteries (2006/66/EC), WEEE (2012/19/EU), and packaging (94/62/EC). Recent developments, such as the Circular Economy Action Plan (2020) and the Ecodesign for Sustainable Products Regulation (ESPR 2024), further strengthen the legislative framework by integrating circularity and product durability considerations.

From a theoretical standpoint, EPR is situated at the intersection of environmental economics, product stewardship, and regulatory innovation. It is based on the "polluter pays" principle and offers a market-based approach to internalising environmental costs across product life cycles. The literature outlines (EEA 2023; OECD 2016; UNEP 2018) two primary goals of EPR: first, to improve the environmental performance of products by incentivising eco-design, and second, to ensure efficient collection and treatment of waste through dedicated systems operated by producers or third-party organisations.

In urban contexts, however, EPR's relationship with logistics is less developed in policy and practice. While the regulations set collection and recycling targets, they often leave the logistical execution, including first-mile collection and reverse logistics flows, to be interpreted and implemented by producer responsibility organisations (PROs), local authorities, or third-party logistics providers. This regulatory gap presents both a challenge and an opportunity for integrating EPR schemes more explicitly into urban logistics planning.

2.2 *Urban Logistics and Circular Economy*

Urban logistics is undergoing a profound transformation, driven by the rise of e-commerce, increased urbanisation, and growing pressure to decarbonise freight transport. Literature in the fields of transport geography, supply chain management, and sustainable urban planning has documented the growing complexity of last-mile delivery systems, the environmental impact of freight vehicles, and the spatial challenges of urban freight infrastructure.

In this context, circular economy models are increasingly recognised as an opportunity to redesign urban logistics systems. Reverse logistics, i.e. the flow of products and materials back from consumers to producers, has been highlighted as a key enabler of circularity in cities. While reverse flows are traditionally associated with returns in e-commerce or faulty goods, circular models extend this to include take-back of end-of-life products and materials.

The integration of EPR take-back systems into urban logistics offers a strategic pathway to operationalise circularity. Yet the literature reveals that this integration is still at a nascent stage. While some studies point to the potential for shared logistics infrastructure and decentralised collection hubs, few systematically examine the role of EPR schemes in optimising reverse logistics within urban areas. There is also a lack of empirical data on how PROs design their collection networks and the extent to which these align with municipal waste systems or urban freight policies.

2.3 Current Practices: Fragmentation and Innovation

The practical implementation of EPR schemes across Europe varies widely. In many Member States, PROs are responsible for organising the logistics of collecting waste products from retailers, municipal collection points, or directly from consumers. However, these systems are often fragmented, lacking coordination with local urban logistics plans and sustainability goals.

In the battery sector, for example, collection points are frequently limited to retail locations or municipal facilities, leading to inconsistent coverage in densely populated urban neighbourhoods. Similarly, WEEE take-back systems often rely on bulky item collection or civic amenity sites, which are not always accessible or convenient for city residents.

That said, examples of innovation are emerging. In France, the non-profit organisation ecosystem has piloted mobile collection events for small electronics in urban areas, partnering with local authorities and mobility providers. In the Netherlands, the textiles PRO Sympany has developed centralised return stations in shopping centres to encourage citizen participation. Meanwhile, in Sweden, reverse vending machines, traditionally used for bottle deposit returns, are being adapted to collect used textiles under EPR initiatives.

These emerging practices show the potential for more integrated, user-friendly, and data-driven take-back systems. However, they also highlight the need for clearer roles, responsibilities, and infrastructure planning between producers, PROs, municipalities, and logistics providers.

2.4 Gaps in Knowledge and Research Needs

Despite the increasing policy emphasis on EPR and the circular economy, the role of urban logistics remains underexplored. Academic research has yet to fully address how EPR systems intersect with urban freight systems and what logistical models could best support circularity goals in dense urban settings. Several key research gaps have been identified:

- **Lack of data on reverse flows:** While there is considerable data on outbound deliveries (e.g. parcel volumes, vehicle trips), little is known about the spatial and temporal dynamics of reverse logistics driven by EPR schemes.
- **Limited integration with urban mobility planning:** Urban logistics is often siloed from broader mobility and land-use planning, leading to inefficiencies, and missed opportunities for circular integration.
- **Inconsistent stakeholder collaboration:** Effective EPR schemes depend on coordination across producers, municipalities, retailers, and logistics providers. However, such collaboration is rarely institutionalised.
- **Unclear environmental impacts:** While EPR is intended to improve environmental performance, few studies quantify the emissions or resource savings from integrating EPR-driven reverse logistics into urban freight systems.

Addressing these gaps will be essential for designing urban logistics systems that are not only efficient and responsive but also aligned with the environmental and social objectives of the circular economy.

3 Case Studies

3.1 Examples of EPR-Schemes

To understand the evolving intersection of EPR and urban logistics in the context of the circular economy, this section presents a selection of case studies from across Europe, each offering unique insights into how take-back systems are operationalised in urban environments.

Battery Collection in Norway (RENAS)
Norway's battery take-back system is jointly managed by RENAS and Batteriretur, which specialises in battery recycling. Together, they operate a nationwide EPR system that includes a strong focus on urban logistics for both portable and industrial batteries.

In cities like Oslo and Bergen, the system uses a hub-and-spoke model for efficient urban collections. Small collection boxes for portable batteries are placed in supermarkets, electronics retailers, and public buildings, while larger units for industrial and automotive batteries are serviced from regional consolidation hubs. The scheme collaborates with logistics providers to optimise collection routes using GPS and scheduling software, ensuring cost-effective and low-emission urban pick-ups.

This case demonstrates how a collaborative, safety-conscious logistics approach, supported by digital planning tools and specialised infrastructure, can enhance EPR compliance in complex urban environments.

WEEE Take-Back in France (Ecosystem)

France has pioneered innovative approaches to WEEE take-back through its non-profit PRO ecosystem. In Paris, the ecosystem has launched mobile collection events in collaboration with local governments and waste companies. Citizens are notified via app or SMS about nearby collection events, often located at transit hubs or local markets.

This flexible approach complements the existing network of fixed drop-off points. It helps to address participation barriers in dense urban areas where access to civic amenity sites is limited. The initiative also gathers granular data on collection volumes by neighbourhood, enabling tailored outreach and optimisation of resource allocation.

This case underlines the value of agile, consumer-friendly collection strategies, especially in complex urban environments. It also demonstrates how digital engagement and modular logistics models can significantly enhance EPR effectiveness.

Textile Return Hubs in the Netherlands (Sympany)

In the Netherlands, the textile PRO Sympany has developed a network of return hubs in retail locations, including shopping malls and public transport stations, to make clothing donation and recycling more convenient.

Sympany coordinates with local authorities and private waste companies to ensure collected materials are sorted into categories for reuse, recycling, or energy recovery. The urban logistics model combines fixed routes with dynamic pick-up schedules based on weather, foot traffic, and seasonal trends in textile disposal.

This model exemplifies how EPR for textiles can be embedded into everyday urban infrastructure and consumer behaviour. It also shows that take-back systems can drive high reuse rates when convenience and visibility are prioritised.

These examples show that decentralised systems with higher consumer accessibility (e.g. retail or mobile collection points) tend to achieve higher collection rates but incur slightly higher logistics costs. However, these systems also result in fewer emissions per tonne due to reduced illegal dumping and increased reuse rates. The findings support a hybrid model where convenience and coverage are balanced through intelligent routing and digital tools. The examples also highlight several common themes:

- The importance of consumer accessibility and awareness in boosting participation in take-back systems.
- The value of partnerships between PROs, municipalities, and private logistics providers.

- The role of digital technologies, such as fill-level sensors, apps, and GIS, in improving logistics planning.
- The potential trade-off between system convenience and cost-efficiency, which can be optimised through careful design.

3.2 Repair and Reuse as Cornerstones of Circular Urban Logistics

Repair and reuse are essential strategies in the hierarchy of waste management, situated above recycling in terms of environmental benefit. Within the EPR framework, these practices help extend product lifespans, reduce demand for virgin resources, and limit the volume of waste requiring collection and treatment. Yet, they are often underutilised in current urban systems, where take-back infrastructure and producer obligations are primarily focused on end-of-life management rather than waste prevention. Integrating repair and reuse into EPR schemes offers cities the opportunity to reduce costs, create local jobs, and encourage more sustainable consumer behaviour.

The introduction of eco-modulated EPR fees is a powerful mechanism to promote design for durability, repairability, and upgradability. When producers are financially rewarded for more sustainable products, they have stronger incentives to rethink product architecture and material choices. Policy frameworks should also mandate the availability of spare parts, repair manuals, and software updates for a minimum period after sale. Additionally, reduced EPR fees could be offered for producers who support take-back systems that include certified repair services or refurbishment pathways.

Repair and Reuse: iFixit and the Right to Repair Movement (EU/Global)

iFixit, an online repair community and toolkit provider, has become a central actor in the global right to repair movement. Originally launched in the United States, iFixit has expanded its influence in Europe through partnerships with EU institutions and NGOs. The platform empowers consumers and small businesses to repair electronic products by offering open-access repair manuals, spare parts, and tools. In collaboration with the European Right to Repair campaign, iFixit has contributed to the European Commission's development of ecodesign and durability requirements for smartphones and tablets.

iFixit's model demonstrates how open-source repair knowledge and community engagement can help extend product lifespans and reduce e-waste. It also highlights the role of civil society and grassroots innovation in complementing policy instruments like EPR. While not a producer-led EPR scheme, the success of iFixit illustrates how repair-oriented infrastructures and business models can flourish when supported by transparency, legislation, and consumer demand, underscoring the potential of integrating repair more directly into formal EPR take-back frameworks.

Urban areas are well-suited to hosting decentralised repair infrastructure due to their dense populations, existing retail ecosystems, and proximity to waste generation. Municipalities and EPR schemes can support the creation of local repair hubs and facilitate training programmes for technicians. Collaboration with existing repair initiatives, such as repair cafes or social enterprises, can scale grassroots efforts and make repair a more accessible and affordable option for consumers. Tax incentives and public procurement standards can further boost demand for refurbished products.

Despite its benefits, repair still faces significant barriers, including high labour costs, limited consumer awareness, and product designs that impede disassembly. A shift in culture is needed to normalise reuse and repair as preferred alternatives to disposal. Urban EPR schemes should include public awareness campaigns, collection systems that distinguish reusable goods from waste, and digital platforms that connect consumers with repair providers. These efforts can enhance material efficiency while empowering citizens to participate actively in the circular economy.

3.3 Urban Logistics Hubs for Waste Take-Back

Urban logistics hubs are emerging as critical components in the infrastructure needed to implement EPR take-back schemes efficiently. These hubs serve as decentralised centres for the aggregation, sorting, and temporary storage of collected products, enabling faster, more cost-effective reverse logistics. Strategically located within or near urban centres, they reduce transport distances and emissions, helping cities meet their climate and circularity goals. By integrating such hubs into existing urban logistics frameworks, cities can optimise the flow of both goods and returned materials.

Effective urban logistics hubs must be adaptable to handle a range of product types and waste streams, including electronics, textiles, batteries, and packaging. Modular designs allow flexibility in space allocation, while digital tools can support real-time tracking, inventory management, and coordination with downstream recyclers or refurbishers. Co-location with retail centres or parcel delivery depots can generate synergies and help lower operational costs. Additionally, safety and environmental compliance, particularly for hazardous waste, must be built into hub design.

The operation of urban hubs may involve diverse stakeholders, including PROs, logistics providers, municipalities, and retailers. Public–private partnerships can leverage the strengths of each actor, combining regulatory oversight with market-driven efficiency. Business models may include shared infrastructure funded through EPR fees, or pay-per-use arrangements that allow smaller players to participate. Integration with digital product passports and return platforms can further enhance efficiency and transparency.

Urban Logistics Hub for Waste Take-Back: Paris's "Recycl'Hub" Pilot
In 2023, the City of Paris launched the "Recycl'Hub" initiative, a pilot network of urban logistics hubs designed to improve waste take-back and recycling flows for small electronics, batteries, and textiles. Located in underutilised urban spaces such as parking garages and former warehouses, Recycl'Hubs act as micro-distribution centers where citizens, retailers, and logistics providers can drop off or sort end-of-life products for collection by authorised PROs.

Each hub integrates digital tracking and QR-code-based registration to ensure transparent waste flows and data sharing between PROs, municipalities, and logistics operators. Designed to support local EPR compliance and reduce last-mile transport emissions, the hubs also provide storage and sorting infrastructure for repairable or reusable items. Through collaboration with PROs like Ecologic and ecosystem partners such as La Poste, the project has improved collection rates in dense urban areas while raising citizen awareness of take-back options.

The Recycl'Hub model offers a compelling example of how EPR logistics can be embedded into urban infrastructures, helping close local material loops and reduce environmental impact in complex city environments.

To ensure long-term viability, urban logistics hubs must be scalable and replicable across different urban contexts. Standardised planning guidelines and zoning support are key to enabling rapid deployment. Pilot programmes can help cities and PROs test layouts, technologies, and business models. Benchmarking against circular economy KPIs can guide continuous improvement. Ultimately, the widespread implementation of urban logistics hubs for waste take-back will be essential to building the decentralised infrastructure needed for a functioning circular economy.

3.4 The Role of Reverse Logistics in Urban Circular Systems and EPR Schemes

As urban areas become hubs of both consumption and waste generation, reverse logistics has emerged as a critical enabler of circular economy models. In the context of EPR, reverse logistics refers to the infrastructure, operations, and coordination required to collect, transport, sort, and reintegrate post-consumer goods into recycling or reuse processes. Efficient reverse logistics systems are essential for achieving the objectives of EPR schemes and reducing the environmental footprint of urban logistics. This chapter explores how reverse logistics supports circularity in urban settings, focusing on its operational models, integration with waste take-back schemes, and the barriers and opportunities facing cities and producers.

Reverse logistics differs from traditional, forward logistics by addressing the flow of used, discarded, or obsolete products back through the supply chain. In a linear system, products flow from manufacturers to consumers and then to disposal. In contrast, reverse logistics aims to close the loop by facilitating the recovery of materials and products for reuse, remanufacturing, or recycling. Within EPR frameworks, reverse logistics is not an ancillary activity but a core mechanism by which producers fulfil their legal and environmental responsibilities.

In urban environments, reverse logistics enables localised collection of used products, supports decentralised sorting and pre-processing, and can reduce transport emissions when aligned with sustainable logistics networks. It also connects waste collection with product recovery systems, forming a backbone for circular flows of materials.

Reverse Logistics for Batteries in the Netherlands

The Dutch organisation Stibat operates a national take-back scheme for portable batteries under an EPR model. Funded by battery producers, the system integrates reverse logistics into daily urban flows via over 25,000 collection points, including supermarkets, schools, and retail outlets.

Collected batteries are retrieved using low-emission vehicles, sorted by chemistry, and processed at certified facilities for material recovery. Digital tracking tools ensure real-time monitoring, compliance reporting, and transparent data for policymakers and producers. Public awareness campaigns and school partnerships have helped maintain collection rates above 50%, exceeding EU targets.

Stibat's success demonstrates how well-designed reverse logistics can align with urban mobility, improve collection efficiency, and support circular economy goals, offering a replicable model for other EPR waste streams.

Reverse logistics in urban EPR systems present several opportunities for improvement and innovation:

- **Design for reverse logistics:** Products should be designed not only for durability and recyclability, but also for ease of disassembly, transport, and collection.
- **Public-private collaboration:** Stronger partnerships between cities and producers can align waste management with local sustainability goals.
- **Digital integration:** Platforms for data sharing, consumer engagement, and real-time logistics coordination can increase system efficiency and transparency.
- **Policy support:** Harmonised regulations, funding mechanisms, and incentives for reverse logistics investment are essential for scaling up.

Reverse logistics plays a foundational role in operationalising EPR and advancing circularity in urban contexts. It bridges the gap between end-of-use products and material recovery, ensuring that the vision of a circular economy becomes a practical reality. While challenges remain, reverse logistics offers a powerful lever for reducing environmental impact, enabling reuse, and recycling, and building more resilient urban systems. When supported by enabling policy, infrastructure, and collaboration, it can help cities transition to a truly circular urban logistics model.

4 Analysis

This section presents a cross-case analysis of the EPR take-back systems discussed previously, highlighting key operational, regulatory, and environmental dimensions. The goal is to identify patterns, success factors, and challenges in implementing circular logistics solutions for waste streams in urban contexts, particularly under the pressures of digitalisation, consumer behaviour, and regulatory expansion.

4.1 Operational Efficiencies and Constraints

A comparative analysis of the case studies reveals varying levels of operational efficiency, shaped by local urban density, logistical models, and the degree of system integration with municipal waste services.

In cities such as Berlin and Paris, where consumer density and collection points are high, take-back systems benefit from economies of scale but must contend with logistical fragmentation. High-frequency, low-volume pickups from dispersed retail outlets or mobile collection points demand agile routing and strong coordination with third-party logistics providers.

Technologies such as fill-level sensors (e.g. GRS Batterien) and demand-responsive pickup models (e.g. ecosystem's mobile WEEE units) have been instrumental in improving route optimisation and reducing unnecessary vehicle mileage. However, integration across waste streams and digital interoperability between systems remain limited, often siloed by waste type, funding model, or administrative boundaries.

In contrast, decentralised return models, such as textile collection hubs in the Netherlands, offer high consumer convenience but tend to incur higher per-tonne collection costs unless subsidised or combined with commercial reuse channels. This trade-off between cost and accessibility is a recurring challenge for urban EPR logistics.

4.2 System Design and Circularity Outcomes

The alignment between EPR take-back logistics and broader circular economy outcomes is heavily influenced by system design, especially in relation to sorting and reuse.

The case studies illustrate that systems which prioritise reuse and material recovery through upstream sorting (e.g. Sympany's textile categorisation) achieve better circularity outcomes than those focused solely on recycling or end-of-life disposal. Reuse-centric models also tend to generate lower net emissions, provided transport logistics are well-managed.

However, the influx of ultra-fast fashion and low-quality consumer goods, especially in textiles and electronics, poses a serious barrier to high-value circularity. These products are often unsuitable for reuse or economically unviable to recycle. PROs are increasingly burdened with managing such waste, raising concerns about cost distribution and eco-design enforcement.

The effectiveness of eco-modulation, where producer fees are adjusted based on environmental performance, remains limited by inconsistent implementation across Member States and product groups. Without stronger links between design standards and collection system funding, EPR risks reinforcing linear disposal patterns.

4.3 Governance and Regulatory Coherence

From a governance perspective, the role of PROs as intermediaries between producers, consumers, and waste operators is central, but also contested.

In all cases, PROs are required to balance compliance enforcement with innovation, often under tight administrative and financial constraints. Where regulatory frameworks are harmonised and PROs enjoy policy stability (e.g. France), they can play a transformative role. Where legal uncertainty or fragmentation exists (e.g. emerging textile EPR systems in Germany), progress is slower and often reactive.

The analysis also highlights a critical gap in the coordination between EPR and broader urban planning and climate strategies. Rarely are take-back systems fully integrated into city-level waste, mobility, or emissions planning. This siloed approach limits potential synergies, such as shared collection infrastructure, last-mile consolidation, or co-investment in reverse logistics depots.

4.4 Environmental and Social Impacts

Although EPR take-back schemes are primarily designed around material recovery targets, their broader environmental and social effects are increasingly relevant. The reduction of illegal dumping, the creation of green jobs (particularly in sorting and repair), and consumer education benefits are well documented in systems with high visibility and engagement, such as mobile WEEE units or urban textile hubs.

Lifecycle assessment (LCA) studies aligned with the cases show that urban collection systems with strong reuse components and optimised transport logistics can significantly outperform traditional disposal-based models in terms of CO_2 emissions and resource intensity. However, unintended consequences persist. Exported waste under the guise of reuse (particularly in textiles and electronics) undermines circularity goals and shifts environmental burdens to lower-income countries. Additionally, social inclusion elements, such as fair labour in reuse or repair, are unevenly addressed across schemes.

4.5 Key Challenges Identified

From the comparative analysis, several recurring challenges stand out:

- **Economic viability:** Ensuring financial sustainability of urban take-back systems without overburdening producers, particularly SMEs.
- **Quality of materials:** Dealing with rising volumes of low-quality, hard-to-reuse products that disrupt reuse and recycling markets.
- **Digital fragmentation:** Lack of common data standards and interoperability between take-back systems limits optimisation and policy evaluation.
- **Consumer participation:** Ensuring regular and informed consumer engagement in urban settings with competing disposal channels.
- **Regulatory divergence:** National variation in EPR implementation hinders cross-border harmonisation and innovation transfer.
- **Integration with urban logistics:** Missed opportunities to connect EPR logistics with broader city transport, waste, and emissions strategies.

In summary, while EPR schemes hold significant promise for advancing circular urban logistics, their success depends on thoughtful system design, cross-sector collaboration, and regulatory alignment. These findings form the analytical foundation for the recommendations and conclusions presented in the next section.

5 Conclusion and Recommendations

5.1 Findings

As cities face mounting pressure to reduce waste, carbon emissions, and resource dependency, the role of EPR in urban logistics has gained increasing relevance. This section has examined how EPR take-back schemes can become central enablers of circularity in urban settings, provided they are thoughtfully designed, effectively governed, and strategically integrated with broader logistics and sustainability efforts.

The analysis presented throughout this section underscores a few key conclusions. First, EPR systems are already facilitating the transition toward circular economy principles in several critical waste streams, such as batteries, electronics, textiles, and packaging. Second, these systems function not only as compliance tools but increasingly as levers for innovation in sustainable logistics, reuse infrastructure, and reverse supply chain design. However, without more harmonised regulatory frameworks, coordinated urban planning, and effective producer engagement, EPR take-back schemes risk falling short of their transformative potential:

- EPR has proven effective in improving material recovery and increasing producer accountability, but its ability to drive upstream circular design and urban logistics integration varies widely across countries and sectors.
- Urban contexts pose unique challenges, including high consumer density, complex logistics needs, and regulatory fragmentation, which must be addressed for EPR schemes to deliver consistent environmental and economic outcomes.
- PROs play a pivotal role, but their performance and influence are constrained by inconsistent regulation, limited digital tools, and lack of integration with municipal and city logistics systems.
- Digitalisation, consumer behaviour, and product design are crucial leverage points. Without adequate data infrastructure, clear eco-modulation rules, and public engagement strategies, even the most well-intentioned systems risk inefficiency or underperformance.
- A systems perspective is essential. Coordination between producers, PROs, logistics providers, municipalities, and regulators is key to creating urban EPR models that are cost-effective, scalable, and environmentally beneficial.

5.2 *Discussion*

(a) **Strategic Role of EPR in Circular Urban Logistics**

This study highlights the growing importance of EPR schemes as a key mechanism for promoting circular economy principles within urban logistics systems. EPR enables the internalisation of environmental costs by assigning responsibility to producers for the end-of-life phase of their products.

In the context of urban logistics, this translates into more organised and efficient systems for the take-back, sorting, reuse, and recycling of discarded goods, especially for waste streams such as textiles, electronics, and batteries. By formalising reverse logistics and facilitating better resource recovery, EPR aligns producer incentives with circular objectives while helping cities reduce the environmental impacts of consumption and waste.

(b) **Policy and Governance Implications**

To unlock this potential, regulatory clarity and coherence across the European Union are crucial. The analysis in this section shows that fragmented national approaches to EPR, each with differing timelines, targets, and governance models, have led to uneven outcomes. Harmonising EPR rules, including definitions, reporting requirements, and the scope of financial responsibility, would allow for more consistent performance across Member States and provide a level playing field for producers and take-back operators. Urban and regional authorities should also be more directly involved in EPR implementation, ensuring that the specific needs and infrastructures of cities are reflected in national and EU-level policy frameworks.

(c) **Infrastructure and System Design**

Well-functioning EPR systems require robust infrastructure and service design tailored to the challenges of urban environments. Dense cities often face logistical constraints such as limited storage space, traffic congestion, and fragmented waste collection responsibilities. Addressing these barriers calls for the integration of reverse logistics planning into broader urban mobility and infrastructure strategies.

Collection points must be accessible, scalable, and adapted to different consumer behaviours and product types. Investment in sorting and pre-treatment infrastructure within or near urban areas is equally critical to reduce transport emissions and improve material recovery rates. Pilot projects and public-private partnerships can play a valuable role in testing new models that support decentralised, efficient take-back solutions.

(d) **Digital Tools and Innovation**

Digitalisation offers major opportunities for strengthening EPR systems in urban logistics. Tools such as Digital Product Passports, QR-based collection systems, and real-time tracking technologies can support data-driven decision-making, enhance transparency, and improve operational coordination. For producers and compliance organisations, digital tools also facilitate more accurate reporting and auditing, which in turn supports compliance and enables continuous system optimisation.

Furthermore, consumer-facing innovations such as mobile apps for product return or repair services can simplify participation in EPR schemes and increase collection rates. However, these tools must be developed with attention to interoperability, user experience, and data security to ensure widespread uptake and effectiveness.

(e) **Design Incentives and Circular Business Models**

Eco-modulation of EPR fees, whereby producer contributions are adjusted based on the environmental performance of their products, has emerged as a promising tool to encourage circular design. When implemented transparently and effectively, eco-modulated fees can incentivise the production of more durable, repairable, and recyclable goods.

These design changes, in turn, make waste collection and sorting more efficient and cost-effective for PROs and municipalities. In parallel, EPR frameworks can support emerging circular business models such as repair services, product-as-a-service models, and second-hand retail by providing infrastructure and financial mechanisms for the reverse flow of goods.

(f) **Collaboration and Inclusivity**

The success of EPR systems in urban settings depends on close collaboration between a wide range of actors, including producers, PROs, municipalities, waste managers, logistics providers, NGOs, and consumers. Inclusive dialogue and stakeholder engagement are essential for ensuring that systems are tailored to real-world needs and constraints.

Moreover, special attention must be paid to including SMEs, social enterprises, and informal actors, particularly in sectors such as textiles, where existing reuse and repair ecosystems can complement formal take-back systems. Support measures, such as capacity-building, grants, or access to digital tools, should be targeted to enable participation and innovation across diverse actor groups.

5.3 Recommendations

Building on these findings, the following recommendations are proposed to enhance the effectiveness of EPR in urban logistics and support the broader circular economy agenda:

- **Design EPR schemes to incentivise circularity, not just compliance:** Eco-modulated fees should be standardised and linked to tangible design improvements, such as ease of repair, recyclability, or material durability. Urban take-back systems should prioritise reuse and refurbishment alongside recycling.
- **Integrate EPR logistics with city-level sustainability and mobility strategies:** Take-back schemes should be coordinated with municipal services and urban freight planning to improve efficiency, reduce emissions, and maximise synergies with existing infrastructure.
- **Support digital transformation of EPR systems:** Investment in shared data platforms, product registries, and tracking technologies will allow for better collection forecasting, policy monitoring, and consumer transparency.
- **Promote harmonisation across Member States:** To ensure a level playing field and avoid market distortion, the EU should advance common standards for reporting, eco-modulation, and producer obligations across waste streams.
- **Strengthen the role of PROs through capacity-building and governance clarity:** PROs should be empowered to innovate, collaborate, and influence policy, supported by clear mandates, transparent oversight, and access to shared best practices across sectors.
- **Ensure inclusive stakeholder engagement:** All actors, particularly SMEs, local authorities, and social enterprises involved in reuse, must be actively included in the design and implementation of urban EPR schemes.
- **Tackle low-quality waste at the source:** Producers of low-grade, ultra-fast fashion and non-repairable electronics should be held financially accountable through fee structures that reflect the environmental cost of their products.

5.4 *Future Research Directions*

The future of sustainable urban logistics lies in collaborative, circular approaches that align environmental goals with practical, data-driven systems. EPR take-back schemes are among the most promising tools in this transformation. They can help close material loops, reduce environmental burdens, and foster producer responsibility, if designed and managed to address the unique complexities of urban environments.

By rethinking logistics as a shared circular infrastructure and leveraging the collective potential of PROs, municipalities, industry, and civil society, EPR can become a foundation for cleaner, smarter, and more resilient cities across Europe and beyond. As EPR schemes evolve in scope and complexity, further research is needed in several areas:

- Comparative studies on EPR performance in urban versus rural logistics contexts
- Lifecycle and emissions analysis of reuse versus recycling models under EPR
- Socioeconomic impacts of urban take-back systems on employment and equity
- Policy evaluation of eco-modulation effectiveness on product design innovation
- Development of circularity indicators specific to EPR logistics networks

As cities seek to reduce waste, cut emissions, and transition to more circular models of production and consumption, EPR systems stand out as a powerful instrument to align environmental and economic goals. When integrated into urban logistics strategies, EPR can help build the infrastructure and incentives needed to close material loops and promote sustainable value chains.

Yet, realising this potential requires harmonised regulation, strategic investment, and inclusive governance. By addressing systemic barriers and promoting digital and organisational innovation, policymakers and industry stakeholders can transform EPR into a cornerstone of circular urban logistics and of Europe's green and competitive future.

Box 1 Battery Collection in Norway (RENAS)
RENAS, 2024. Sustainable Recycling in Practice: Batteries. [online] Available at: https://renas.no/en [Accessed 20 May 2025].

Box 2 WEEE Take Back in France (Ecosystem)
Ecosystem, 2023. Mission Committee Report 2023. [online] Available at: https://pro.ecosystem.eco/en/ [Accessed 20 May 2025].

Box 3 Textile Return Hubs in the Netherlands
City of Amsterdam, 2022. Circular Amsterdam – Urban return logistics strategy. [online] Available at: https://www.amsterdam.nl [Accessed 20 May 2025].

Box 4 Repair and Reuse: iFixit and the Right to Repair Movement (EU/ Global)
R.U.S.Z, 2023. Repair is the future – Annual Impact Report. [online] Available at: https://rusz.at [Accessed 20 May 2025].

Box 5 Lindholmen - Reverse logistics Project (cargo/waste co-freight)
Gothenburg European Office, MOVE21 – turning cities into smart, zero-emission nodes for mobility and logistics, 2022. Circular economy in e-commerce. [online] Available at: https://goteborg.se/wps/portal/enhetssida/gothenburg-european-office/projects-in-gothenburg/move21--turningcities-into-smart-zeroemissions-nodes-for-mobility-and-logistics [Accessed 20 May 2025].

Box 6 Reverse Logistics for Batteries in the Netherlands
Stichting OPEN, 2024. Annual Report 2024. [online] Available at: https://www.stichting-open.org/en/about/ [Accessed 20 May 2025].

References

Bocken NMP, de Pauw I, Bakker C, van der Grinten B (2016) Product design and business model strategies for a circular economy. J Ind Prod Eng 33(5):308–320

European Commission (2020) A new circular economy action plan: for a cleaner and more competitive Europe (COM(2020) 98 final). [online] Brussels: European Commission. Available at: https://eur-lex.europa.eu/legal-content/EN/TXT/?uri=CELEX%3A52020DC0098 [Accessed 20 May 2025]

European Commission (2040) Regulation (EU) 2024/1781 of the European Parliament and of the Council of 13 June 2024 on ecodesign requirements for sustainable products, OJ L 178/1, 28 June 2024. [online] Brussels: European Commission. Available at: https://eur-lex.europa.eu/legal-content/EN/TXT/HTML/?uri=OJ:L_202401781

European Environment Agency (EEA) (2023) Textiles and the environment: The role of digital technologies in Europe's circular economy. [online] EEA. Available at: https://www.eea.europa.eu/publications [Accessed 20 May 2025]

Lindhqvist T (2000) Extended producer responsibility in cleaner production: policy principle to promote environmental improvements of product systems. PhD. International Institute for Industrial Environmental Economics (IIIEE), Lund University

OECD (2016) Extended producer responsibility: updated guidance for efficient waste management [online]. OECD Publishing, Paris. Available at: https://doi.org/10.1787/9789264256385-en [Accessed 20 May 2025]

Rizos V, Tuokko K, Behrens A (2017) The circular economy: a review of definitions, processes and impacts. CEPS Research Report. [online] CEPS. Available at: https://www.ceps.eu/ceps-publications/circular-economy-review-definitions-processes-and-impacts/ [Accessed 20 May 2025]

United Nations Environment Programme (UNEP) (2018) Redefining value—The manufacturing revolution: Remanufacturing, refurbishment, repair and direct reuse in the circular economy. [online] UNEP. Available at: https://www.resourcepanel.org/reports [Accessed 20 May 2025]

From Circular Economy to Circular Society: The Role of Logistics

Jeevan Kamsagara Maheshwarappa, Annika Hoppe, Denis Daus, and Ani Melkonyan-Gottschalk

1 Foundations and Key Concepts

A CE contrasts sharply with the linear "take-make-dispose" model. It promotes closing material loops through reuse, repair, remanufacturing, recycling, and other "Rs." By design, CE strategies seek to reduce resource extraction and waste while maintaining product value. The European Parliament, for example, defines CE as "a model of production and consumption that involves sharing, leasing, reusing, repairing, refurbishing, and recycling existing materials and products as long as possible", thereby extending product life cycles. Commonly cited principles are "designing out waste and pollution, keeping products and materials in use, and regenerating natural systems." In practice, CE policies and business models include eco-design, industrial symbiosis, extended producer responsibility (EPR), remanufacturing, and service-based models.

However, critics note that many CE narratives are technocratic and economy-centered, giving insufficient attention to social equity, ecological boundaries, and political power. The circular society concept broadens the focus to explicitly include social and cultural dimensions. As an umbrella term, it encompasses eco-social visions and alternative frameworks that emphasize democracy, equity, and well-being. In a circular society, cycles of resource use are interwoven with cycles of knowledge, care, and power, and activities are re-embedded in ecological contexts. For example, Calisto Friant et al. (2024) propose seven key socio-ecological cycles (biogeochemical, ecosystem, resource, power, wealth, knowledge, care) that must be addressed for truly circular systems.

J. K. Maheshwarappa (✉) · A. Hoppe · D. Daus · A. Melkonyan-Gottschalk
Institute of Geotechnology and Mineral Resources, Clausthal University of Technology, Clausthal-Zellerfeld, Germany
e-mail: jeevan.kamsagara.maheshwarappa@tu-clausthal.de

A. Melkonyan-Gottschalk, Y. Li (eds.), *Urban Logistics Transformation*, Greening of Industry Networks Studies 15,
https://doi.org/10.1007/978-3-032-08387-6_18

1.1 Principles and Key Players of Circular Societies

Circular society proponents advocate a mix of strategies that go beyond technical fixes. These include shrinking flows (simpler lifestyles, low-tech design to reduce overall throughput), slowing flows (extending product life through reuse, leasing, maintenance), shortening flows (repair, remanufacture, cascading reuse), and closing flows (recycling, composting). Crucially, societal values drive these strategies: for instance, prioritizing social well-being over continuous growth, or fostering care and reciprocity rather than mere consumption. In sum, the foundations of a circular society fuse CE's material loops with social justice, participatory governance, and regenerative ecology.

Stakeholders: Industries, Citizens, and Institutions
Building a circular society is inherently multi-actor. **Industries and businesses** adopt circular practices by redesigning products, their value, supply, and logistics chains. Companies may use sustainable materials, design for durability and recyclability, offer product-service systems, and invest in remanufacturing or recycling infrastructure. For example, in manufacturing sectors (textiles, electronics, automotive), firms set targets for recycled content and product longevity. Industrial symbiosis projects connect industries in regional clusters (where waste from one process feeds another), while shortening the supply and logistics chains. Large corporations often respond to policies and consumer demand by innovating new circular business models.

Citizens and communities play a central role besides industry, both as consumers and co-producers. Their everyday choices shape demand: opting for repairable, second-hand, or sharing economy options, reducing resource use. Grassroots initiatives community tool libraries, repair cafés, clothing swaps, and urban gardens, demonstrate circular practices at the local level. For instance, citizens can adopt circular lifestyles by recycling and upcycling, support green businesses, and start community initiatives that facilitate the sharing of goods and materials. Urban communities have formed makerspaces and tool libraries ("libraries of things") where neighbors borrow equipment rather than buying new. Policy advocacy is another lever: informed citizens lobby for EPR legislation, waste reduction targets, or local procurement of circular products. In short, citizens are not passive recyclers but can be active change agents in shaping norms and pushing for system change.

Institutions and governments provide enabling frameworks that guide industries and the actions of citizens. National and local authorities adopt policies, standards, and incentives for circularity. The EU's Circular Economy Action Plan (2020) exemplifies this: it integrates CE into the European Green Deal, setting measures across product life cycles and sectors with high resource use. The Action Plan explicitly states that CE transition is key to reducing resource pressure, creating sustainable growth and jobs, and meeting climate and biodiversity goals. In practice, institutions also fund research and demonstration and facilitate multi-stakeholder platforms. For example, the Ellen MacArthur Foundation's Circular Economy 100 network convenes companies and governments to co-develop solutions. In Asia, governments like Japan and China have national CE strategies. Sub-national governments (cities, regions) craft CE roadmaps, organize living labs or hackathons, and integrate circularity into urban planning. International bodies (UN, OECD) also promote global guidelines for sustainable consumption and circular transitions.

Logistics and E-Commerce

Supply Chains and Logistics are undergoing transformation as businesses respond to environmental pressures, policy shifts, and the growing demand for circular practices. Companies increasingly design supply chains that support product life extension, reuse, remanufacturing, and material recovery (Mishra et al. 2023). Reverse logistics systems used to collect, sort, and reintegrate products and materials after use are becoming central to circular operations. Localized and decentralized supply networks are emerging, reducing transport distances and supporting more resilient, community-based economies. For example, some manufacturers are shifting from global to regional supply models to better align with circular principles and minimize carbon footprints. Urban consolidation centers, shared delivery platforms, and electric last-mile logistics further illustrate how logistics can be restructured for sustainability (van Woensel et al. 2019). In addition, digital tools like IoT, blockchain, and AI are being used to track product lifecycles, optimize routes, and enable transparency across the value chain. Just as consumers shape circular demand, logistics providers play a pivotal role in enabling the practical flow of goods and materials in circular systems linking producers, users, and re-users in closed-loop networks.

E-commerce is rapidly evolving in response to sustainability concerns and shifting consumer expectations, offering both challenges and opportunities for circularity. Online retailers are beginning to incorporate circular business models such as product-as-a-service, rental platforms, and take-back schemes into their operations (Nanayakkara et al. 2022). Digital marketplaces for second-hand goods, refurbished electronics, and clothing resale are gaining popularity, reducing the demand for new resources and extending product lifecycles (Cano et al. 2022). Some companies offer repair services or promote modular product design to support reuse and maintenance. Packaging remains a critical issue; however, innovations such as reusable packaging, compostable materials, and minimal-waste delivery solutions are emerging in response. E-commerce also enables local and peer-to-peer exchanges, connecting consumers directly with circular goods and services in their area. Algorithms and recommendation systems can nudge consumers toward sustainable options, while transparent sustainability ratings can empower informed choices (Cano et al. 2022). Overall, e-commerce platforms are more than transaction points—they are evolving into ecosystems that can either reinforce or help transform consumption norms toward a circular society (Li et al. 2021).

In summary, circular society transformation is co-produced by (i) industrial innovation and business-model change, (ii) citizen behavior and community initiatives, (iii) institutional policies and governance, and (iv) adjusted new supply and logistics chains. Successful cases often involve partnerships: firms collaborating with local government and communities to pilot new solutions, or transnational consortia fostering knowledge exchange. Examples include industrial parks with shared recycling facilities, urban programs that engage residents in co-designing waste solutions, and multi-level governance aligning standards and goals. The fashion e-commerce platform **Zalando** partnered with **RePack**, a reusable packaging provider, to pilot circular packaging solutions in urban deliveries across select European cities. Customers could return the empty packaging via local postal networks, enabling reuse cycles that significantly reduce single-use waste. To optimize last-mile logistics and emissions, Zalando also integrated urban micro-hubs and bicycle couriers in dense areas such as Berlin and Amsterdam. This model demonstrates how e-commerce businesses can embed circular principles into packaging, logistics, and consumer interaction supporting the goals of circular cities by reducing waste, emissions, and congestion in urban environments. Each group has different incentives and constraints, so deliberate coordination is needed. For instance, governments may support circular business incubators to help SMEs overcome entry barriers. Policies like tax benefits for recycled materials can make circular practices more competitive for industry. Likewise, educating and empowering citizens helps create the cultural shift toward valuing circular behaviors.

1.2 Gaps in Circular Society Transformation

Despite growing momentum, significant gaps remain in realizing circular societies. These span **policy, research,** and **practice** dimensions:

- **Policy gaps**, despite growing momentum, persist in realizing circular societies. Although many governments endorse CE, implementing coherent circular society policies is difficult. Coordination across sectors (waste, industry, social policy) is often weak, leading to fragmented efforts. Extended Producer Responsibility laws may exist for packaging but not for textiles or electronics, for example. Equity and just-transition considerations are frequently under-addressed: many policies assume uniform benefits, whereas in reality, vulnerable communities can be left out without safeguards. Inertia and vested interests also slow change; industries used to linear models may resist stricter circular regulations. Often time, logistics chains are not incorporated into the solutions. Moreover, stakeholder engagement in newly designed logistics ecosystems for circularity transition is merely neglected by the policy-makers. In short, there is a policy implementation gap: a need for stronger, cross-cutting frameworks that explicitly integrate social equity. The Green Economy Coalition highlights these needs, recommending policies that support social equity, provide worker retraining, and involve marginalized groups in planning.
- **Research gaps** are evident as literature often remains fragmented and mostly limited to waste management. Major gaps include: geographically seen much work focuses on Europe and China, while Africa, Latin America, and many

Asia-Pacific countries are understudied; social dimensions like behavioral, cultural, gendered aspects of circular practices are less explored compared to technical aspects and interdisciplinarity, few studies integrate economics, sociology, ecology, and engineering together, though a circular society requires just such integration. There is also a shortage of evaluations of systemic impacts—for example, how would a nationwide circular procurement policy affect employment patterns? Moreover, many studies assume certain imaginaries without critically examining alternatives. Therefore, we call for more holistic and contextual analyses: including long-term case studies of communities striving for circularity, comparative international research, and participatory action research methods that co-produce knowledge with stakeholders. Moreover, we highlight the urgent need to integrate stakeholder perspectives from the logistics sector, while proposing integrated forward and reverse-logistics flow for circularity transition at the infrastructural level, in particular within complex urban systems.

- **Practice gaps** reveal on the ground, as many circular solutions struggle to scale due to a lack of infrastructural solutions, particularly in urban regions. Business uptake can be slow, especially among small and medium enterprises (SMEs) that lack capital or know how to redesign products and design appropriate infrastructure for implementing their businesses. Consumers often have limited awareness or incentives to change habits. Only 20% of global waste is recycled worldwide. Crucial infrastructure remains missing in many regions: for example, many cities lack adequate facilities and logistics infrastructure for material recycling or composting, and digital systems are nascent. Finance is also a gap: banks and investors are just starting to consider circular business models. Finally, citizen engagement in practice can be tokenistic if not intentionally cultivated; many initiatives fail to maintain community involvement over the long term.

Addressing these gaps requires integrated action. Policymakers must strengthen and coordinate regulations (e.g., by updating waste laws to include new materials and by designing just-transition measures). Academics should target underexplored and work across disciplines to understand circular societies and their role within complex systems, such as logistics or circular cities and regions. Practitioners need support for scaling innovation: incubators for circular startups, platforms for knowledge exchange, and awareness campaigns to shift consumer behaviour. Initiatives like the EU's Circular Cities Mission and the UN's Sustainable Development Goals provide frameworks, but translating them into equitable, context-sensitive action, while integrating the existing structures or designing new (logistics) infrastructure, remains a major challenge worldwide.

1.3 Circular Principles in Regional Contexts

The **European Union (EU)** has been a leader in articulating CE policy. Its 2020 Circular Economy Action Plan (and its 2015 predecessor) set a "whole lifecycle" agenda: influencing design, production, consumption, waste management, and secondary raw material markets. The Plan's objectives include making sustainable products the norm, empowering consumers, focusing on high-impact sectors, and "making circularity work for people, regions and cities." Many member states have translated this into national strategies. For example, the Netherlands and Finland include social innovation and citizen engagement in their CE roadmaps. The EU also measures progress via indicators. At the city/regional level, initiatives like the European Circular Cities and Regions Initiative (CCRI) encourage local governments to set targets (e.g., Amsterdam aims for a fully circular city by 2050). In the Southeast Lower Saxony region, a CCRI fellow is focusing on advancing topics such as circular battery production, CO_2-reduced steel production, and urban mining. These initiatives are pivotal for driving sustainable industrial practices and innovations in the region. The EU approach is largely policy-driven but also supports niche innovation and citizen involvement in design (e.g., stakeholder consultations in drafting legislation).

Also **Asia** exhibits diverse CE trajectories. In Japan, the concept of "Sound Material-Cycle Society" (SMCS) dates back to the 1990s and is enshrined in law (Basic Act on Establishing a SMCS). Recent research shows that Japan's CE discourse remains technocratic, focusing on recycling and technology while "revisiting ecological thinking" is recommended for greater inclusivity. Nonetheless, Japan's policy (4th Fundamental Plan for Establishing SMCS) includes ambitious recycling and reuse targets, and the government often promotes eco-design and small-scale reuse initiatives. China has adopted a CE model as part of its national development plans. Recent analysis of China's 14th Five Year Plan (2021–2025) highlights tasks like "building a resource recycling industry system," "fostering a recycling-oriented society." In Southeast Asia, countries like Indonesia and Singapore are developing national CE roadmaps, often spurred by international frameworks (e.g., ASEAN's 2021 Framework for Circular Economy). The ASEAN region's policy momentum was noted with the 2022 APEC leaders' Bangkok goals on the Bio-Circular-Green (BCG) economy. In sum, Asian contexts feature strong governmental leadership and growing industrial investment in recycling technology, while citizen engagement is at an earlier stage compared to Europe.

Many cities in Europe and Asia have explicitly adopted CE plans. **Circular cities** serve as focal points for circular experimentation due to their high resource flows and innovation capacity. These plans often emphasize urban recycling infrastructure, circular procurement by city governments, eco-districts, and local sharing platforms. For instance, Amsterdam has set targets to halve raw material use by 2030 and be fully circular by 2050, though its policies currently emphasize waste reduction and energy efficiency. (OECD, 2025) In practice, cities support urban living labs, circular urban planning, and citizen co-creation workshops. Industrial symbiosis parks are being piloted within or near cities, where urban waste streams (food waste, municipal solid waste) feed into bioenergy or composting facilities. Public awareness campaigns and education in schools are common to mobilize residents. Some municipalities involve citizens in decision-making, for example, through participatory budgeting of circular projects or citizen advisory boards on waste policy.

Besides cities, rural areas, referred to as **circular regions**, engage with circular principles primarily through the closure of resource loops. Small-scale industries in rural areas may practice remanufacture or repair strategies. While citizens in rural areas may have less influence on formal policy, local cooperatives and NGOs often drive circular projects. Government programs sometimes support rural circularity with grants for biomass energy or circular farming practices. The dynamics in rural contexts often differ from urban ones: networks tend to be looser, and informal reuse is common. Thus, enabling local innovation networks and extending CE networks beyond cities is a current challenge.

Logistics within circular cities and regions play a crucial role in enabling the transition towards sustainable practices, which emphasize sustainability through resource efficiency, reuse, and waste minimization. Traditional linear logistics systems based on take-make-dispose models are being reimagined to support circular flows of materials, energy, and goods. This shift necessitates a transformation in urban logistics strategies, focusing on reverse logistics, urban mining, local production, and last-mile delivery optimization (Bocken et al. 2016). In circular cities, logistics must support closed-loop supply chains where products are returned, repaired, refurbished, or recycled. Efficient reverse logistics systems are critical in managing the flow of end-of-life products and waste materials back into productive use (Genovese et al. 2017). Moreover, the localization of production through urban manufacturing and 3D printing reduces transport distances, thus cutting emissions and congestion (Bressanelli et al. 2017). Digitalization is a key enabler in circular logistics. Smart logistics platforms using IoT, AI, and big data analytics improve the coordination of materials flows, enhance traceability, and support sharing-based business models, such as product-as-a-service (Geissdoerfer et al. 2017). For example, platforms that connect surplus food donors with recipients, or sharing systems for tools and appliances, are increasingly integrated into urban logistics networks. Finally, city planners must align logistics infrastructure with circular goals by creating urban consolidation centers, integrating multimodal transport, and designing zoning regulations that encourage reuse and recycling hubs within neighborhoods. As circular economy principles become embedded in urban development, logistics will act not just as a support function but as a strategic driver of sustainability.

2 E-Commerce and Logistics in a Circular Society

While the previous sections have outlined the foundational principles of circular societies, it is in specific economic sectors that exhibit both benefits and potential tensions of circularity in action. In particular, the rapid growth of e-commerce platforms and the extensive logistics networks to serve them present a fertile ground for circular-economy innovation, but can also be explored regarding the deeper meaning of a circular society.

E-commerce is increasingly recognized as a valuable component of the CE, as it offers various platforms and models that facilitate sustainable consumption and production systems. One major challenge posed by the rapid growth of e-commerce is its environmental impact, due to high levels of waste generation and resource depletion. The integration of logistics and technology in rural e-commerce considerably contributes to the CE. For instance, it enables farmers to sell their products directly to consumers through online platforms, which minimizes intermediaries and consequently reduces waste (Huang et al. 2022; Liu et al. 2023). This model supports the financial viability of smallholder farmers and also enhances their roles in the local economic ecosystem, which is a fundamental aspect of a circular society. A CE is also facilitated by e-commerce through the deployment of blockchain and IoT technologies. Such technologies enable more efficient tracking of products throughout their lifecycle, fostering a closed-loop system with minimal waste (Prajapati et al. 2022).

In this view, e-commerce heavily relies on logistics. In particular, reverse logistics encompasses all operations related to the reuse of products and materials. It plays an important role by managing the flow of materials back through the supply chain, enabling recycling, remanufacturing, and upcycling processes within a closed-loop system. This way, it enhances resource efficiency and supports the overall goals of the CE (Van Buren et al. 2016). As e-commerce transitions to such sustainable models, it improves not only the environmental responsibility of companies but also encourages consumer engagement in sustainable practices. When examining the CE and logistics from a societal viewpoint, it is essential to consider how logistics and the circular society interconnect. Logistics plays a key role in this framework by serving as the orchestrator of materials and ensuring resources are circulated within the economy rather than disposed of as waste. Such activities require technological advancements and cultural shifts that encourage participation among individuals, businesses, and governments (Kocaoğlu and Bulut 2024; Murray et al. 2017), but also according infrastructure, especially in highly dense urban regions.

However, while e-commerce and related logistics have significant potential in advancing CE goals, critical reflections on their role reveal essential challenges, particularly in developing regions. Issues such as unequal access to digital platforms and variations in logistical capabilities (Nazir and Roomi 2020; Reardon et al. 2021) can hinder the inclusivity that a circular society strives for.

Transitioning to a CE also presents challenges regarding logistical infrastructure and societal readiness. Effective reverse logistics systems must manage complex flows of materials and stakeholders' responsibilities along the entire supply chain, while ensuring that the necessary regulatory frameworks and cooperation are in place (Adriana Hofmann Trevisan 2022). Barriers such as resistance to change, lack of awareness among consumers, and insufficient governance can hinder the implementation of effective circular logistics strategies. Engaging stakeholders is crucial to overcoming these barriers, as their participation can foster innovation and streamline the integration of circular practices into existing linear models (Rizos et al. 2016). The role of policy and regulatory frameworks is also critical because it creates an enabling environment that promotes sustainable logistical practices and incentivizes circular business models (Zarbakhshnia et al. 2023).

Even though e-commerce has the potential to elevate sustainable practices and contribute to a circular society, it can also deepen economic gaps within societies. That's why it's vital to engage with society throughout supply chains to ensure inclusivity and fair conditions for everyone.

3 Case of Lower Saxony

For example, the Circular Region Southeast Lower Saxony, Germany, home to 1.15 million residents across an area of 5100 km^2, has long been recognized for its prowess in raw materials, recycling, and high-tech production, particularly in the steel, automotive, and critical raw materials (CRM) sectors. The region includes urban centers like Braunschweig and Wolfsburg, along with peri-urban and rural areas in the surrounding districts. Historically, this region has been deeply rooted in resource extraction, especially through mining and metallurgy, which today translates into strong competencies in recycling and circular economy practices. (Goldmann and Mrotzek-Blöß 2024).

Currently, Southeast Lower Saxony is transitioning into a circular economy (CE) model, focusing on "Circular Production and Use." Key industrial areas include mobility, steel production, and resource extraction through metal recycling. A central component of Southeast Lower Saxony's shift toward a circular economy is supporting production systems in transitioning to sustainable circular economy models. This includes the development of closed-loop battery production and the conversion of steel production to hydrogen-based processes, while managing excess material storage and designing recyclable product structures (Goldmann and Melkonyan-Gottschalk 2025).

The circular economy initiatives in Southeast Lower Saxony have direct implications for local logistics infrastructure. By promoting circular economy structures, logistics must be optimized, specifically concerning the collection and reintegration of products into the economic cycle, particularly for electronic products. This necessitates an efficient logistical infrastructure to ensure the transport and processing of materials with minimal environmental impact. In this context, innovative concepts for decentralized e-waste logistics and digital tools for tracking, sorting, and reintegrating materials are being developed and implemented. A newly established Circular Economy Lab explores data-driven solutions for sustainable logistics systems. ("6RLogistics" 2024) In the e-commerce context, circular economy practices open up new opportunities for virtual marketplaces, particularly for second-hand platforms, the sharing of products, and the promotion of recycled products. Digital platforms play a crucial role in extending product lifecycles and facilitating the return of materials. They also encourage sustainable consumption models by easing access to reused or recycled products, thereby raising awareness of the importance of resource conservation.

4 Conclusion

Transitioning to circular societies demands integrated efforts: industry innovation, citizen-driven actions, and enabling governance must co-evolve. The shift involves not just new technologies (niches), but altered business regimes and cultural expectations (imaginaries), guided by supportive policy landscapes. Urban and rural contexts each play unique roles: cities as hubs of pilot projects and consumer demand, rural areas as sources of biomass and stewardship of resources. Across regions, local context matters—for example, Europe's regulatory approach differs from Asia's state-led industrial planning—but the underlying principles of closing loops and prioritizing sustainability remain common.

Recommendations Finally, ongoing research is essential. Scholars should continue examining the "social" in circularity (justice, labor, culture) as proposed in key literature. Empirical studies comparing regional approaches (EU vs. Asia, urban vs. rural) can reveal what policies and practices work where. As CE gains traction, a circular society framework can guide more holistic innovation: one that loops in not just materials, but also values, knowledge, and care.

References

6RLogistics [WWW Document] (2024) Entsorgungslogistik von Elektroaltgeräten ganz neu gedacht. https://www.tu-clausthal.de/universitaet/einrichtungen/presse-und-oeffentlichkeitsarbeit/pressemitteilungen/artikel/entsorgungslogistik-von-elektroaltgeraeten-ganz-neu-gedacht (accessed 6.12.25)

Adriana Hofmann Trevisan DG (2022) Stakeholders' responsibilities in reverse logistics of product-service systems: towards circular economy in the healthcare industry 19, 0–0. https://doi.org/10.4322/pmd.2021.010

Bocken NMP, De Pauw I, Bakker C, Van Der Grinten B (2016) Product design and business model strategies for a circular economy. J Ind Prod Eng 33:308–320. https://doi.org/10.1080/21681015.2016.1172124

Bressanelli G, Perona M, Saccani N (2017) Reshaping the washing machine industry through circular economy and product-service system business models. Proc CIRP 64:43–48. https://doi.org/10.1016/j.procir.2017.03.065
Calisto Friant M, Vermeulen WJV, Salomone R (2024) Transition to a sustainable circular society: more than just resource efficiency. Circ Econ Sust 4:23–42. https://doi.org/10.1007/s43615-023-00272-3
Cano JA, Londoño-Pineda A, Rodas C (2022) Sustainable logistics for E-commerce: a literature review and bibliometric analysis. Sustainability 14(19):12247. https://www.mdpi.com/2071-1050/14/19/12247
Geissdoerfer, M., Savaget, P., Bocken, N.M.P., Hultink, E.J., 2017. The circular economy—a new sustainability paradigm? J Clean Prod 143, 757–768. https://doi.org/10.1016/j.jclepro.2016.12.048
Genovese A, Acquaye AA, Figueroa A, Koh SCL (2017) Sustainable supply chain management and the transition towards a circular economy: evidence and some applications. Omega 66:344–357. https://doi.org/10.1016/j.omega.2015.05.015
Goldmann D, Melkonyan-Gottschalk A (2025) Meet the CCRI stakeholders: Southeast Lower Saxony (CCRI Fellow) | Circular Cities and Regions Initiative
Goldmann, Mrotzek-Blöß (2024) Broschüre - Circular Region SüdOstNds
Huang J, Su L, Huang Q, Liu X (2022) Facilitating inclusive ICT application and e-commerce development in rural China. Agric Econ 53:938–952. https://doi.org/10.1111/agec.12728
Kocaoğlu B, Bulut M (2024) Circular supply chain network design for e-commerce. Gazi Univ J Sci 37:840–852. https://doi.org/10.35378/gujs.1234548
Li Y, Feng C, Liu H (2021) Package-free returns: a trend in green ecommerce. In: Advances in Digital Marketing and eCommerce (DMEC 2021). Springer, Cham, pp 330–334. https://doi.org/10.1007/978-3-030-76520-0_34
Liu M, Shi P, Wang J, Wang H, Huang J (2023) Do farmers get a greater return from selling their agricultural products through e-commerce? Rev Dev Econ 27:1481–1508. https://doi.org/10.1111/rode.12968
Mishra A, Dutta P, Jayasankar S, Jain P, Mathiyazhagan K (2023) A review of reverse logistics and closed-loop supply chains in the perspective of circular economy. Benchmark Int J 30(3):975–1020. https://doi.org/10.1108/BIJ-12-2021-0786
Murray A, Skene K, Haynes K (2017) The circular economy: an interdisciplinary exploration of the concept and application in a global context. J Bus Ethics 140:369–380. https://doi.org/10.1007/s10551-015-2693-2
Nazir MA, Roomi MA (2020) Barriers to adopting electronic commerce for small and medium-sized enterprises in emerging economies. EMAJ Emerg Market J 10:43–55. https://doi.org/10.5195/emaj.2020.203
OECD (2025) The circular economy in cities and regions of the European Union, OECD urban studies. OECD Publishing. https://doi.org/10.1787/e09c21e2-en
Prajapati D, Jauhar SK, Gunasekaran A, Kamble SS, Pratap S (2022) Blockchain and IoT embedded sustainable virtual closed-loop supply chain in E-commerce towards the circular economy. Comput Ind Eng 172:108530. https://doi.org/10.1016/j.cie.2022.108530
Reardon T, Belton B, Liverpool-Tasie LSO, Lu L, Nuthalapati CSR, Tasie O, Zilberman D (2021) E-commerce's fast-tracking diffusion and adaptation in developing countries. Appl Econ Perspect Policy 43:1243–1259. https://doi.org/10.1002/aepp.13160
Rizos V, Behrens A, Van der Gaast W, Hofman E, Ioannou A, Kafyeke T, Flamos A, Rinaldi R, Papadelis S, Hirschnitz-Garbers M, Topi C (2016) Implementation of circular economy business models by small and medium-sized enterprises (SMEs): barriers and enablers. Sustainability 8:1212. https://doi.org/10.3390/su8111212
Van Buren N, Demmers M, Van der Heijden R, Witlox F (2016) Towards a circular economy: the role of Dutch logistics industries and governments. Sustainability 8:647. https://doi.org/10.3390/su8070647

van Woensel T, Rubio S, Rodrigues V (2019) Reverse logistics and urban logistics: making a link. Sustainability 11(20):5684. https://doi.org/10.3390/su11205684
Zarbakhshnia N, Govindan K, Kannan D, Goh M (2023) Outsourcing logistics operations in circular economy towards to sustainable development goals. Bus Strateg Environ 32:134–162. https://doi.org/10.1002/bse.3122

Conclusions and Recommendations

The Role of Public Authorities in Fostering the Transition Towards Sustainable Urban Logistics

Andrea Lorenzini and Giorgio Ambrosino

1 Urban Logistics and Its Challenges in European Cities

Urban logistics is a fundamental component of contemporary urban life and plays a central role in shaping the efficiency, liveability, and sustainability of cities. It encompasses a wide range of activities, including the delivery of goods purchased online, the distribution of food and beverages to hotels, restaurants, and cafés (HoReCa activities), the provision of essential supplies to hospitals and schools, and the transport of construction materials for both public infrastructure and private housing developments and renovations. In parallel, reverse logistics operations—such as the return of goods and the collection and disposal of waste—are increasingly acknowledged as essential pillars of the circular economy.

As cities across Europe and beyond undergo significant urban regeneration efforts, the role of freight transport in these processes becomes increasingly evident. Urban logistics is not merely a technical issue but a multidimensional challenge, interwoven with spatial planning, economic development, environmental policy, social equity, and urban mobility governance. Freight transport is also a major contributor to greenhouse gas (GHG) emissions. In 2020, global freight activities were responsible for approximately 3 billion tonnes of CO_2 emissions, representing around 9% of total global carbon emissions (International Transport Forum 2021).

Zooming in on the urban scale, city logistics in particular is responsible for a disproportionate share of environmental externalities. Urban freight accounts for over 25% of transport-related CO_2 emissions in cities, and between 30% and 50% of harmful particulate matter (PM) and nitrogen oxides (NOx) emissions (ALICE 2015). It is also a significant contributor to road congestion, with freight vehicles

A. Lorenzini (✉) · G. Ambrosino
MemEx Srl, Livorno, LI, Italy
e-mail: andrea.lorenzini@memexitaly.it

A. Melkonyan-Gottschalk, Y. Li (eds.), *Urban Logistics Transformation*, Greening of Industry Networks Studies 15,
https://doi.org/10.1007/978-3-032-08387-6_19

making up approximately 15% of urban traffic. Furthermore, the energy efficiency of urban logistics remains suboptimal, as evidenced by low average load factors—often below 50%—for delivery vehicles.

As demonstrated by several European initiatives—such as ENCLOSE, LIFE ASPIRE,[1] and the ongoing MED-COLOURS[2] project—the challenges and impacts associated with urban freight delivery, as well as with broader logistics activities (including waste collection, reverse logistics, and packaging recovery), are particularly acute in historic and touristic cities. These challenges stem from a series of distinctive urban and regulatory characteristics that intensify the complexity of logistics operations in such contexts. Among the most prominent features are:

- The physical constraints of the urban environment and infrastructure, such as narrow streets, irregular road networks, and the general absence of dedicated loading and unloading platforms.
- Increasingly stringent access and parking regulations, often designed to preserve the historic fabric and reduce traffic congestion.
- The presence of architectural and cultural heritage assets, which limit opportunities for infrastructural interventions or street redesign.
- The coexistence of pedestrian flows and vehicular traffic in shared spaces, raising concerns about safety and accessibility.

These factors generate a range of negative externalities, including increased emissions, noise pollution, and traffic congestion, while simultaneously driving up the operational costs of logistics activities. Moreover, the structural rigidity of the built environment and the regulatory landscape offer limited flexibility for stakeholders to adapt or innovate their delivery schemes.

In recent years, the rise of e-commerce has expanded the scope of freight-related challenges beyond the traditional boundaries of historic city centers. Many small towns have begun to observe similar pressures in adjacent districts, where increased residential deliveries are introducing new demands on local infrastructure and mobility solutions. This spatial shift calls for a more comprehensive urban logistics strategy that addresses not only the core but the broader urban area as a whole.

Adding further complexity, seasonal tourism flows place substantial strain on urban logistics systems in many historic and touristic towns (Hącia et al. 2023). During peak tourist seasons, the temporary population surge results in increased demand for retail goods, hospitality services, and waste management, all of which elevate the volume and frequency of freight activities. At the same time, the intensification of pedestrian flows and the prioritisation of the visitor experience often led to tighter delivery windows, restricted access zones, and greater competition for limited urban space. These seasonal fluctuations require adaptive logistics solutions that can reconcile the dual needs of supporting local economies and preserving the liveability and heritage value of urban environments.

[1] LIFE ASPIRE Project, https://www.life-aspire.eu/

[2] Interreg EURO-MED MED-COLOURS Project, https://medcolours.interreg-euro-med.eu/

Based on the experience of cities such as Lucca, Zadar, and Stockholm's historic peninsula area, the LIFE ASPIRE project has identified a series of shared objectives and planning criteria that can support the development of effective urban freight strategies tailored to historic and touristic areas. A central aim of this work is to maintain accessibility and ensure regulatory compliance among last-mile transport operators, while avoiding the simple displacement of logistical pressures from the historic core to surrounding districts. To this end, the project proposed several guiding principles and measures:

- Redesign selected streets to enhance pedestrian safety while ensuring efficient delivery routes.
- Introduce incentives to encourage sustainable behaviours among transport operators, including the adoption of clean "last-mile" delivery solutions.
- Revise zoning policies to define differentiated freight distribution rules in urban districts beyond the inner center.
- Pilot alternative delivery schemes in suitable districts, enabling the testing of innovative logistics models under real-life conditions.
- Support the uptake of electric vehicles through financial incentives and the deployment of adequate charging infrastructure.
- Promote coordination with neighbouring municipalities within the same functional urban area to harmonize regulations and foster cross-jurisdictional consistency.

Taken together, these strategies underscore the need for a context-sensitive, integrated, and forward-looking approach to urban logistics in historic and touristic towns—one that addresses spatial, temporal, and institutional dimensions and aligns with broader sustainability and quality-of-life goals.

In this context, the governance of urban logistics emerges as a crucial factor in managing and mitigating its impact. The logistics sector is predominantly driven by private operators—ranging from global freight companies to local delivery service providers—who contribute substantially to economic growth and employment. However, the sector operates within public regulation and spaces, relying heavily on urban roads, curbside areas, and logistics hubs, often in direct competition with other urban functions such as passenger mobility, public space usage, and residential needs.

This inherent tension between private operations and public space necessitates a coordinated and strategic governance approach. Public authorities, while not direct operators of logistics services, hold a critical role in shaping the regulatory, spatial, and operational frameworks that enable more sustainable urban logistics systems. Policy instruments such as Sustainable Urban Mobility Plans (SUMPs), Sustainable Urban Logistics Plans (SULPs), access regulations, urban freight zones, low-emission areas, and incentives for zero-emission vehicles and consolidation centres are among the tools that cities can employ to steer the transition.

As urban logistics continues to evolve—driven by the rapid growth of e-commerce, the adoption of digital technologies, and the push for circular and climate-neutral urban economies—the involvement of public authorities becomes even more pivotal. The following chapters explore these dynamics in detail, highlighting policy roadmaps, governance models, and planning frameworks that can support a more integrated, efficient, and sustainable approach to urban freight management.

2 Regulation and Legislation in EU Transport Sector

The European Commission's Sustainable and Smart Mobility Strategy[3] outlines a clear roadmap for transforming the transport sector into a more sustainable, intelligent, and resilient system. As part of this strategy, cities are identified as key actors in leading the transition toward sustainable mobility. The Communication emphasizes that, by 2030, all urban areas should develop and implement Sustainable Urban Mobility Plans (SUMPs). Recognized as a mature and well-established strategic planning tool at the EU level, SUMPs are designed to address the mobility needs of both individuals and businesses in urban and peri-urban contexts. Their overarching objective is to foster inclusive, efficient, and low-emission transport systems that contribute to broader environmental, social, and economic sustainability goals.

Building upon the broader strategic framework provided by SUMPs, Sustainable Urban Logistics Plans (SULPs) serve as a complementary planning instrument specifically focused on the freight and logistics dimension of urban mobility. Urban logistics presents distinct governance challenges due to its hybrid nature: while the use of public infrastructure and space is regulated and influenced primarily by local and regional authorities, the organization and operation of logistics services are largely managed by private actors. This structural duality necessitates a careful balancing of often diverging interests among a wide range of stakeholders, including municipal administrations, freight operators, retailers, residents, and other urban users.

In this context, SULPs offer a structured approach to developing targeted strategies that address the unique operational, environmental, and spatial impacts of urban freight (Ambrosino 2015). The most up-to-date European reference for SULP development is found within the SUMP Guidelines 2.0, and more specifically, the accompanying Topic Guide on Sustainable Urban Logistics Planning, published by the European Commission in June 2019 (Aifandopoulou and Xenou 2019). This document builds on and refines the original Sustainable Urban Logistics Plan (SULP) Guidelines, developed by MemEx under the European IEE ENCLOSE project.[4] Together, these guidelines provide cities with a coherent methodology for

[3] COM(2019) 640 final

[4] Energy efficiency in City Logistics Services for small and mid-sized European Historic Towns (ENCLOSE) Project, https://trimis.ec.europa.eu/project/energy-efficiency-city-logistics-services-small-and-mid-sized-european-historic-towns

integrating freight considerations into urban mobility planning, promoting more efficient, sustainable, and inclusive logistics systems at the local level.

For SUMPs and SULPs to be effectively embedded within broader regional and national policy frameworks, they must be developed through a lens of integration and coordination (Lorenzini 2024; Matusiewicz 2019). This requires aligning them not only with existing local territorial and transport planning instruments but also with broader planning and governance strategies at both horizontal and vertical levels. In this sense, SUMPs and SULPs should not function as isolated sectoral tools, but rather as interfacing instruments—capable of interacting with and supporting overarching objectives from various policy domains.

A well-conceived SULP must be rooted in a multi-scalar planning framework that takes into account the following dimensions (as illustrated in Fig. 1):

- European and national-level frameworks, including directives and strategic plans in sectors such as transport, land use, infrastructure, digitalization, and energy.
- Regional strategies, encompassing spatial development, environmental protection, and infrastructure planning.
- Local and municipal planning frameworks, where alignment is required not only with EU-level societal goals—such as environmental sustainability and social equity—but also with local cross-sectoral strategies (e.g., Smart City Strategies, Digital Transformation Agendas, Local Development Plans, Environmental and Air Quality Plans, and Sustainable Energy and Climate Action Plans).

Given these interdependencies, SULPs should be regarded as dynamic planning tools—closely/mutually linked to SUMPs and to local environmental policies, particularly those addressing air quality and emissions. From a governance perspective, municipalities must approach SULP development as an iterative and adaptive process (Ambrosino and Liberato 2014). This involves not only strategic planning but also the identification and mobilisation of the necessary governance capacities, technological tools, and institutional resources to ensure effective implementation and long-term management of urban logistics systems.

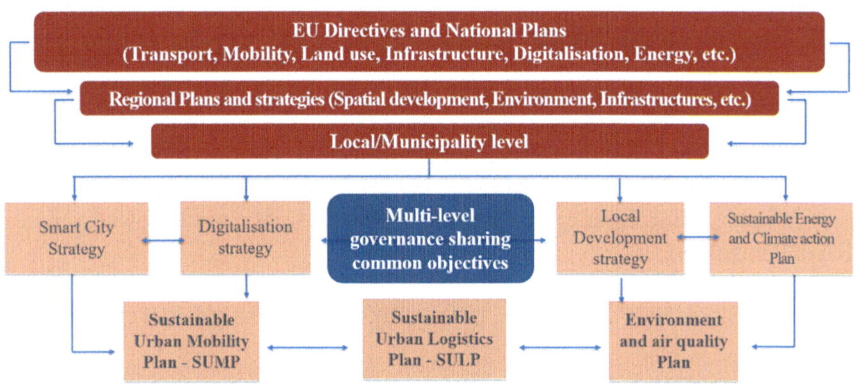

Fig. 1 Framing city logistics within different urban planning strategies. Source: Own elaboration

3 Policy Frameworks at Local Level

The experiences of Lucca, Lisbon, and Zadar have been selected as representative case studies due to their active engagement in European projects and initiatives focused on sustainable urban logistics.[5] Despite their differing sizes and urban structures, all three cities share a common commitment to addressing the challenges posed by freight distribution in historic and touristic environments. Their participation in projects such as ENCLOSE, LIFE ASPIRE, and MED-COLOURS has led to the development and testing of innovative strategies, as illustrated in Table 1, for managing last-mile deliveries, reducing emissions, and improving accessibility—making them valuable examples for other cities facing similar challenges.

Lucca: Heritage Preservation and Stakeholder-Centered Logistics
Lucca, a city in the Tuscany Region (Italy) of approximately 80,000 inhabitants and a surface area of 185 km^2, stands out for its well-preserved historic core. The city's Roman and medieval origins are still visible today, with the Renaissance-era walls—stretching for 4.2 km—encircling the historic centre. These fortifications have played a critical role in maintaining the city's architectural integrity, making it a major destination for tourists and an important site for cultural heritage.

In this singular context, the Municipality of Lucca integrated urban logistics considerations into its Sustainable Urban Logistics Plan (SULP), aligning it with the broader Sustainable Urban Mobility Plan (SUMP). The SULP promotes a multi-stakeholder governance approach, encouraging continuous dialogue between local administrations, logistics operators, retailers, and citizens.

Key actions include:

- Supporting virtuous logistics practices, such as reducing the number of delivery journeys, improving loading efficiency, and incentivizing off-peak deliveries.
- Promoting zero-emission deliveries, including the use of cargo bikes and pilot efforts for consolidation centres tailored to specific logistical flows.
- Establishing clear access rules within the Restricted Traffic Zone (RTZ), linked to vehicle type and emission levels, in line with the "polluter pays" principle.

The SUMP further envisions progressively stricter access criteria for polluting vehicles and supports policies aimed at expanding the use of clean delivery modes.

Lisbon: Logistics Governance in a Growing Low Emission Zone
Lisbon, the capital of Portugal, is home to over 544,000 residents within the city and nearly 2.8 million in its metropolitan area. As a major European capital, Lisbon faces complex urban logistics challenges, particularly in its historic downtown area (Baixa Pombalina).

The city's logistics policy is currently undergoing a partial revision to ensure alignment with its SUMP. A central pillar of this effort is the planned expansion of

[5] The focus here is on cities located in the EURO-MED programme area: https://interreg-euro-med.eu/en/

Table 1 Urban logistics approaches from EURO-MED Cities experience. Source: Own elaboration

Theme	Lucca	Lisbon	Zadar
Population & Area	~80,000 residents; ~185 km^2	~545,000 in city; ~2.8 million metro area; 100 km^2	~70,000 residents; 25 km^2 (city proper)
Key urban features	Medium-sized town in Tuscany; enclosed by renaissance walls; narrow streets; strong cultural heritage	Capital of Portugal; dense urban core (Baixa) and large metropolitan area	Historic city on the Adriatic; cultural heritage on the peninsula; key ferry hub; tourist hotspot
Commercial vehicles/day in the city Centre	1680 vehicles/day	2100 vehicles/day[a]	370 vehicles/day
Urban logistics focus	SULP integrated with SUMP; focus on stakeholder engagement and low-emission deliveries	Logistics strategy aligned with SUMP and LEZ development	Focus on regulating deliveries in pedestrian zones
Regulation	Restricted traffic zone (RTZ) with vehicle-specific access and pricing based on emissions	Progressive implementation of low emission zone (LEZ) aiming to reduce exempted vehicles	Proposed improvements include differentiated time windows, upgraded vehicle standards, weight limits revised to 3.5 tons, and implementation of a comprehensive control of the vehicle accesses
Measures and initiatives	- RTZ access rules with "polluter pays" pricing - Support for zero-emission delivery (i.e., cargo bikes) - Awarding platform	- Progressive implementation and expansion of LEZ - Stakeholder coordination - Specification and development of a multi-purpose logistical Centre	- Video surveillance to count delivery vehicles -Control system based on RFID antennas to check time spent in pedestrian zones - Regulatory updates for better delivery vehicle management

[a]Data regarding daily access to the Low Emission Zone in Lisbon was retrieved from the presentation in the MED-COLOURS Meeting held in Lyon on May 15, 2025

the Low Emission Zone (LEZ), which is now the cornerstone of the city's logistics strategy. The new LEZ is scheduled for implementation in 2026 in the historic downtown area, and will gradually expand to other zones of the city, aiming to limit circulation to cleaner vehicles and reduce exemptions.

However, implementation requires not only political will and financial investment but also a cultural shift. Like many Mediterranean cities, Lisbon faces challenges in public acceptance and enforcement of regulatory measures. Building consensus among public and private stakeholders is therefore essential to the success of the LEZ and broader urban logistics measures.

Zadar: Historic Peninsula Logistics Management and SULP Development
Zadar is the administrative, cultural, and economic hub of northern Dalmatia, with a population of around 70,000 and a long-standing historical identity as the oldest continuously inhabited city in Croatia. Located on the Adriatic coast, its administrative area includes nearby settlements and seven islands, making it a key node for regional ferry and fast-boat connections—particularly important during the busy tourist season, when smart separation of passenger flows has helped reduce congestion in the historic centre.

Zadar has been actively engaged in improving its urban logistics framework through European initiatives. In 2018, the city developed a feasibility study exploring sustainable freight measures and prepared for the adoption of a Sustainable Urban Logistics Plan (SULP). Within the LIFE ASPIRE project, Zadar laid the groundwork for a more integrated and data-driven urban logistics strategy. This included the development of video surveillance software to count vehicles entering pedestrian zones and the installation of RFID antennas to monitor vehicle dwell times. These tools support evidence-based planning and highlight the externalities associated with delivery operations. The city also began introducing regulatory measures aimed at more efficiently managing delivery vehicles within the historic peninsula, balancing the need for goods movement with the protection of pedestrian areas and the unique heritage of the old town.

4 Policy Recommendations and Way Forward

Building on the considerations outlined in the previous sections, several guiding principles can support local administrations in addressing the challenges of urban freight distribution. From a logistics planning perspective, several policy factors can support the transition toward more sustainable and efficient urban logistics systems:

- Urban logistics must be recognised as an integral component of city functioning, with direct and indirect impacts on mobility, environmental quality, and urban life. It often involves competing interests and priorities that must be carefully balanced.
- Freight distribution should be fully integrated into broader urban and mobility policies, including Sustainable Urban Mobility Plans (SUMPs) and city service plans, particularly regarding spatial and temporal coordination.
- All implemented solutions should contribute to the long-term sustainability of logistics operations and support the transition to zero-emission delivery systems.

Addressing urban logistics requires action across multiple dimensions of city life. This includes regulatory and policy frameworks such as access restrictions and low-emission zones, the active involvement of stakeholders through participatory processes, and the implementation of both physical infrastructure and soft measures designed with a clear understanding of their impacts. It also involves the adoption of effective business and management models, as well as a growing need to consider the implications of e-commerce on traffic flows and urban space. Within this framework, cities—particularly through Sustainable Urban Logistics Plans (SULPs)—should establish policies that promote collaborative logistics (Expert Group on Urban Mobility (EGUM) 2024), encouraging shippers and logistics service providers to share vehicles, assets, and data to optimise resource use and reduce empty runs. In this context, urban logistics hubs can play a central role in better organising deliveries and enhancing overall urban liveability. Sustainable logistics measures should be tailored to the local context, contributing to zero-emission goals while aligning with broader urban planning and mobility strategies.

Once a shared vision among local stakeholders is established, a structured, step-by-step approach should follow, beginning with the identification of appropriate logistics measures and services. These should be assessed not only for their technical feasibility but also in terms of governance models, business viability, key performance indicators (KPIs) related to mobility and energy, and environmental impacts.

A wide array of solutions is available, varying in complexity, level of innovation, and digital integration. However, too often the organisational and operational dimensions are overlooked. Public administrations should therefore be well-informed of the full range of implications—technical, economic, social, and environmental—before moving forward (Liberato et al. 2022).

Drawing on diverse experiences and case studies, six core types of measures have emerged as particularly relevant and recurrent in many European urban logistics strategies, especially in small and medium-sized towns:

- Regulatory frameworks.
- Access control systems.
- Cargo-bike services and mini transhipment hubs.
- Smart control of loading/unloading areas.
- Urban consolidation centers (UCCs).
- Demand aggregation schemes.

Digitalisation cuts across all these areas, enabling better coordination, monitoring, user experience, and data-informed policymaking. It supports a range of functions—from access control and emissions monitoring to dynamic space management and incentive systems—and should be considered a transversal enabler rather than a standalone measure.

Table 2 presents the key features, benefits, and considerations for each of these six solutions, offering insights into their potential transferability and adaptation in other urban contexts.

Public authorities, particularly at the city level, must reclaim a central role in orchestrating sustainable urban freight, balancing regulatory, technological, and collaborative efforts to meet the mobility challenges of the future. The decarbonization transition calls for a clear policy roadmap that guides decision-makers from strategic planning and regulations to implementation, monitoring, and scaling. The stepwise approach outlined below offers a proposed roadmap to support cities in tackling the transition toward more sustainable and efficient logistics systems.

1. Establish strategic policy foundations.

   Local authorities should incorporate urban logistics within broader sustainable mobility and climate plans, by:

Table 2 Key features and opportunities for public-enabled solutions

Public-enabled solutions	Concept	Opportunities
Regulatory frameworks	Definition and enforcement of access and parking rules (e.g., UVARs, ZTL, LEZ) based on static and dynamic parameters to regulate commercial vehicle operations, integrated with broader mobility and urban space management strategies	Establish stable, flexible regulations that support fleet renewal, enable controlled access, regulate space use, encourage low-emission vehicles and higher load factors, and promote data exchange between operators and authorities
Access control systems	Implement ICT-based access control systems to monitor and manage vehicle entry/exit in urban areas using static and dynamic criteria, supporting enforcement, reduced violations, policy compliance, and integration with EU UVAR standards and NAP systems	Enable monitoring of compliance and vehicle dwell times, reduce roadside enforcement, support fines and rewards, and integrate with EU UVAR standards and NAP systems
Cargo-bike services and mini hubs	Shared zero-emission cargo bike services enabling freight, shopkeepers, and citizens to deliver goods within restricted zones during off-hours, often supported by app-based multi-station systems	Ensure guaranteed access for last-mile operators, extend use to shopkeepers and citizens, support occasional client use, and integrate with reward schemes to promote sustainable deliveries
Smart control of loading/ unloading areas	ICT-based monitoring of vehicle presence and parking duration in loading/unloading bays to ensure compliance, reduce misuse, and improve availability	Real-time stall status info for transporters, enhanced enforcement of parking rules, reduced traffic from searching for free bays, and the need to review bay layout and signage
Urban consolidation centres (UCCs)	A transhipment hub where goods from long-haul transport are consolidated and delivered last-mile by low-emission vehicles, optimizing deliveries and reducing traffic and pollution	Lower traffic and emissions, potential for additional services (reverse logistics, pick-up points), remote warehouse options for shopkeepers, fresh supply chain handling, and the importance of public procurement and alignment with access rules
Demand aggregation schemes	Local authorities promote agreements among shopkeepers and HoReCa to coordinate deliveries, share assets, and encourage low-emission transport, improving load efficiency and reducing environmental impact without relying on new technology	This approach can cut vehicle numbers, boost load factors, and strengthen shopkeepers' influence, supporting clean vehicle use. Success requires strong municipal leadership, managing stakeholder resistance, and fitting local regulations

- Defining long-term visions with measurable targets for reducing emissions.
- Developing Sustainable Urban Logistics Plans (SULPs) aligned with SUMP and other urban development strategies.
- Creating multi-stakeholder governance structures to ensure inclusive planning and shared responsibilities (Lorenzini 2024).

2. Adopting resilient regulatory and operational frameworks.

   Municipalities should set up clear, enforceable rules and ensure the infrastructure and digital tools to support compliance:

   - Establishing regulatory frameworks (e.g., RTZ, LEZ, UVARs) governing time, space, and vehicle types for deliveries.
   - Facilitating access and sharing of digital infrastructure, including ICT tools, across stakeholders.
   - Develop and implement impact monitoring systems based on KPIs (congestion, emissions, safety, etc.).

3. Accelerate the uptake of clean and innovative logistics solutions.

   Support the shift to zero-emission and shared logistics models through:

   - Incentive schemes for electric and cargo bike adoption, and the use of micro-hubs.
   - Promotion of cooperative models (e.g., van-sharing, demand aggregation), especially in historic centers.
   - Collaboration with regional, national, and EU bodies for funding and policy alignment.

4. Monitor, learn, network, and scale.

   Establish continuous and dynamic learning and fine-tuning loops:
   - Use real-time data to adapt rules, better allocate space (prioritizing collective transport and safe freight delivery zones), and refine service design.
   - Share results and lessons with peer cities to build collective knowledge and accelerate replication, transferability, and scalability of target solutions.

In summary, a successful transition requires municipalities to balance regulation, innovation, and coordination, embracing a new policy framework and leadership role that is proactive, data-driven, and inclusive. With a clear roadmap, cities can shape freight systems that support liveability, economic growth, and environmental goals, meeting the mobility challenges of the future.

References

Aifandopoulou G, Xenou E (2019) Sustainable urban logistics planning; Topic Guide. Europäische Kommission

ALICE (2015) Urban freight research roadmap. European Road Transport Research Advisory Council, Brussels

Ambrosino G (2015) SULP PORTFOLIO, Sustainable Urban Logistics Plan developed by 9 European cities for enhancing the sustainable urban mobility, including the methodology/guidelines for developing the SULP in your city. ENCLOSE Project Publication

Ambrosino G, Liberato A (2014) Sustainable mobility governance in small and medium historic towns: the LIFE+ PERHT project approach - TRA 2014, Transport solutions: from research to deployment, Paris, 14–17 April 2014

Expert Group on Urban Mobility (EGUM), subgroup 4, Sustainable Urban Logistics Planning (SULP), Recommendations on Urban Logistics, Adopted by the EGUM Plenary on 05.12.2024, https://transport.ec.europa.eu/document/download/d1580f49-1332-4401-97f8-9595ea33092e_en?filename=EGUM_Recommendations_SG4_D2_DataSharing.pdf

Hącia E, Wagner N, Łapko A (2023) The importance of City logistics for urban tourism development: searching for a new research field. Energies 16(1):175. https://doi.org/10.3390/en16010175

International Transport Forum (2021) ITF transport outlook 2021. OECD Publishing

Liberato A, Lorenzini A et al (2022) Policy recommendations for implementing sustainable freight distribution and last-mile delivery in small and mid-sizes historic towns. LIFE ASPIRE Project Publication

Lorenzini A (2024) Navigating the long-term transition toward net zero cities. Cutter Consortium. Retrieved from AMPLIFY J, 37 (3)

Matusiewicz M (2019) SULP (Sustainable Urban Logistics Plan) as a tool for shaping sustainable urban logistics: a review of European projects supporting the creation of SULP. Transp Econ Logist:Article 84. https://doi.org/10.26881/etil.2019.84.06

Mainstreaming the *Twin Transition* in Urban Logistics and Planning

Paola Cossu

1 The EU Challenge of Decarbonising Urban Logistics *by Digital* and *of Digital*

We are placed in the middle of a new era, branded by advanced and disruptive innovations and trends, predominantly digital (e.g. e-commerce, AI), twinned with an increasing pressure to reduce considerably Greenhouse Gas (GHG) emissions from all industries and economies, incorporating the ones generated by transport and logistics. Urgent actions are needed to implement adaptive legislative frameworks so that all industrial and economic sectors can move on sustainable paths. Policies and regulations have the challenge to encourage the development and implementation of innovative, interoperable and affordable solutions, prioritizing and incentivizing their scale-up.

'Climate logistics' is the collective term for the multiple roles that logistics will play in this battle (McKinnon 2018).

Both the green and the digital transitions are strategic priorities of the European Commission, delineating the shape of our future planet in the long term. While these two synchronised, or 'twin', transitions has potential to empower each other in diverse sectors, they are not naturally on the same path. For example, digital technologies could have stronger environmental footprints that need to be reversed if compared to net-zero economies' strategic goals by 2050. Therefore, a proactive and integrative approach to managing the twin transitions is important to ensure their successful implementation.

The European Commission is tackling the challenges through the European Green Deal, setting targets towards Europe's net-zero economy by 2050. The green

P. Cossu (✉)
FIT Consulting, Rome, Italy
e-mail: cossu@fitconsulting.it

A. Melkonyan-Gottschalk, Y. Li (eds.), *Urban Logistics Transformation*, Greening of Industry Networks Studies 15,
https://doi.org/10.1007/978-3-032-08387-6_20

and digital transitions are two main trends that will shape the future of the European Union. Both are at the top of the political agenda. Therefore, it is indispensable to anticipate their interaction and the possible *pressure* points or *gaps*.

According to Accenture analysis (2022) and in collaboration with the World Economic Forum, digital technologies, if suitably and sustainably scaled across industries, have huge potential to generate up to 20% of the reduction needed to comply with the International Energy Agency net-zero paths by 2050, impacting the energy, materials and transport, logistics and mobility industries. Therefore, these industrial markets can reduce emissions by 4–10% if they rapidly adopt efficiently and affordably available digital technologies. The 'twin transitions' refer to uniting the digital and green transitions, which could accelerate necessary changes and bring societies closer to the level of transformation needed. To succeed with the green and digital transitions, a better understanding of the interdependencies is fundamental, especially when it comes to understanding priorities. The fast growth of e-commerce has altered consumer prospects and behaviours, leading to a surge in deliveries. Although this leads to economic and consumer benefits and opportunities, it generates challenges such as congestion, emissions and safety issues. With a business-as-usual scenario, all urban delivery traffic is estimated to generate an increase of carbon emissions by 60% by 2030.

While digital and green transitions have the potential to cooperatively transform societies and economies, they express diverse dynamics. The green transition is predominantly guided by climate neutrality and sustainability targets and the need to reach them swiftly. It entails a clear and forward-looking political and societal commitment. The digital transition process is still ongoing and determined by the technology evolution, in which the industrial and business sectors are the main players. Consequently, regulation and policy support are key ingredients to guarantee that the digital transition can be a leverage for achieving a just and sustainable green transition, harnessing digital technologies, empowering both people and businesses. Successful accomplishment of the green and digital "twin" transitions is the foundation for delivering a sustainable, fair and competitive future. To unlock potential and anticipate possible adverse effects, the green and digital transitions require a proactive and integrated management. The benefit of a twin transition is an efficient urban logistics system where interventions to reduce GHG emissions will also improve the efficiency and reduce the social cost of transport and logistics systems to society.

The EU Commission is thus moving fast by achieving the Sustainable and "Smart Mobility Strategy" goals (COM/2020/789) and "A Europe fit for the digital age" targets and to establish conditions for setting a common European Mobility Data Space—EMDS (COM/2023/751). The aim is to accelerate the digital and green transformation of the EU's mobility and transport sector, empowering efficiency, safety, sustainability, resilience and agility. One of the main challenges to achieve fast is to decrease the level of fragmentation in the supply chain, promoting and enabling exchange of mobility and transport data between public and private actors within a trusted, inclusive and harmonised ecosystem. Particularly, the EMDS can generate cross-sectoral values due to synergies with other data spaces (e.g. tourism, health, circular economy). It relies on common principles and building blocks that can be adopted in other sectors and domains without exacerbating existing solutions.

Urban logistics, being positioned at the core of the EU strategic goals, is strongly contributing to reducing GHG emissions by 90% by 2050. Data-driven revolution is needed to make multimodal urban logistics more efficient and greener. In urban logistics, the green and digital transitions can reinforce each other, but they do not necessarily always align. For instance, cities are responsible for approximately. Seventy-five percent of CO_2 emissions.

Smart cities can leverage the twin transitions in a holistic and more systemic way. ICT-based solutions can reduce GHG emissions by 10–15%, while the adoption of Digital Twins can enhance city planners' competence in simulating prospective scenarios or modelling the potential impact of implemented measures and policies (e.g., in SULPs), fostering a more innovative and competitive logistics market.

For instance, digitalised urban logistics would support LSPs to find backloads for their empty trips.

However, there are domains where the two transitions can impede each other. The development of digital infrastructure needs to be aligned with the long-term goal of the green transition, especially when dealing with energy consumption and the environmental footprint generated by the digital infrastructures themselves.

2 Spotlight on Coupling Digital and Green Transition in Urban Logistics

According to ALICE (BOOSTLOG project 2021), Urban logistics ensure that shops and businesses are stocked, people access goods (including all forms of e-commerce delivery), equipment is installed and repaired, buildings are supplied, returns are collected and waste is removed. Every place of activity requires deliveries/collection and servicing, logistic demands proper infrastructure and resources planning by companies and cities to make logistics serve people, planet and profit in an efficient and sustainable way.

The recent physical disruptions and economic shockwaves caused by the Covid-19 pandemic, also impacted by worldwide ongoing geopolitical events, increased the relevance for urban logistics players to boost digital transformation.

With the implementation of the EMDS, city planners will be able to adequately and smartly analyse urban freight data (e.g. volumes, vehicle and routes, loading/unloading operations) to drive informed planning, decision making/policies and sustainable investments in infrastructure. Urban infrastructure-related data exchange (e.g. curbside, dynamic vehicle access, parking space and consolidation, real-time traffic monitoring and e-charging infrastructure) will support Logistics Service Providers in increasing their operational efficiency, facilitating collaborative models and solutions, especially to removing barriers due to space scarcity or conflicting mobility operations. Market players can seize new business opportunities, form partnerships and integrate services for seamless coordination. Real-time information makes private and public actors' operations more efficient, ensuring higher level of compliance with the regulation in force.

Rethinking decarbonised deliveries is supported by promoting virtuous models enabled by technology but driven by sustainability in logistics planning. For instance, some of them are inspired by the Physical Internet paradigm for structural collaboration, implying strategic and collaborative use of assets, optimizing use of existing capacity (e.g. infrastructure, vehicles), with more consolidated flows, adopting and combining transport modes and logistics assets smartly and seamlessly (ALICE Roadmap on PI 2020).

A structured methodology for evaluating the urban logistics ecosystem, defining urban logistics innovations transition paths and creating city-specific Physical Internet operational schemes has been developed by the DISCO project, co-funded by the EU Commission (2024). It emphasises the importance of data availability, supportive policies and regulations for the successful adoption of urban logistics innovations. Additionally, the introduction of the DISCO Urban Freight Data Space and the Meta Model Suite—a set of tools and frameworks that synchronises spatial and multimodal freight transport logistics planning—guides cities in selecting appropriate solutions based on their capacities and requirements. The open data-sharing space will provide city planners with tools that can help them optimally manage, monitor and dynamically predict city freight flows.

3 Twin Transition-Led Enablers for Paradigm Change in Urban Logistics

Digital technologies deliver functions that can boost the green transition in urban logistics. For example, monitoring and tracking can provide real-time information and be a catalyst for optimal urban space management and dynamic access control. Technologies like Artificial Intelligence (AI) are key enablers in addressing climate change. AI technologies facilitate efficient resource allocation through data-driven decision-making (Olawade et al. 2024). Some examples of functions for digitalising urban logistics are provided below.

Simulation and forecasting can improve efficiency, for example in the form of Digital Twins that enables data-driven decision making real-time view of the logistics network, e.g. vehicle locations, traffic conditions and delivery status, as well as changes in traffic patterns, weather conditions, or delivery demands, to assess the impact on logistics operations and optimise strategies for improved decision making and logistics planning. Digital Twin can be instrumental in developing SULPs by providing city authorities with a comprehensive understanding of the existing logistics ecosystem and enabling them to make data-driven decisions. Public administrations can be provided with a smarter and affordable way to improve not just urban logistics but overall sustainability of the last mile of the supply chain (LEAD project 2023 Policy Recommendations, LEAD (2022) Digital twins for last-mile low-emission logistics).

Predictive analysis relies on a diverse array of data sources to forecast future trends accurately in urban logistics. It can enable the identification of the optimal infrastructure and fleet operation configurations for last-mile logistics services, based on a multitude of service and system sustainability and efficiency indicators. The provision of this service relies on a joint but modular prediction-optimisation-simulation modelling solution framework which enables: Day-to-day parcel demand prediction; Infrastructure and operations optimisation; Fleet operations simulation in realistic traffic conditions (GREEN-LOG project 2024).

Virtualisation has the potential to change and reduce the environmental footprint of deliveries. Extended reality technologies, such as augmented reality and metaverses, have the potential to move more services, production, or consumption activities online. Carmigniani et al. define augmented reality (AR) as "a real-time direct or indirect view of a physical real-world environment that has been enhanced by adding virtual computer-generated information to it". Virtualisation covers trends such as increasing online shopping. Taking retail as an example, the demand for physical shops can be reduced, for example, with the so-called "virtual fitting room" supported by a 3D avatar." More importantly, the return of goods can be limited as well. For example, Zalando is piloting a virtual fitting room, allowing customers to try on clothing by creating a 3D avatar and inputting their height, weight and gender details. The retailer offers customers personalised size recommendations based on their purchase and return history, as well as information that customers can add to their size profile. For those items where Zalando provides size advice, size-related returns have decreased by 10%. Therefore, a sound virtualisation is based on the motivation of individual customers to change behaviour and be adaptive to advanced sustainable solutions.

Monitoring and tracking for accurate real-time information, for example, to enable the monitoring of emissions for enhanced supply chain carbon footprint visibility. This involves using technology to collect and retrieve real-time data, with marginal delay between data gathering and further analysis. Other examples are tracking vehicle locations in fleet management or tracking goods in transit in a supply chain (Kumari A, Gaikwad KK (2025).

Information and communication technologies enable high levels of interaction and modern ICT leads to higher potential to positively affect user behaviour. For instance, if we consider labels and smart packaging, they can both deliver the environmental footprint so that the "full cost" of an item can be substantiated. Being in the Internet of Things era, packaging is more than a cover for items but a means for information exchange on the item between shipper and customer (Kumari and Gaikwad 2025). For this reason, "packaging" is also acknowledged as a "silent seller." Smart packaging embeds technology for additional functionality, thinking beyond conventional packaging. This can refer to sensors, RFID, or QR codes retrieving information to track and trace a product's condition and communicate with the final buyer, with a greater efficiency and sustainability in the supply chain and circularity. Smart packaging could reveal temperature changes, guaranteeing food quality or also provide immediate feedback when the item is scanned by a device.

Societal governance is fundamental for the success of the twin transitions. A just twin transition can enable extensive acceptance and adoption of green-digital solutions. For instance, social responsiveness could limit extreme consumption and bulk buying. The sector necessitates new skills and competences, and onboarding of small enterprises is fundamental for a competitive and innovative market of green-digital solutions, as for small carriers, creating conditions for inclusiveness and competitiveness of the urban logistics sector, supported by a source of financing for innovation able to reduce environmental footprint. Open partnerships and balanced governance between corporates, start-ups, technology providers, lenders and public authorities will create potential for scaling investments, de-risking technologies and enabling trusted data exchange. It is thus important that the digital transformations that boost the green transition in urban logistics are also inclusive and sustainable, so the benefits belong to all. The living lab approach (Nesterova and Quak 2016) ensures that the stakeholders are involved much earlier in the planning and implementation processes, and that the proposed city logistics implementation is revised and continuously improved to meet stakeholder needs and obtain maximum impact for a long time.

4 Data–Driven Collaborative Operating Models for Net-Zero Economies

The operating models outlined in the paragraph include advanced solutions and hence showcase diverse context-specific challenges adopted by authorities and private players. In some interventions, the emphasis is on strategic land-use planning with mobility strategies, able to generate efficient urban logistics consolidation. Some cities have already implemented *zoning* and *regulations* to promote the use of low-emission vehicles with determined areas dedicated to logistics operations. For instance, the Logistiek 010 community in the city of Rotterdam (NL) involves almost 3000 logistics companies, government, knowledge institutions and educational bodies, having the common challenge of removing inefficient and unnecessary journeys and mitigating congestion. They have jointly designed and implemented a Zero-Emissions Zone for Logistics vehicles by 2025.

A recent study in Brussels evidenced the expected operational benefits of cargo bikes when compared to conventional vans. The results showed that the smaller vehicles can park nearby to the last delivery point (i.e. less than 30 m), travel faster in congested conditions (16 kph versus 11 kph) and reduce related costs (kale ai 2023). It was estimated that adopting cargo bikes could bring 60% reduction in time spent cruising for parking and go by walking to the destination. Combining cargo bikes with hubs, a gradual reduction in distances and hours travelled can be achieved, "up to a density of about 150 deliveries/km^2" (Dalla Chiara et al. 2020).

Other interventions are oriented to manage the demand growth for urban logistics hubs, mapping strategic locations, incentivising mixed-use development, retrofitting and repurposing existing buildings and infrastructures, filling empty parking lots, event areas and unused spaces in cities. In some cities, public authorities have also introduced curbside management to govern urban freight traffic effectively and encourage the establishment of micro-hubs in the inner city, which can serve nearby deliveries.

Some of the most relevant examples of innovative models are reported below.

Public–private cooperation for dynamic, data-driven and automated communications, incentivising effective and zero emission operations in last-mile deliveries. Within the Horizon Europe project DISCO (Cossu P, Astegiano P (2025)), the city of Ghent (Belgium) implemented the Urban Access Control (UAC), with the contribution of a local partner. It is a tool that integrates urban vehicle access regulations (UVARs) into logistics planning. The tool optimises deliveries by ensuring compliance and suggesting collaborations with Green Last Mile operators when they are cost-effective. By enabling real-time emissions and cost calculations, the UAC helps compare conventional and green delivery options, supporting Ghent's broader goals for sustainable urban logistics. Initial findings already demonstrated that the UAC system has a significant effect on safety and stress-free planning. The city of Ghent also prioritises innovation through data-driven logistics. Ghent created an interconnected logistics ecosystem, enabling seamless communication between public and privately owned databases and systems without costly integrations by establishing an urban freight dataspace.

Multipurpose, multi-tenant and temporary use of buildings as a logistics hub is a successful solution implemented in the Greek city of Thessaloniki (DISCO project 2024). The WareM&O platform (Parodos et al. 2022), is a digital marketplace where warehouse owners can list available spaces and tenants can easily search, compare and book the capacity they need. It integrates e-signatures and blockchain-based smart contracts for secure, transparent transactions. By offering real-time space spare capacity, the platform enables efficient and collaborative logistics operations (i.e. the capacities are updated dynamically on the platform). TIF-HELEXPO's underused pavilions, shops and offices have been retrofitted to logistics hubs adopting the WareM&O platform. Real-time capacity tracking, e-signatures and blockchain-based smart contracts enabled a smooth collaboration between TIF-HELEXPO and the courier ACS, which resulted in the relocation of ACS's principal depot, reducing congestion in Thessaloniki's inner city. Infrastructure upgrades—such as RFID access and around-the-clock operations—further streamlined processes, showcasing how strategically located large event venues can be repurposed for more flexible and sustainable urban freight solutions. Also, the city of Paris (OECD 2024) offers underused spaces for logistic operations, adopting "calls for projects." For instance, they use an area under the ring road at Porte de Pantin retrofitted by a public-private real estate developer, Sogaris, into a logistics space, leased under a public domain occupation agreement.

Digital twins for decision making have been demonstrated in the Danish city of Copenhagen (DISCO Project 2024) to simulate prospective logistics scenarios. The outcomes of the simulations yielded a range of Key Performance Indicators (KPIs), including economic indicators, which were adopted to evaluate the feasibility and potential impact of implementing different logistics measures (e.g. zero-emission zones, location of micro-hubs, etc.). Central to the success was ensuring data security and confidentiality. Rigorous data protection protocols have been put in place to guarantee the full confidentiality of the data provided by the LSPs (3 weeks of historical routing information were needed to build the scenarios). In parallel, synthetic data were generated to reproduce the global logistical flows representative of the local demand. A comprehensive logistics activity in each area is determined, embedding additional data over time improving the level of representativeness, robustness and acceptance. The results of this analysis have been introduced in the newcoming SULP, currently under political approval.

Hub and Spoke delivery model supported by Digital Twins has been demonstrated successfully in the Greek city of Thessaloniki (URBANE project 2023). The city Living Lab introduced a macro and micro-consolidation delivery system, leveraging data-driven optimisation and collaborative logistics. Extensive operational and socioeconomic data informed facility location modeling, ensuring strategic placement of micro-hubs in public spaces. Using Digital Twins technology and Agent-Based Modeling (ABM), the system simulated green last-mile solutions, integrating Urban Consolidation Centers (UCC) to upgrade delivery efficiency. Three scenarios explored varying degrees of logistics operator collaboration, demonstrating that an *Alliance Locker Network* significantly enhances sustainability KPIs and urban space utilisation. A centralised UCC further optimises distribution, reducing travel distance while improving network performance.

Smart Micro-Hubs: Enhancing Collaboration Among Logistics Providers. Has been installed in Bologna (Italy). A Nearby Delivery Areas (NDA) has been introduced as a key measure in the Sustainable Urban Logistics Plan (SULP) to improve last-mile delivery efficiency within the city's Limited Traffic Zone (LTZ). This initiative combined zero-emission vehicles with unnamed micro-hubs, strategically positioned at key entry points to facilitate the transfer of freight from conventional vans to light electric vehicles. A blockchain-based system ensured transparent tracking of logistics events, providing proof of delivery and strengthening collaboration among logistics operators. By sharing infrastructure, competing LSPs improved operational efficiency while significantly reducing congestion in Bologna's historic center. The NDA proved an enhancement of sustainability, accessibility and service reliability, demonstrating how automation and digital integration can optimise urban freight operations while reducing environmental impact.

Pick-up and drop-off (PUDO) networks are digital solutions and collaborative operating models that drive operational efficiencies and minimise travel distances and decouples courier and customer routing plan. They represent a new form of consolidation and reduce the amount of unsuccessful first-time deliveries. PUDO points can reduce the number of delivery trips by up to 15% and reduce congestion by up to 2%. They can reduce the cost of deliveries (up to 15%). Major companies such as Amazon, DHL, FedEx have their own lockers. Nevertheless, there are "agnostic parcel locker networks", like those with labelled ones coming from InPost and Posten.

Curbside management: a recent study of commercial vehicle traffic in the US city of Seattle evidenced that cruising for a parking space enhances an average of 2.3 min to each delivery. Curbside management optimally allocates space for increasing the efficiency in its use by different mobility players. It includes dedicated zones for last-mile operations, PUDO, waste collection and bike parking. Smart curbside management enhances this by dynamically rethinking curb space on demand. In New York, smart curb technology can satisfy the increasing demand for goods deliveries, waste collection with cleaner modes such as e-bikes and scooters.

Key considerations for implementing smart curbside embrace data integration, empowering compliance, technology infrastructure costs and managing conflicts in the use of space.

White-label deliveries, combining urban consolidation centres and micro-hubs, can substantially reduce the number of trips. White-label deliveries can reduce the number of journeys by 30% and cut total delivery costs by 51% when combined with the use of e-bikes and other zero-emission micromobility modes.

Cross-industry collaboration and data exchange require adequate coordination among public entities, business and research for joint urban logistics efforts. For instance, ColisActiv', a French programme, aims at financing bike and foot last mile deliveries if they exchange routing data, driving local infrastructure investments. They are requested to exchange real-time traffic data and delivery routes, so that stakeholders can optimise use of space for higher delivery efficiency. City of Rotterdam has devoted programs based on data-driven private-sector recommendations, including a zero-emissions zone.

5 Filling the Gaps Towards the Twin Transition in Urban Logistics

Urban logistics players are requested to embrace *data & digital assets*, *forward-looking mindsets* and *collaboration* as the three key levers enabling real change. In this way, urban logistics can significantly give impulse (rather than harming) city liveability and global well-being.

This chapter has highlighted the critical role that digitalisation will play in the transformation of the European road freight market towards a zero-emission future Planet. It presents insights gathered directly from the current literature and recent EU research and innovation projects and studies, aims at improving the understanding of how digitalisation and carbon neutrality can support cities in Climate City Contracts. It will include a journey to climate neutrality, with initiatives established together by the cities, national and regional authorities, delivering their ambitious climate neutrality targets with investors and citizens.

Every city's revolution journey is exclusive, as players have evolved to different layers of readiness and maturity in terms of digitalisation and transport infrastructure for logistics and planning. Therefore, the top policies and strategic decision making depend on urban-specific contexts. The recommendations in this section are presented as a checklist to be considered for city governments' assessments, considering their unique ecosystem. City governments should take an active role, but each player in the ecosystem needs to play its part.

To reach a positive shift towards carbon-neutral urban logistics, it will be crucial for all actors in this sector to be truthful when declaring their level of engagement and commitment to collaborate and exchange, and in the willingness to redress traditional business as usual practices and behaviours. Local authorities have an important role.

Solutions like PUDO pick-up deliveries have the potential to limit the number of vehicles circulating and the overall distance to the customer, implying education to the responsibility of e-consumers.

EU strategies and several innovative solutions, technology enablers are analysed, all having potential to boost local authorities and LSPs alignment of their freight transport decarbonisation trajectory needed to reach 2030 and 2050 targets.

Recommendations refer to the importance of adopting technologies supporting the reduction of unnecessary kms, improving the adoption of low carbon transport modes, smarter loading of freight vehicles, delivering higher energy efficiency of goods movement, creating conditions for strategic use of space, data exchange and collaboration for informed decision making and smarter Sustainable Urban Logistics Planning.

All recommendations are derived from a high-level topic that impacts across mixed multiple solutions and measures for cities and service providers.

High-level topic	Recommendation for LSPs
Collaboration models	Enable a deeper understanding of public-sector needs on data and collaborate with local authorities. Promote open innovation to drive urban solutions. Collaborate with local research and innovators and be active in cross-domain living labs. Establish partnerships with real estate owners for micro consolidation. Cooperate with city authorities to provide fair urban access and optimise use of space dynamically. Adopt a data-driven (informed) approach to locate micro-hubs, EV charging stations, parcel lockers/PUDO and loading bays.
Operating models	Balance demand for road space from last-mile operations vs availability. Select suppliers with a sustainable footprint and green policy embracing sustainability as a contractual requirement, with support and incentive systems. Investigate models focused on sustainability and city liveability, addressed to manage and satisfy consumer needs.

High-level topic	Recommendation for LSPs
Carbon Footprint	Prioritise carbon-neutral delivery vehicles. Promote visibility and transparency of their carbon footprint. Adopt advanced technology for route optimisation and real-time vehicle tracking for enhanced fleet management. Explore tax incentives or affordable funding programmes embracing green investments plans.
Multimodality	Prioritise micromobility with cargo bikes and on-foot delivery. Incentivise the adoption of EV layouts that better support modularisation. Introduce micro-hubs for cross-docking, consolidation and charging of micro-fleets. Offer out-of-home consolidated delivery options prioritizing unnamed proprietary or agnostic PUDO networks, embracing proximity logistics for nearby deliveries.
Incentives	Retrieve information to retail suppliers on resource-efficient and sustainable delivery choices. Promote incentive schemes for consolidated and off-peak deliveries and collaboration.

High-level topic	Recommendation for city authority
Regulation and standardisation	Develop clear, standardised and fair logistics regulation and governance, for the last-mile, curb-side management and micromobility, involving manufacturing in the political debate. Promote a collaborative ecosystem for innovation and investment by creating an appropriate framework. Allocate resources to support the private sector in navigating regulations and expediting planning processes, especially for emerging technologies. Share global best practices to speed up the implementation of initiatives. Use other cities' experiences and allocate dedicated resources in support of this effort.
City-led policy	Integrate logistics into urban planning, prioritising common needs, guided by sustainable urban logistics plans (SULPs). Plan and optimally locate urban public space for unnamed micro-hubs and EV charging and generate incentives schemes that inspire collaboration. Professionalise logistics employment with training, accreditation, career advancement and job enrichment, increasing the appeal to the younger generation. Deliver and promote a clear and long-term transition strategy and vision, e.g. climate City contract and dialogue with logistics players and industries, e.g. living labs, to facilitate the adoption and scaling up of innovations in urban logistics. Understand different hub typologies, their interrelation with supply chains and logistics movements in urban areas and develop strategies and plans for diverse logistics hub establishment.

High-level topic	Recommendation for city authority
Collaboration	Map and allocate adequate urban public space for logistics, e.g., for EV charging, lockers and micro-hubs. Establish collaboration models between the city, LSPs and real estate for optimal and strategic locations and multipurpose use of existing (and often underused) infrastructures (multitenant). Guarantee larger scale network of EV charging stations, e.g. collaborating with charging-infrastructure operators. Establish the right ecosystem enabling data exchange between urban logistics players, favouring data availability, real time information, deconflicting relationships among different curb-side community players and guaranteeing prioritised use of infrastructure and city access. Data spaces should be promoted among players, as a decentralised platform guaranteeing data sovereignty, neutrality and trust. Establish the right layers leveraging innovation and pilots. Co-creation approach in living labs.
Carbon emissions and circularity	Create incentives for micromobility and on-foot modes. Consider designating low-traffic neighbourhoods and zero-emission zones, limiting vehicle access and providing sustainable alternatives. Guaranteeing that zero-emission fleet adoption is financially viable and affordable. Road-user charging incentives, purchase subsidies and favourable loans. Including plans for the efficient disposal and recycling of waste generated during the delivery and servicing process.
Safety	Ensure safe environments for on-foot and micromobility deliveries. Providing clear and enabling rules on the use of public space. Assure high safety standards for road users with adequate legislation and the involvement of governments. Enhance safety standards for commercial vehicles, e.g., implementing a safety accreditation system.
Infrastructure	Promote curb-side management for more efficient operations and better match delivery needs. Mixed use of space with additional loading capacity, supported by digital platforms managing real-time demand, making it easier the use zero-emission vehicles. Promote inland waterways and air deliveries, introducing adequate measures and regulations for LSPs. Provide data on charging infrastructure, e.g. with a citywide open-source platform to map available infrastructure and generate fair tariffs and to satisfy EV charging infrastructure supply and demand. Ensure visibility of the last mile. Consider using a data platform to process urban deliveries, and focused investments in infrastructure.
Technology	Optimise freight transport network operations adopting digital twins to predict traffic flows and scenarios, with geospatial data supporting last-mile operators, especially in case of road impediments (e.g. road work).

In conclusion, although the optimistic effect of all such initiatives, their impact at scale will finally depend on the city's local ecosystem, including its political, geographical, cultural and economic features. Every successful measure should be generated by a co-design approach and prioritised according to local expectations so that they can be accepted, and available resources allocated. City governments and local stakeholders should take care and understand local contexts and dynamics before prioritizing measures that can concretely and systematically accelerate the Twin Transition of urban logistics.

References

Abel S et al (2021) Dynamic curbside management: keeping pace with new and emerging mobility and technology in the public right-of-way. The National Academies Press

ALICE-ETP, 2020 (2020) Roadmap to the Physical Internet, funded under the SENSE UE project

CIVITAS (2020) Smart choices for cities: making urban freight logistics more sustainable. CIVITAS Forum Network, s.l.

Cloud Report on Urban Logistics (2022) BOOSTLOG project. Funded by the EU Commisison

Cossu P, Astegiano P (2025) Enhancing urban logistics with a physical internet-led approach and integrated data spaces. INUF reimagining the future of urban freight Los Angeles

Dalla Chiara G, Alho AR, Cheng C, Ben-Akiva M, Cheah L (2020) Exploring Benefits of Cargo-Cycles versus Trucks for Urban Parcel Delivery under Different Demand Scenarios. Transportation Research Record 2674(5):553–562

Fried T, Goodchild A, Browne M, Sanchez-Diaz I (2024) Seeking equity and justice in urban freight: where to look? Transp Rev 44(1):191–212

ITF (2024) Urban logistics hubs: summary and conclusions, ITF Roundtable Reports, No. 165. OECD Publishing, Paris

JRC's Strategic Foresight Report (2022) Towards a green and digital future

Kumari A, Gaikwad KK (2025) Data carriers for real-time tracking and monitoring in smart, intelligent packaging applications: a technological review

McKinnon A (2018) Decarbonizing logistics. Kogan Page, London. ISBN: 978-0749483807

Montreuil B, Meller RD, Ballot É (2013) Physical internet foundations. In: Studies in computational intelligence, pp 151–166

Nesterova N, Quak H (2016) A city logistics living lab: a methodological approach

Olawade DB, Wada OZ, David-Olawade AC, Fapohunda O, Ige AO, Ling J (2024) Artificial intelligence potential for net zero sustainability: Current evidence and prospects

Parodos L, Tsolakis O, Tsoukos G, Xenou E, Ayfantopoulou G (2022) Business model analysis of Smart City logistics solutions using the business model canvas: the case of an on-demand warehousing E-marketplace

Policy Recommendations, LEAD (2022) Digital twins for last-mile low-emission logistics. Funded by the EU Commission

Zhang Y, Li Y, Liu Y (2021) A literature review of smart warehouse operations management. Int J Prod Res 59(5):1387–1403

Collaborative Urban Logistics: The Transformative Potential of AI, Data Spaces, and Living Lab Ecosystems Towards Zero Emission Urban Logistics

Ioanna Fergadiotou

1 Challenges and Future Outlook

Urban logistics and delivery systems, integral components of the European freight Transport and Logistics (T&L) industry, have been put under serious strain over the last 10–15 years. Beyond the rise of on-demand logistics and the negative consequences of the 'uberisation of logistics (Stępnicka et al. 2025), cities are confronted with the necessity to quickly reform urban space, infrastructure and regulation to accommodate new service models and operators' needs, for example, installing parcel lockers or charging infrastructure in public spaces. Urban planners are tasked with an ambitious mission. On the one hand, to adapt to the fast-paced market developments, serving businesses and consumers' needs for delivery speed and flexibility. On the other hand, to promote environmental sustainability and improve cities' liveability, meeting eventually the green transition decarbonisation goals set by the Fit for 55 Package (EC COM 2021). Mission impossible? Maybe not, if bold and coordinated measures are taken to support the systemic integration of policy, technology and private-public partnerships. In any case, balancing these often-competing objectives also requires rethinking and speeding up the traditional planning lifecycle of urban governance, disrupting the slow and linear planning phases with more agile, adaptive models that can keep up with the speed of change in the urban logistics landscape.

Turning competition into coordination—leveraging technology to transform fragmented operations into a shared, adaptive and efficient urban ecosystem towards zero emission urban logistics

I. Fergadiotou (✉)
Inlecom Innovation, Athens, Greece
e-mail: ioanna.fergadiotou@inlecomsystems.com

A. Melkonyan-Gottschalk, Y. Li (eds.), *Urban Logistics Transformation*, Greening of Industry Networks Studies 15,
https://doi.org/10.1007/978-3-032-08387-6_21

Living in an era shaped by digital solutions, automation, and advanced technological applications, one should consider their adoption would by default lead to optimised urban logistics operations. Technology and automated processes have transformed the way urban deliveries are planned and executed. Automation in warehousing combined with robotics has already begun to enhance service reliability while reducing errors, costs, and often emissions. Equipped with advanced sensors, telematics, and communication technologies, operational systems enable real-time tracking, predictive maintenance, and intelligent route planning. Thousands of papers have been written on route optimisation, and millions of code lines of admittedly sophisticated algorithms have been developed for such systems.

However, the rapid pace of innovation also presents a set of structural challenges that threaten to fragment rather than unify urban logistics systems. Many technological solutions have been adopted independently across cities, operators, and sectors, often without interoperability, standard protocols, or integration strategies. As a result, urban logistics networks risk becoming a patchwork of isolated systems that fail to communicate effectively. Infrastructure investments—such as urban or peri-urban consolidation centres, micro-hubs, or digital platforms—may be underutilised or be abandoned if they are not planned according to a holistic strategy driven by cities. These efforts must be aligned within a coherent, system-level approach that integrates logistics operations but also considers urban energy systems and spatial planning. Logistics providers, local authorities, and infrastructure operators need to work in sync toward harmonised levels of automation and shared operational goals, enabling smoother flows of goods and better utilisation of limited urban space.

According to Nagy and Szabolcs (2024), collaborative logistics can offer a compelling response to the structural fragmentation and inefficiencies in urban delivery systems by enabling all participants in the logistics chain to make the best use of available resources while reducing idle times and excess inventories. These efforts can contribute to meeting sustainability goals through joint resource usage and optimised transportation. This approach is built on the principle of sharing (Fig. 1) and is about aligning stakeholders—public and private, large organisations and SMEs—around a common vision and strategy, along with shared resources—shared infrastructure, data and operational processes—to unlock efficiencies and reduce urban burdens.

In a context where multiple delivery providers operate within the same geographical areas, collaborative models enable joint planning and exploit unused network capacity. Consolidation of goods flows with deliveries from multiple suppliers being bundled reduces redundant trips, improves vehicle fill rates, and minimise traffic congestion. The shared use of assets such as micro-hubs, charging points, locker networks, which can serve multiple operators instead of being siloed for exclusive use, reduces the demand on limited curb and warehouse space and contributes to more equitable use of urban space in general. Shared vehicles' capacity can lead to fewer delivery trips, consequently resulting in lower emissions. The urgency of climate goals acts as a powerful incentive for actors to work together rather than compete in isolation, aligning individual business models with collective

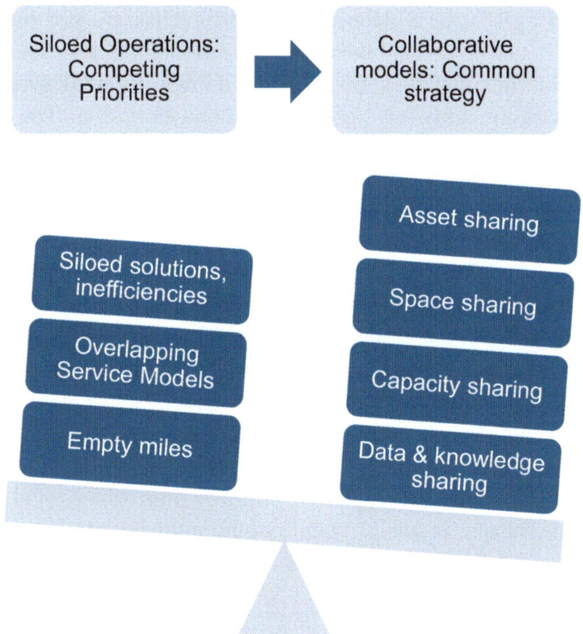

Fig. 1 Operational dimensions of collaborative logistics

urban and planetary well-being. A number of collaborative initiatives across Europe have put these ideas into practice with varying degrees of success.

2 Real Life Case Studies

A number of practical examples across Europe show how collaboration in urban logistics can make a difference, but also reveal where things can go wrong. In 2016, the Chapelle International Urban Logistics Hotel (Baron and Khodadad Saryazdi 2019), a multi-storey facility built in the centre of Paris, operated as an intermodal urban terminal, demonstrating how public-private partnerships and shared infrastructure can reduce congestion, improve delivery efficiency, and make better use of urban space. It included freight rail connections, space for logistics operators, offices and housing. It's a good example of how logistics can be integrated efficiently into urban life, supporting night-time deliveries and reducing the need for trucks to enter the city during the day. The facility remains active as a mixed-use logistics hub, offering shared warehousing and supporting sustainable last-mile deliveries. The rail component, even if central in the initial design, is not operational anymore due to cost barriers. The KoMoDo Project in Berlin (Found at: https://www.behala.de/en/komodo-the-sustainable-delivery-traffic-solution/; URBANE Project 2023a) established collaboration among, five major parcel delivery companies—DHL, DPD, GLS, Hermes, and UPS. They worked together using a shared micro-hub and e-cargo bikes for last-mile deliveries. The results were impressive considering that they managed to deliver 160,000 parcels in a low-emission zone, cutting down on traffic, noise, and pollution. The initiative was sponsored by the Federal Ministry of Transport and Digital Infrastructure (BMVI).

While there are promising examples, not all initiatives have gone as planned, offering valuable lessons about implementation pain points and practical challenges in scale-up that can be extrapolated to other contexts. There are cases where shared consolidation centres failed to gain traction, e.g. in Vicenza, Italy (Ville et al. 2012), or city-backed white-label delivery platforms envisioning to unify urban deliveries under a neutral delivery system failed to take off. In these cases, the business models struggled due to regulatory and judicial pressures, the hesitance of retailers or LSPs to join the platforms, competitive tensions, the lack of participation incentives or governance rules and the absence of a neutral trustee to coordinate operations. Without a clear business case or shared vision, these models didn't move forward. These examples show that collaboration requires more than good intentions; it needs strong governance, financial viability and solid foundations that guarantee stakeholder alignment (Nagy and Szabolcs 2024).

3 Enabling Change: Integrating Digital Tools and Business Innovation in Collaborative Logistics

New prominent innovative solutions are emerging from European research projects, based on the foundations of collaborative and sustainable logistics. Multidisciplinary teams are working together to co-design green solutions, implement interventions with significant impact and deliver innovative tools for the replication of successful models. Digital enablers such as AI, Digital Twins, blockchain and Data Spaces are being gradually integrated into real-world living labs environments (European Network of Living Labs et al. 2025). These technological advancements support the broader transition toward the Physical Internet (PI), where logistics is envisioned as a seamless, interconnected, and modular ecosystem that can adapt dynamically to business and consumer needs. The integration of the PI framework within city logistics planning presents a forward-looking strategy to reshape urban freight systems (Matusiewicz 2024; Montreuil et al. 2012). In a simplified interpretation, the PI can be understood as a framework for organising collaboration and asset sharing, particularly among Logistics Service Providers (LSPs), to reduce redundancies and overlaps between their service models. Further, it provides the essential systems-thinking approach to ensure interoperability across physical, operational, and digital layers.

The URBANE project (GA No: 101069782) is set to provide evidence of the impact of PI's introduction in real-world last-mile deliveries, using Digital Twin models (Rizopoulos et al. 2023) to support planning and operations. It introduces a novel approach for replicating and scaling up successful collaborative logistics solutions across different urban contexts. A project key building block is the transferability platform, comprising:

(a) An open, extendible library of containerised models that 'solve' specific problems; they vary from computational models for e.g. calculating energy consumption of a mixed delivery fleet, to Agent Based Models (ABM) for simulating behavioural responses to specific interventions and measures, to Cost-Benefit Analysis models (CBA) for investments and operations costs evaluation, to Artificial Intelligence (AI) models for route planning and optimisation, etc.;
(b) A domain-agnostic data ingestion infrastructure and an orchestrating engine that facilitates the sequential execution of selected models employed within a specific what-if scenario. Example use cases include determining the optimal placement of parcel lockers, estimating the required number of e-bikes and e-vans for specific delivery volumes in a targeted area, or simulating emissions impact under different policy regimes;
(c) An impact assessment radar to support strategic planning and the calculation of externalities of freight urban distribution;
(d) An intuitive suite of applications (CitIQore suite) that supports the digital representation of different testing scenarios and decision-making for urban planners, policy makers and business stakeholders—it currently supports experimentation with i) collaborative delivery models between last-milers, using shared lockers as transhipment points and different types/sizes of electric vehicles, ii) logistics operations that combine autonomous delivery vehicles with e-bikes, iii) Comparison of planned with actual deliveries with e-bikes.
(e) A blockchain-based smart contract generator (URBANE Project 2023b) for smart governance of collaborating parties, enabling the automated management of service agreements. It ensures data integrity, immutability and tamper-proof peer-to-peer business transactions.

The URBANE transferability platform is a ready-to-use facility that empowers collaborating urban and peri-urban logistics communities to use data-driven simulation to understand market and city change needs, work out best response strategies and monitor the impact of new policies and solutions, while accounting for consumers' preferences. Its development started in 2020 in the LEAD project (GA No: 861598), where the technology research focused on the need to containerise the models to make the platform open, modular and scalable. As explained above, models come in a variety of types, programming languages, frameworks, and technologies. This heterogeneity made it imperative the decision to adopt a containerisation approach. The platform, therefore, supports the abstraction over the implementation specifics of each model and manages them in a uniform, standardized way. A Dynamic Data-Driven Application System (DDDAS) built on Apache Airflow queries the model library to access models' properties and metadata, and dynamically feeds the parameters and outputs of every execution as inputs to the next model.

This way, any model can be executed as part of a scenario, and models are reusable and adaptive.

The implementation and evolution of the platform have met several challenges beyond technology constraints. A notable example here is the communication between transport modellers and the engineering teams, who had to work together on the digital model integration and the realisation of a common decision support tool. Even if transport modellers and software engineers have both an engineering background, computer scientists and transport engineers use different fundamental laws of science, resulting in different solution approaches to similar problems. Furthermore, designing a model to be used in a parametrised and dynamic system, comes with its own set of challenges and may differ from the workflow used by model developers, who traditionally do not structure their models to be part of a sequenced scenario. Early collaboration between domain experts and technical teams establishes a mutual understanding of such projects' objectives, hurdles, and requirements.

Besides decision-support simulation environments, data governance and semantic interoperability are critical enablers for urban logistics transformation. The DISCO project (GA No: 101103954), an initiative that started in 2023, establishes the Urban Freight (UF) Data Space, a digital enabler for the deployment of data-driven collaborative planning services. It provides a trustworthy framework for sharing data related to transportation, delivery, space use, and supply chain operations in urban areas across decentralised networks, whilst introducing semantic interoperability. The latter is particularly critical for establishing a common understanding of data meaning and context across diverse sources and logistics applications. The UF Data Space aims to ensure data sovereignty, transparency and fairness, creating the ground for stakeholders' onboarding into collaborative ecosystems. Data connectors facilitate seamless data exchange services while incorporating robust data security enforcement mechanisms. A catalogue of key reusable services is currently being developed to support the development of sustainable urban logistics strategies. Data sovereignty is a fundamental principle of the UF Data Space framework, supported by both centralized and decentralized identity management mechanisms. However, nearly all existing open-source dataspace connector implementations exclusively support centralized identity management, presenting significant challenges for cross-Data Spaces interoperability. This limitation forces organisations to either isolate their data spaces or compromise their security by relying on an external authority controlling the single Dynamic Attribute Provisioning Service (DAPS). Recent research (Protopapas 2025) presents a proof-of-concept extension to the Sovity Connector (https://sovity.de/en/connect-to-data-space-en/), addressing the identity management limitations. This approach proposes a federated architecture that enables connectors to operate across multiple DAPS, facilitating interoperability across independently managed Data Spaces. This development is critically important for urban planning, since it encourages cross-domain collaboration. Traditional urban planning manages mobility, energy, land use, and industry in silos. Even integrated approaches like Sustainable Urban Logistics Plans (SULPs) do not incorporate energy and circular economy considerations, limiting systemic impact. Moving forward, urban and logistics operations planners need to link transport demand forecasting with grid energy management and potentially waste-to-energy systems. AI-based decision-making tools tapping into cross-sectoral data spaces can drive synergies between mobility, energy, circular economy, and governance, and promote wider impactful public-private cooperation.

Another promising development planned for the near future involves capitalising on the datasets available through the UF Data Space—such as warehouse and depot locations, parcel locker networks characteristics, and daily delivery-to-customer address patterns—to support the extension of the CitIQore suite. In addition to leveraging existing datasets, these applications will also enable users to generate synthetic data when needed, providing greater flexibility for simulation, scenario testing, and planning in data-sparse contexts. A non-exhaustive list of potential use cases includes (i) synthetic data generation of realistic delivery point datasets using AI-trained distributions, (ii) Comparing the performance of shared versus non-shared parcel locker systems across logistics providers, (iii) Evaluating key operational and environmental trade-offs, or (iv) designing optimal locker placement strategies by combining real and synthetic delivery data.

4 Promising Results from European Living Labs

As mentioned above, living labs have become important real-life testbeds and experimentation environments across Europe, where stakeholders co-create, trial, refine, and showcase innovative urban logistics solutions. The collaborative models developed through European projects demonstrate how multi-actor cooperation, supported by digital technologies and data-driven planning, can help cities advance towards more sustainable and resilient logistics systems. A few interesting examples are presented below.

The LEAD project, one of the first to use digital twinning in urban logistics, delivered substantial advancements in sustainable last-mile logistics through six living lab pilots (GA No: 861598) deployed across major European cities. Electrification and micro-hubs consistently showed environmental benefits but required infrastructure and policy support for full scalability.

More specifically, Madrid focused on a two-echelon distribution model using an Urban Consolidation Centre (UCC) in Plaza Mayor's underground parking, with hybrid vans for the first leg and electric three-wheelers for the last mile. This model reduced mileage by up to 33% and service time by up to 28%. CO_2 emissions were also cut significantly. The key to success was the close collaboration between CityLogin and Empresa Municipal de Transportes de Madrid (EMT).

The Hague tested crowdshipping via Nimber start-up, a public parcel locker network, and an integrated model combining both. Even if site selection and partner engagement were challenging, the pilot showed potential for 9% carbon reduction and 35% cost savings on crowdshipped deliveries. Testing highlighted the importance of administrative agility and collaboration across actors.

Lyon explored UCC use combined with electric cargo bikes and lockers for B2B and B2C deliveries. Digital twin models showed UCC use could reduce GHG emissions by 78% for local deliveries.

Budapest created the digital twin of a hub near the Western Railway Station and simulated multiple delivery configurations combining electric vehicles, LNG trucks, and mobile cross-dock facilities. Simulation results indicated potential reductions in traffic and parking space use (up to 76% savings) and considerable emissions reductions.

Oslo tested four value cases, evolving from direct e-van delivery to crowdshipping with micro-hubs. Significant environmental benefits were measured (GHG -99%, PM2.5 -100%). One of the important key takeaways for future scaling was the need for early integration of landowners and robust micro-hub planning.

Porto's pilot focused on B2C deliveries led by retailer SONAE. Three scenarios tested electrified delivery networks, optimised charging infrastructure, and same-day rescheduling via electric motorcycles. The pilot sprints confirmed substantial energy (1.4 MJ/delivery) and emissions savings (57 gCO_2e/delivery) and demonstrated the value of digital twins for network optimisation.

URBANE, as a follow-up innovation project, has demonstrated the impact of collaborative urban logistics again through six living labs (four lighthouses and 2 twinnings), each developing tailored solutions. Further, six early adopter cities are conducting feasibility studies to explore the potential implementation of similar measures, supported by the digital tools described earlier. The impact assessment results are promising.

In Bologna, three automated micro-hubs called Nearby Delivery Areas (NDAs), were installed in public spaces, so that logistics companies TYP and Due Torri could share them, using them as a transhipment node of parcels from vans to electric vehicles. The pilot combined public and private infrastructure, reduced CO_2 emissions by 52% and improved delivery reliability. Smart contracts were used to govern the collaboration agreements.

Thessaloniki's collaborative locker network and dynamic routing system also demonstrated operational gains. 16 ACS lockers were strategically installed in high-demand areas to improve accessibility. A facility location model analysed urban density and demand to determine optimal locker placements. Projections on the use of AI-based logistics planning estimate reduced CO_2 emissions by up to 68.2% and delivery distances by 82%. ACS achieved a 10.2% reduction in parcel pickup time and improved locker fill rates from 4.5% to 29%.

Helsinki tested Autonomous Delivery Vehicles (ADVs) and cargo bikes in a three-sprint pilot. Customer satisfaction reached 88.7%, delivery safety was 100%, and CO_2 emissions were reduced by 3%. Iterative feedback loops ensured service adaptation to user needs, highlighting the importance of establishing living lab communities for the co-design of logistics innovations.

Valladolid piloted AI-based monitoring of loading/unloading zones and electric cargo vehicles powered partly by solar energy. Findings showed that sustainable delivery models match the performance of traditional combustion vans. The living lab also evaluated in-trunk delivery models linked to park-and-ride lots, supporting logistics efficiency and reduced urban congestion.

Barcelona's pilot combined RFID tracking, blockchain smart contracts, and a digital twin for real-time optimisation of cargo-bike operations. This integration ensured delivery traceability and assisted in warehouse organisation by aligning parcel placement with delivery sequences.

Finally, Karlsruhe explores the integration of autonomous mobile robots with public tram infrastructure to enable intermodal, sustainable, and efficient parcel delivery. The solution involves an autonomously navigating robot to designated tram stops, boarding the tram, travelling alongside passengers, and disembarking to complete the final delivery leg directly to the recipient.

More impact assessment results are expected from the DISCO project, which is geared to improve the knowledge of city authorities in sustainable urban logistics planning. Further, the IKIGAI project (GA: 101202912), building on aforementioned assets and results, will focus on smart and synchromodal hubs for hyperconnected urban logistics, establishing standard processes for data and asset sharing, mobile hub (land and inland waterways) operations, and volume-pooled logistics, to reduce kms in parcel delivery.

5 Conclusion and Future-Looking Solutions

Collaborative logistics can serve as the foundation for an integrated, city-wide delivery ecosystem, with demonstrated operational and environmental benefits. The conceptual diagram (Fig. 2) illustrates the elements that enable effective and sustainable collaborative logistics. It captures four critical domains—multimodal transport networks, logistics operations, decarbonisation goals and roadmaps, and digital enablers. Their systemic consideration in urban and logistics planning drives innovative, system-wide collaborative logistics solutions.

At the centre of the diagram lies the convergence zone where interoperability—physical, operational, and digital (Pan et al. 2021)—is achieved, enabling scalable and resilient logistics systems. Collaboration in logistics can be achieved by digital innovation alone, nor by regulation or infrastructure improvements in isolation. It lies in the integration of physical infrastructure, harmonised operations, digital intelligence, and a common sustainability agenda. Through neutral, open platforms and shared governance frameworks, cities can facilitate cooperation among traditionally competitive actors, ensuring that operational decisions align with broader urban mobility and sustainability goals. Living labs and public–private partnerships can play a key role in this transformation, offering real-world testbeds to prototype, validate, and scale new collaborative solutions. Living labs support the technical deployment of shared services but also help build trust among all parties toward long-term success, paving the way for wider adoption and scalable impact. The path ahead calls for continued experimentation combining simulation and real-life testing, knowledge sharing, and the establishment of standards and governance models

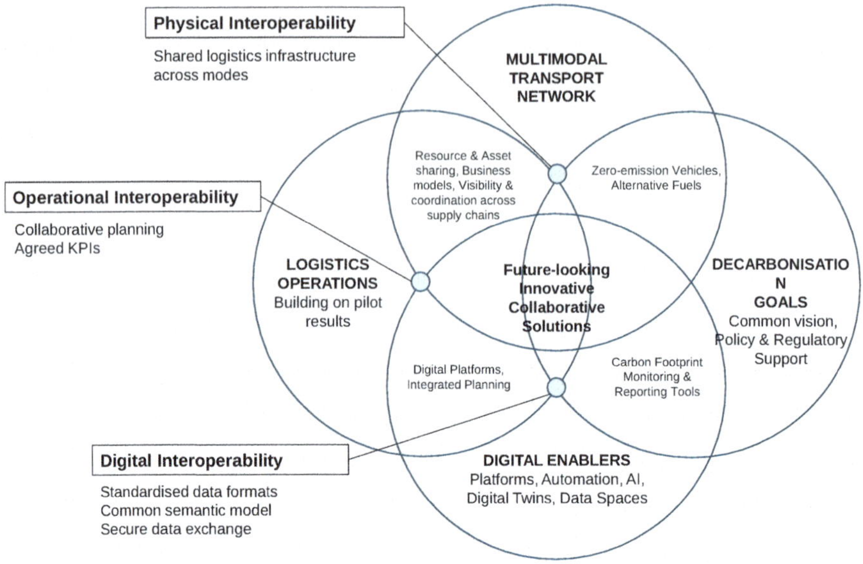

Fig. 2 Main elements of innovative solutions in collaborative logistics

that enable collaborative logistics to move from pilots to mainstream urban operations.

The adoption of collaborative urban logistics models can support the transition path to decarbonisation. Through digital technologies, e.g. Blockchain-enabled smart contracts and optimization tools, companies can consolidate deliveries by using shared urban consolidation centres or micro-hubs, allowing fewer vehicles to serve multiple destinations, which reduces congestion and emissions. Public and private actors can collaborate to share data on traffic flows, curb availability, or low-emission zones, enabling smarter routing and scheduling decisions. Infrastructure such as parcel lockers, depots, or last-mile delivery vehicles can be jointly used, avoiding redundancy and making better use of scarce urban space.

Moreover, collaboration supports joint planning and governance. Through mechanisms like living labs or public–private partnerships, stakeholders can co-design and test new logistics solutions, pilot innovative business models, and implement urban policies that balance commercial needs with public interests.

References

Baron M-L, Khodadad Saryazdi A (2019) Public and public-private cooperation in building resilient urban logistics: the case of "La Chapelle International", Paris. https://doi.org/10.4018/978-1-5225-8160-4.ch001

CitIQore applications suite presentation, https://www.youtube.com/watch?v=mwi3Mth5YEg

DISCO (GA No: 101103954), Data-driven, Integrated, Syncromodal, Collaborative and Optimised urban freight meta model for a new generation of urban logistics and planning with data sharing at European Living Labs

EC COM (2021) 550 final, Fit for 55′: delivering the EU's 2030 Climate Target on the way to climate neutrality

European Network of Living Labs, Schuurman D, DeLosRíos-White, MI, Desole M (2025) Living lab origins, developments, and future perspectives. Zenodo

IKIGAI (GA: 101202912), Physical Internet and the Key steps to Innovation-driven supply chain transformation towards Green, Affordable, scalable and collaborative Zero-EmIssions Freight Transport solutions

LEAD (GA No: 861598), 'Low-Emission Adaptive last mile logistics supporting on Demand economy through digital twins' has received funding from the European Union's Horizon Europe research and innovation programme

LEAD (GA No: 861598), Deliverable D3.2 Living Labs monitoring, refinements and conclusions

Matusiewicz M (2024) Framework for physical internet deployment in cities. Urban Plan Transp Res 12(1):2303341. https://doi.org/10.1080/21650020.2024.230334

Montreuil B, Rougès J-F, Cimon Y, Poulin D (2012) The physical internet and business model innovation. Technol Innov Manag Rev 2:32–37. https://doi.org/10.22215/timreview/566

Nagy G, Szabolcs S (2024) Collaborative logistics: An innovative strategy to address future logistics challenges. Adv Logist Syst Theor Pract 18:83–95. https://doi.org/10.32971/als.2024.031

Pan S, Trentesaux D, Mcfarlane D, Montreuil B, Ballot E et al (2021) Digital interoperability in logistics and supply chain management: state-of-the-art and research avenues towards physical internet. Comput Ind 128:103435

Protopapas P (2025) Toward interconnected dataspaces: implementing decentralised identity management via DAPS, ESWC Conference, Portoroz

Rizopoulos D, Niavis H, Kampa M, Zafeiropoulou A, Franklin R, Mygiakis A, Kakouris A, Alexopoulos A, Fergadiotou I (2023). Digital Twinning platforms as an enabler for the ex-ante evaluation of PI-inspired interventions to last-mile logistics networks. https://doi.org/10.35090/gatech/7980

Stępnicka N, Monastyrskyi G, Sadowska B, Walasek R, Wiączek P, Zimon G (2025) Economic consequences of the phenomenon of uberization on the example of the services of selected carriers in the transport industry

URBANE (GA No: 101069782), 'Upscaling innovative green urban logistics solutions through multi-actor collaboration and PI-inspired last mile deliveries', has received funding from the European Union's Horizon Europe research and innovation programme

URBANE Project (2023a) Deliverable D1.1: operating models for innovative urban logistics. Horizon Europe

URBANE Project (2023b) Deliverable D3.1 – Collaboration governance ledger and smart contracts

Ville S, Gonzalez-Feliu J, Dablanc L (2012) The limits of public policy intervention in urban logistics: lessons from Vicenza (Italy). Eur Plan Stud 21(10):1528–1541. https://doi.org/10.1080/09654313.2012.722954

The manufacturer's authorised representative in the EU is Springer Nature Customer Service Centre GmbH, Europaplatz 3, 69115 Heidelberg, Germany. If you have any concerns regarding our products, please contact ProductSafety@springernature.com

Printed and bound by CPI Group (UK) Ltd, Croydon, CR0 4YY
21/09/2026
02220009-0010